PQQ AND QUINOPROTEINS

PQQ and Quinoproteins

Proceedings of the First International Symposium on PQQ and Quinoproteins, Delft, The Netherlands, 1988

edited by

J. A. JONGEJAN and J. A. DUINE

Biotechnology Centre,
Delft University of Technology,
Delft, The Netherlands

KLUWER ACADEMIC PUBLISHERS

DORDRECHT / BOSTON / LONDON

Library of Congress Cataloging in Publication Data

International Symposium on PQQ and Quinoproteins (1st : 1988 : Delft,
Netherlands)
    P Q Q quinoproteins : proceedings of the First International
Symposium and PQQ and Quinoprotein, Delft, the Netherlands, 1988 /
edited by J.A. Jongejan and J.A. Duine.
        p.   cm.
    Includes indexes.

    1. PQQ (Biochemistry)--Congresses. 2. Quinoproteins--Congresses.
I. Jongejan, J. A., 1944-   . II. Duine, J. A. III. Title.
QP801.P654I67 1988
574.19'25--dc20   89-7956

    ISBN-13: 978-94-010-6920-5   e-ISBN-13: 978-94-009-0957-1
    DOI: 10.1007/978-94-009-0957-1

Published by Kluwer Academic Publishers,
P.O. Box 17, 3300 AA Dordrecht, The Netherlands.

Kluwer Academic Publishers incorporates
the publishing programmes of
D. Reidel, Martinus Nijhoff, Dr W. Junk and MTP Press.

Sold and distributed in the U.S.A. and Canada
by Kluwer Academic Publishers,
101 Philip Drive, Norwell, MA 02061, U.S.A.

In all other countries, sold and distributed
by Kluwer Academic Publishers Group,
P.O. Box 322, 3300 AH Dordrecht, The Netherlands.

printed on acid free paper

The first international Symposium on PQQ and Quinoproteins was held in Delft from 5-7 September, 1988. It attracted over 100 participants from many countries, from universities as well as industrial companies. In view of the late announcement, this is a surprising large number. Also the enthousiasm with which the proposal was received to make preparations for ISQ-2 (most probably to be held in Japan), gave the impression that there exists a need for symposia specially devoted to this topic.

The initiative is indeed justified in the light of the growing importance of this novel field of enzymology. While originally quinoproteins seemed restricted to some specialised microbes, nowadays it appears that for many well known mammalian enzymes which once were thought to be simple metalloproteins or pyridoxoproteins (enzymes containing pyridoxal phosphate), PQQ functions as a cofactor. Since the large part of these enzymes are involved in the degradation or biosynthesis of bioregulators, it is already clear that an impact can be expected in the field of pharmacology and nutrition. First indications for this can already be found in this volume, presenting the formal presentations, either as a lecture or as a poster, given during the symposium. Also the variety of contributions nicely reflects the involvement of several different disciplines in this topic.

May ISQ-1 form the beginning of a long standing tradition.

October, 1988   J.A. Duine.
     Delft.

The organizing Committee wishes to acknowledge the following companies for their financial aid:

Fluka Chemie, Buchs, Switzerland
Ciba-Geigy, Basel, Switzerland
Hoffmann-La Roche, Basel, Switzerland
Boehringer Mannheim, Mannheim, FRG
Unilever Research Laboratories, Vlaardingen, The Netherlands
Merrell Dow, Strasbourg, France
Gist-Brocades, Delft, The Netherlands
Andeno, Venlo, The Netherlands
Hewlett Packard Nederland, Amstelveen, The Netherlands
Sandoz, Basel, Switzerland
AVEBE, Veendam, The Netherlands

TABLE OF CONTENTS

PARTICIPANTS OF THE 1st INTERNATIONAL SYMPOSIUM ON PQQ AND QUINOPROTEINS, ISQ 1, DELFT, SEPTEMBER 1988

QUINOPROTEINS IN C_1-DISSIMILATION BY BACTERIA

C.Anthony, Biochemistry Department, University of Southampton, Southampton, SO9 3TU, UK.

INTRODUCTION

This review will first briefly summarise the role of quinoproteins in the physiology of methylotrophic bacteria growing on methane, methanol and methylamine; the oxidation of methylamine will then be considered in detail with special emphasis on its mechanism and the nature of its electron acceptor; finally the nature of methanol dehydrogenase and its interaction with its specific electron acceptor will be considered.

I shall give a 'historical' introduction to each area of interest, concentrating on those aspects about which there is most debate, and including brief summaries of some of our recent work at Southampton.

ENERGY TRANSDUCTION DURING METHANOL AND METHYLAMINE OXIDATION

Bacteria growing on methanol, methane or methylamine depend on the oxidation of these substrates to formaldehyde which is then assimilated into cell material or further oxidised to carbon dioxide to provide energy.

The dehydrogenases for methanol and methylamine are both quinoproteins; they are especially important in the physiology of methylotrophic bacteria because every molecule of growth substrate is metabolised by one or other of these enzymes, including the molecules that are assimilated into cell material. As a result of this, between 50 and 90% of the oxygen consumed by methylotrophs is by way of electron transport chains involving the dehydrogenases for methanol and methylamine rather than the usual flavoprotein dehydrogenases (Anthony, 1982, 1986). Because methanol and methylamine are highly reduced, not all of the energy available from their oxidation needs to be harnessed as ATP and it is partly because of this that methylotrophs have unusual electron transport chains. They terminate in membrane-bound oxidases of the aa_3-type or the co-type, but all other components are soluble, periplasmic proteins (for reviews see Anthony, 1986,1988). Electrons do not pass from the dehydrogenases to the lower potential ubiquinone or cytochrome bc_1 complex but are passed directly from the periplasmic electron transport proteins to the oxidases (Fig. 1). In this respect the routes for oxidation of methanol and methylamine differ from those for all other organic compounds, being more like the routes for oxidation of many inorganic ions by chemolithotrophs (Ferguson, 1988; Wood, 1988).

In the summary scheme in Fig.1 it is shown that electrons move across the membrane from the periplasmic side, reducing oxygen and consuming protons in the cytoplasm. Protons are released on the opposite side of the membrane in the periplasm during reaction of the dehydrogenases with their electron acceptors. Operation of this 'protonmotive redox arm' leads to a protonmotive force and hence, by way of a proton-translocating ATP synthase, to ATP synthesis. This scheme is consistent with measurements of proton

1

2

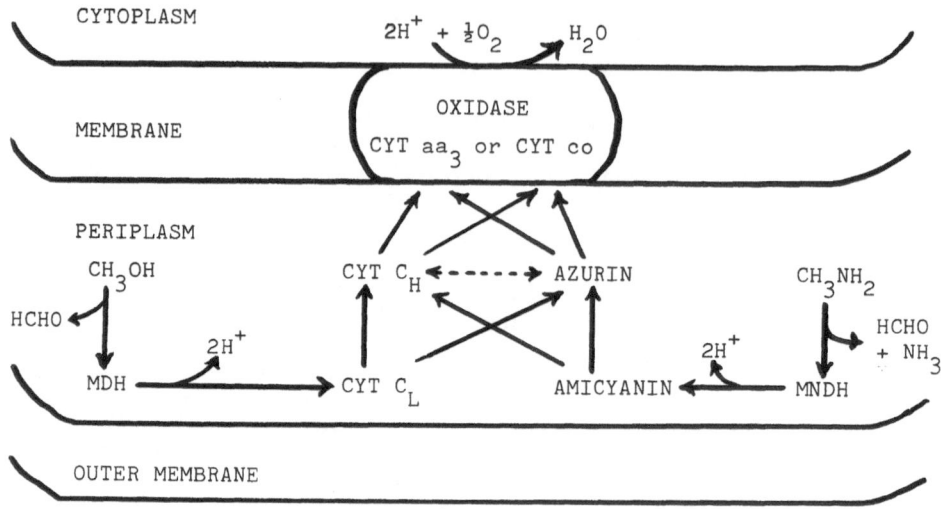

Fig.1. Electron transport during methanol and methylamine oxidation

release, membrane potential and ATP synthesis during respiration which have demonstrated that the oxidation of methanol (and almost certainly methylamine) is coupled to the synthesis of only one ATP (Anthony, 1986, Anthony & Jones, 1987).

THE OXIDATION OF METHYLAMINE

Methylamine dehydrogenase

Methylamine dehydrogenase (MNDH), induced during growth on methylamine, catalyses oxidation of this substrate to formaldehyde and ammonia in the presence of PMS or its physiolgical electron acceptor, amicyanin (for reviews, see Anthony, 1982, 1988). MNDH consists of two light subunits (about 13,000 daltons), each of which carries one molecule of prosthetic group, plus two heavy subunits (about 40,000 daltons) (Matsumoto *et al.*, 1978, 1980; Matsumoto & Tobari, 1978). It is located exclusively in the periplasm (Burton *et al.*, 1983), and it has been suggested that it is the lack of a periplasmic space in Gram-negative bacteria that has led to the alternative copper-containing amine oxidase system in these methylotrophs (Duine *et al.*, 1987b).

When MNDH was first described by Eady & Large (1968, 1971), it appeared possible that its novel prosthetic group might be the same as that of methanol dehydrogenase, but they showed that it was covalently-bound and so not readily removed from the enzyme for direct comparison; and this has remained a problem in the study of its prosthetic group. They showed that MNDH has an absorption peak due to the prosthetic group at 430nm and a shoulder at 460nm. When reduced by methylamine or borohydride the 430nm peak diminishes and a new peak at 325nm appears; the spectrum of the oxidised enzyme returns after addition of PMS. The spectrum is also affected by reaction with carbonyl reagents. The reduced form of the enzyme is fluorescent with an excitation maximum at about 330nm and an emission maximum at 380nm. Consideration of these properties led to the suggestion that the prosthetic group is either a completely novel compound or a previously undescribed pyridoxal derivative. It has now been shown that the prosthetic group of MNDH is covalently-bound PQQ or, perhaps, a closely related derivative of it (de Beer *et al.*, 1980; Kenney & McIntire, 1983), but the suggestion that a Schiff's base may be involved in the mechanism (Eady & Large, 1971) remains a possibility (see below). Amino acid sequence studies initially showed that the PQQ is bound to two amino acid residues (Ishii *et al.*, 1983) and this has been confirmed by McIntire & Stults (1986) who have concluded that a modified PQQ (lacking the three carboxyl groups) is bound to the enzyme by a cysteine thio-ether (via a methylene bridge) plus a serine ether linkage. These conclusions contradict, however, those derived from studies of MNDH from *Thiobacillus versutus;* in this enzyme typical PQQ could be released by treatment with pronase, indicating that it was bound by peptide or ester bonds (van der Meer *et al.*, 1987). The recent crystallization of MNDH by Vellieux *et al.* (1986) holds out exciting prospects for future study of the relationship between the structure and mechanism of quinoproteins.

Investigations of the kinetics of MNDH (Eady & Large, 1971) showed that catalysis occurs by way of a Ping Pong mechanism: methylamine combines with the enzyme to produce a modified enzyme and formaldehyde is released; PMS then oxidises the enzyme leading to release of ammonia and production of the original form of the enzyme. This evidence, together with the spectral evidence quoted above led to the following postulated reaction sequence:

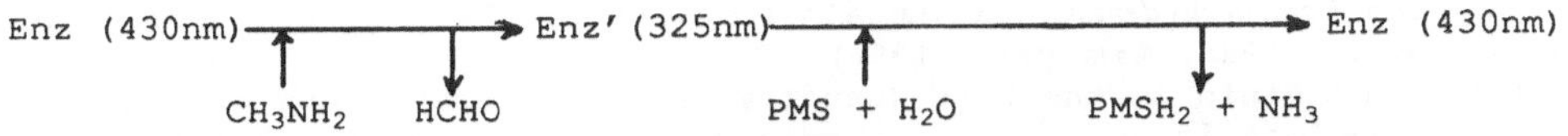
Enz (430nm)
Enz' (325nm)
Enz (430nm)
CH3NH2
HCHO
PMS + H2O
PMSH2 + NH3

When it was shown that the prosthetic group is probably PQQ (de Beer *et al.*, 1980), it became possible to combine this knowledge together with the spectral and kinetic work of Eady & Large (above) and to propose a mechanism (Anthony, 1982); this is reproduced in Fig. 2. As then pointed out, reaction with methylamine might occur at position 4 or 5. Since then the basic observations leading to the proposal of this mechanism have been repeated in the obligate methylotroph bacterium W3A1 (Kenney & McIntire, 1983), the

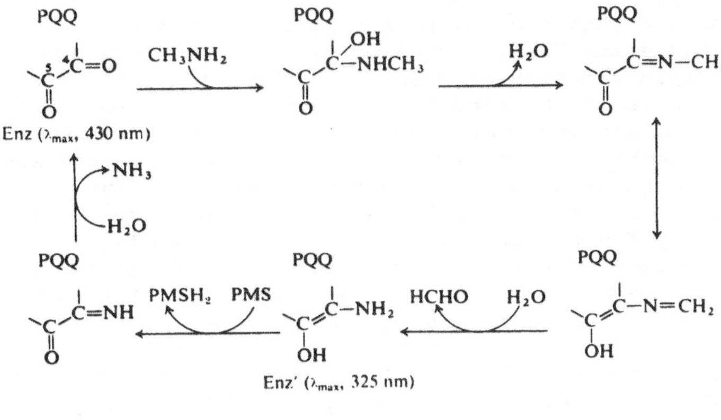
PQQ
CH3NH2
H2O
Enz (λmax, 430 nm)
NH3
H2O
PMSH2
PMS
HCHO
Enz' (λmax, 325 nm)

Fig. 2. A mechanism proposed for methylamine dehydrogenase (from Anthony, 1982).

mechanism has been shown to be consistent with work with model compounds (Eckert & Bruice, 1983; Duine *et al.*, 1987a), and subsequent work with the enzyme has provided considerable evidence that this mechanism is, in principle, correct (McWhirter & Klapper, 1987, 1988).

Electron transport by way of methylamine dehydrogenase

The physiological electron acceptor for MNDH is usually the type I blue copper protein, first discovered by Tobari in *Methylobacterium extorquens* AM1 and called amicyanin (Tobari & Harada, 1981; Tobari, 1984). Its function was initially indicated by the observation that it is induced only during growth on methylamine, that the rates of electron transfer to amicyanin were very rapid, and that no significant electron transfer occurred with *c*-type cytochromes or azurin from the same bacteria. It was subsequently shown that amicyanin is located solely in the periplasmic space together with the dehydrogenase and *c*-type cytochromes (Lawton & Anthony, 1985b; Husain *et al.*, 1986) and that the rate of electron transfer from MNDH to amicyanin is sufficient to account for the respiration rate in whole bacteria (Lawton & Anthony, 1985a). The conclusion that amicyanin is the electron acceptor from MNDH is supported by work with at least five very different methylotrophs including the pink facultative methylotroph *Methylobacterium* and the obligate methylotroph *Methylomonas* J (Tobari & Harada, 1981; Tobari, 1984; Ambler & Tobari, 1985), the obligate methylotroph organism 4025 (Lawton & Anthony, 1985a,b) and the facultative autotrophs *Thiobacillus versutus* (van Houwelingen *et al.*, 1985) and *Paracoccus denitrificans* (Husain & Davidson, 1985, 1986; Husain *et al.*, 1986; Gray *et al.*, 1986; Lim *et al.*, 1986).

Although amicyanin is the best (usually only) electron acceptor for MNDH and it is sometimes induced to extraordinarily high concentrations on methylamine (Lawton & Anthony, 1985b), it is not detectable in all methylotrophs growing on methylamine by way of methylamine dehydrogenase. The best example of this is found in trimethylamine-grown *Methylophilus methylotrophus* (Burton *et al.*, 1983). The picture is confused by the large variation in amounts of amicyanin found in those bacteria that do produce it, by variations in amounts of total blue copper proteins brought about by varying copper concentrations during growth, and by variations in the ease of dissociation of copper from the amicyanin when studied *in vitro*.

Methylamine dehydrogenase is sometimes able to react with the typical small cytochrome *c* of methylotrophs (usually called cytochrome c_H) but whether or not this has any physiological significance is not known. In the obligate bacterium W3A1, for example, copper protein was "not looked for" but the rapid reaction of MNDH with cytochrome *c*-552 from this organism suggested that this is the electron acceptor (Chandrasekar & Klapper, 1986). The evidence from studies *in vivo* are, however, equally consistent with amicyanin or cytochrome *c* being the primary acceptor; if amicyanin is present and able to react with cytochrome *c* then addition of methylamine to whole bacteria will, of course, lead to rapid reduction of the cytochrome *c*, as was previously demonstrated in *Methylobacterium* (Anthony, 1975). It has recently been suggested that when grown in copper-deficient conditions cytochrome c_H may replace amicyanin as the usual electron acceptor in this organism (Fukumori & Yamanaka, 1987). We have come to a different conclusion from work with the obligate organism 4025 which usually produces so much amicyanin that it appears blue in colour (Lawton & Anthony, 1985a,b). We have shown that in oxygen-limited continuous culture, copper deficiency led to lower growth densities of bacteria and lower respiration rates with methylamine, and that blue copper proteins were not readily detectable (K.Auton & C.Anthony, unpublished data). It was impossible, however, to demonstrate that cytochrome c_H is an electron acceptor from methylamine dehydrogenase, or to reconstitute from pure proteins a complete electron transport chain consisting of dehydrogenase, cytochrome c_H and the oxidase (cytochrome co). We concluded that amicyanin may be the only electron acceptor for MNDH but are surprised that such low levels of amicyanin may be sufficient to permit growth. When sufficient copper was present to achieve maximum growth, very large amounts of amicyanin and 'azurin' were produced and complete electron transport chains could be reconstituted using methylamine dehydrogenase plus amicyanin and oxidase, with either cytochrome c_H or azurin as the intermediate electron carrier between amicyanin and the oxidase.

In all methylotrophs studied, amicyanin, azurin, and the periplasmic cytochromes c are all able to interact with each other. An apparent exception was reported in *Paracoccus denitrificans* (Gray *et al.*, 1986). In this organism the midpoint redox of amicyanin is 294mV and that of cytochrome c-551i is 190mV; in the experiment described, about 15% of the cytochrome present might have been expected to become reduced if rapid electron transfer could occur, but none was observed. In the presence of MNDH, rapid reduction of the cytochrome c occurred, although direct reduction in the absence of amicyanin did not. However, this does not indicate, as suggested (Gray *et al.*,1986; Husain *et al.*, 1987), that methylamine dehydrogenase significantly affects the midpoint redox potential of amicyanin; it merely reflects the fact that the midpoint redox potential of an intermediate electron carrier (amicyanin) does not have to be intermediate in value between that of the electron donor (methylamine) and the terminal acceptor (cytochrome c).

In summary, the most likely electron transport chain for the oxidation of methylamine in methylotrophic bacteria is as shown below (NB; the cytochrome c referred to is not cytochrome c_L but the typical small cytochrome often called cytochrome c_H):

Methylamine ⟶ MNDH ⟶ amicyanin ⟶ cytochrome c ⟶ oxidase

In bacteria lacking amicyanin (copper-deficient *Methylobacterium* and perhaps *M.methylotrophus* and bacterium W3A1) the following chain may also occur:

Methylamine ⟶ MNDH ⟶ cytochrome c ⟶ oxidase

In some bacteria at high copper concentrations azurin may replace cytochrome c_H (e.g. organism 4025 and perhaps *Methylobacterium*), thus providing an example of an organism in which blue copper proteins completely replace soluble cytochromes:

Methylamine ⟶ MNDH ⟶ amicyanin ⟶ azurin ⟶ oxidase

THE OXIDATION OF METHANOL

Methanol dehydrogenase

The first quinoproteins to be described and shown to have an unusual prosthetic group were the dehydrogenases for methanol (Anthony & Zatman, 1964) and for glucose (Hauge, 1964), and the prosthetic group of MDH was the first PQQ to be isolated and characterised. The absorption spectrum of MDH was clearly different from that of flavoproteins, having an absorption spectrum due to its prosthetic group with a peak at 345nm and a shoulder at about 400nm; denaturation released a reddish-brown, low molecular weight, polar-acidic molecule with a typical green fluorescence, having an excitation maximum at 365nm and fluorescence maximum at 470nm (Anthony & Zatman, 1967a,b). These fluorescence characteristics are typical of pteridines and it was originally concluded that the novel prosthetic group of MDH might be an unusual pteridine. It was more than 10 years later that it was shown to be PQQ, and MDH was recognised as the first of the quinoproteins (Duine *et al.*, 1978, 1979a, 1980; Westerling *et al.*, 1979; Salisbury *et al.*, 1979). MDH constitutes up to 15% of the soluble bacterial protein, its concentration in the periplasm being about 0.5mM. Most MDHs appear to be dimers of identical subunits (60kD); most are basic proteins, and most require ammonia as activator when assayed with phenazine methosulphate or Wurster's blue (For extensive reviews see Anthony, 1986, Duine *et al.*, 1987a). A technically important (and irritating) characteristic of MDH is the presence on it of an unidentifiable endogenous substrate.

The subunit structure of methanol dehydrogenase

When pure MDH is run on SDS-PAGE systems using highly cross-linked gels and stained with fresh staining reagent, a small 'contaminant' is often observed (e.g. see Elliott & Anthony, 1988). When DNA of *M.extorquens* containing methanol oxidation genes was expressed in *E.coli* a similar small band was seen on SDS-PAGE in addition to MDH (60kD) and cytochrome c_L, suggesting that MDH might have a second, previously ignored, subunit (Anderson & Lidstrom, 1988). We have now been able to confirm this (Nunn & Anthony, 1988). Immediately adjacent to the gene for cytochrome c_L (*MoxG*) is a coding region (*MoxI*) which codes for a protein of about 8Kd. We have shown that the deduced protein sequence corresponds to the protein sequence (determined by direct protein sequencing) of the small protein band isolated by SDS-PAGE from pure MDH; this new MDH B subunit is tightly attached and is hydrophilic and highly basic (Nunn, Day & Anthony, unpublished data).

The reaction of methanol dehydrogenase with cytochrome c_L

All methylotrophic bacteria growing on methane or methanol contain at least two periplasmic c-type cytochromes. One, called cytochrome c_H, is small, usually basic and is now known to correspond to the typical soluble c-type cytochromes found in mitochondria and many bacteria; it functions as electron donor to the cytochrome oxidase (see Anthony, 1986, 1988). Cytochrome c_L, by contrast, is unusually large and acidic, but otherwise appears to be a typical c-type cytochrome. That it plays an essential role in the oxidation of methanol has recently been confirmed *in vivo* by the failure of mutants lacking it to grow on this substrate while able to grow well on all other substrates including methylamine (Nunn & Lidstrom, 1986a,b).

Demonstration of direct electron transfer from methanol to cytochrome c_L catalysed by MDH has been much more difficult than might have been expected. When these proteins were incubated together, the cytochrome was always found to be in the reduced state, even in the absence of methanol; and after oxidation of the cytochrome with ferricyanide, methanol-dependent cytochrome reduction could not be demonstrated (Anthony, 1975; O'Keeffe & Anthony, 1980b). This result suggested that either an activator is required or that some component in the system becomes damaged during extraction; and this appears to have been confirmed by using enzyme preparations from *Hyphomicrobium X* (Duine *et al.*, 1979b). Extracts prepared anaerobically contained reduced cytochrome *c*, which, after oxidation with ferricyanide, could be reduced with methanol in the absence of ammonia as activator, and ammonia was no longer required in the dye-linked assay. After exposure to air, ammonia became essential for dye reduction, the cytochrome c became oxidised, and its reduction by methanol could no longer be demonstrated. Similar results to these were sometimes obtained with *Methylobacterium extorquens* AM1, but most preparations differed in some important respects and were more like those obtained with *Methylophilus methylotrophus;* in this organism anaerobic preparations always required ammonia for the dye-linked assay, and aeration did not lead to cytochrome oxidation (Beardmore-Gray & Anthony, 1984). In the presence of methanol, the rate of cytochrome reduction was greater than in its absence, but after aeration this could not be demonstrated. This was not because methanol oxidation had ceased but because the rate of reduction of cytochrome c by endogenous reductant had increased to the same as that in the presence of methanol.

The first conclusion from these studies with *Hyphomicrobium* and *Methylophilus* is that there is not a clear relationship between the loss of methanol-dependent reduction of cytochrome c by MDH and the production of 'classical' ammonia-requiring MDH in all organisms. A second conclusion is that cytochrome c_L is probably susceptible to damage by ferricyanide. The most important development of the studies with *Hyphomicrobium* has been the discovery of a labile 'factor X' which stimulates reduction of cytochrome c_L and which replaces ammonia as activator (Dijkstra *et al.*, 1988a).

A second complication (besides the presence of endogenous substrate) in investigating methanol-dependent reduction of cytochrome c_L is that the autoreduction of cytochrome c_L (normally occurring on raising its pH) appears to be stimulated by MDH to occur at lower pH values. Autoreduction is, by definition, a first order, intramolecular reaction. We suggested that the rapid reduction of cytochrome c when MDH is added to it (even in the absence of methanol) might be due to a change in pK of a group involved in the autoreduction phenomenon. If this is so, then no electron transfer between the proteins need occur, the presence of methanol will clearly make no difference, and the kinetics will be first order with respect to oxidised cytochrome c_L. This was indeed shown to be the case (for review, see Anthony, 1986). Support for this idea also came from work with the acidophilic *Acetobacter methanolicus*, in which autoreduction occurs at pH 7 (instead of pH 10), and MDH stimulates it to occur at pH 4, the growth pH of the organism (Elliott & Anthony, 1988). Against this explanation of the methanol-independent reduction by MDH of cytochrome c_L is the fact that first order kinetics are difficult unequivocally to demonstrate and that MDH-stimulated reduction of cytochrome c can be demonstrated in bacteria (or conditions) in which the cytochrome c_L is not rapidly autoreducible (see Dijkstra *et al.*, 1988b).

In summary, autoreduction is a frequently observed characteristic of cytochrome c_L, indicating (at least) the unusual nature of this cytochrome. It may be invoked to explain some observations and it may reflect part of the reaction mechanism. On balance, however, it is appearing more probable that MDH-catalysed reduction of cytochrome c in the absence of methanol may be due to the endogenous reductant.

Because of the problems outlined above, it was important to find an alternative way to demonstrate electron transfer from methanol to cytochrome c_L catalysed by MDH; for this to be unequivocal it was necessary to show a rate stimulation by methanol, together with production of formaldehyde. This was eventually achieved by coupling the system to a second electron acceptor (horse heart cytochrome c) (Beardmore-Gray *et al.*, 1983; Beardmore-Gray & Anthony, 1984). These experiments demonstrated that, after oxidation of endogenous substrate, further reduction of cytochrome depended on methanol.

The system was specific for cytochrome c_L, had the same affinity for methanol, the same substrate specificity, and the same sensitivity to EDTA. It has been used to demonstrate unequivocally that MDH catalyses electron transfer directly to cytochrome c_L from methanol with concomitant production of formaldehyde in four completely different types of methylotroph: *Methylomonas* (Ohta & Tobari (1981), *Methylobacterium, Methylophilus, Paracoccus* (Beardmore-Gray *et al.*, 1983; Beardmore-Gray & Anthony, 1984) and *Acetobacter* (Elliott & Anthony, 1988). Despite this unequivocal evidence of the specific, direct, reaction of cytochrome c_L with MDH, considerable discussion of the significance of these results has occurred because we calculated that the rates measured were too slow to account for the rate of respiration in bacteria. It will be of great importance to see if addition of 'factor X', described by Dijkstra *et al.* (1988a) is sufficient to increase sufficiently these rates. It should be noted that 'factor X' is not proposed as an electron transport mediator, but as an activator. Unless damaged by oxygen (in the presence of cytochrome c) it replaces ammonia as the usual activator for MDH in the dye-linked assay.

The structure of cytochrome c_L: a novel class of cytochrome c

Having summarised some of the problems of studying the interaction of MDH and cytochrome c_L it is encouraging that we can report some of our recent work on the structure of this unusual cytochrome. Cytochrome c_L is soluble, it has a high midpoint redox potential (256mV at pH 7), it has an absorbance maximum in the reduced state at about 550nm, it is low spin, and it has histidine and methionine as ligands to the single haem (O'Keeffe & Anthony, 1980a,b; Beardmore-Gray *et al.*, 1982). In these respects it appears to be typical of cytochromes c belonging to Class I as proposed by Ambler (1982). That it is not typical of this class, however, has been previously indicated by its novel function, its large size (about 19kD), its reaction with carbon monoxide (albeit slow and incomplete), its rapid autoreduction at high pH and the response of its redox potential to changes in pH (O'Keeffe & Anthony, 1980a,b; Beardmore-Gray *et al.*, 1982).

The protein sequence, deduced from the gene sequence, confirms that this cytochrome is very unusual and that it constitutes a novel class of c-type cytochrome (Nunn & Anthony, 1988). Except for the haem-binding site, the sequence of cytochrome c_L shows no significant homology with any other cytochrome; in particular, none of the conserved features of c-type cytochromes are seen in the sequence of cytochrome c_L. One key feature of special importance in a typical Class I cytochrome c is the 6th ligand methionine which is more than 59 residues towards the C-terminal from the haem-binding histidine. In cytochrome c_L the 3 methionines are all closer to this histidine, and the sequences around the methionines bear no relation to those around the methionines of other c-type cytochromes. This is consistent with the previous observations of cytochrome c_L that have indicated that it may have an unusual environment for the haem (O'Keeffe & Anthony, 1980a,b; Beardmore-Gray *et al.*, 1982), and that the axial methionine ligand has a novel configuration, as directly observed by NMR studies (preliminary results quoted by Santos & Turner, 1988).

The position of many aromatic and lysine residues in the polypeptide chains of Class I c-type cytochromes are highly conserved. The lysine residues arranged around the haem pocket are of particular importance in binding cytochrome c to its electon donors and acceptors (terminal oxidase or photosynthetic reaction centres). That there is no obvious homologous arrangement of lysines in cytochrome c_L, as shown in the present work, is not surprising because this cytochrome is unlikely to be involved in reaction with the oxidase (Anthony, 1986, 1988). Its function is to react with methanol dehydrogenase and with the small basic cytochrome c_H which reacts with the oxidase and which is similar in structure to other Class I cytochromes.

The determination of the structure of cytochome c_L, and the discovery and analysis of a novel subunit for MDH clearly holds out great hope for developing our understanding of electron transfer processes in this important quinoprotein system.

REFERENCES

Ambler, R.P. and Tobari, J. (1985). Biochem. J. 232, 451-457.

Anderson, D.J. and Lidstrom, M.E. (1988). J.Bact. 170, 2254-2262.

Anthony, C. (1975). Biochem. J. 146, 289-298.

Anthony, C. (1982). In "Biochemistry of Methylotrophs". Academic Press, London.

Anthony, C. (1986). Adv. Microbiol. Physiol. 27, 113-210.

Anthony, C. (1988). In Bacterial Energy Transduction (C. Anthony, ed.), pp. 293-316.

Anthony, C. and Jones, C.W. (1987). In Microbial Growth on C_1 Compounds (H.W. van Verseveld and J.A. Duine eds.), pp. 194-202. Martinus Nijhoff, Dordrecht.

Anthony, C. and Zatman, L.J. (1964). Biochem. J. 92, 614-627.

Anthony, C. and Zatman, L.J. (1967a). Biochem. J. 104, 953-959.

Anthony, C. and Zatman, L.J. (1967b). Biochem. J. 104, 960-969.

Beardmore-Gray, M. and Anthony, C. (1984). In "Microbial Growth on C_1-compounds". (Crawford, R.L. and Hanson, R.S., eds.), pp 97-105. American Society for Microbiology, Washington.

Beardmore-Gray, M., O'Keeffe, D.T. and Anthony, C. (1982). Biochem. J. 207, 161-165.

Beardmore-Gray, M., O'Keeffe, D.T. and Anthony, C. (1983). J. Gen. Microbiol. 129, 923-933.

de Beer, R., Duine, J.A., Frank, J. and Large, P.J. (1980). Biochim. Biophys. Acta. 622, 370-374.

Burton, S.M., Byrom, D., Carver, M., Jones, G.D.D. and Jones, C.W. (1983). FEMS Microbiol. Lett. 17, 185-190.

Chandrasekar, R. and Klapper, M.H. (1986). J.Biol.Chem. 261, 3616-3619.

Dijkstra, M., Frank, Jzn., J. and Duine, J.A. (1988a). FEBS Lett. 227, 198-202.

Dijkstra, M., Frank, Jzn., J., van Wielink, J.E. and Duine, J.A. (1988b). Biochem. J. 251, 467-474. Duine, J.A., Frank, Jzn., J. and Jongejan, J.A. (1987a). Adv. in Enzymol., 59, 169-212.

Duine, J.A., Frank, Jzn., J. and Westerling, J. (1978). Biochim. Biophys. Acta 524, 277-287.

Duine, J.A., Frank, Jzn., J. and van Zeeland, J.K. (1979a). FEBS Lett. 108, 43-446.

Duine, J.A., Frank, Jzn., J. and de Ruiter, L.G. (1979b). J.Gen.Microbiol. 115, 523-526.

Duine, J.A., Frank, Jzn., J. and Verwiel, P.E.J. (1980). Eur. J. Biochem. 187-192.

Duine, J.A., Frank, Jzn., J. and Dijkstra, M. (1987b). In Microbial Growth on C_1 Compounds (H.W. van Verseveld and J.A. Duine eds.), pp. 105-112. Martinus Nijhoff, Dordrecht.

Eady, R.R. and Large, P.J. (1968). Biochem. J. 106, 245-255.

Eady, R.R. and Large, P.J. (1971). Biochem. J. 123, 757-771.

Eckert, T.S. and Bruice, T.C. (1983). J. Amer. Chem. Soc., 105, 4431.

Elliott, E.J. and Anthony, C. (1988). J. Gen. Microbiol. 134, 369-377.

Ferguson, S.J. (1988). In Bacterial Energy Transduction (C.Anthony, ed.), pp. 151-182. Academic Press, London.

Fukumori, Y. and Yamanaka, T. (1987). J. Biochem. 101, 441-445.

Gray, K.A., Knaff, D.B., Husain, M. and Davidson, V.L. (1986). FEBS. Lett. 207, 239-242.

Hauge, J.G. (1964). J.Biol.Chem. 239, 3630-3639.

van Houwelingen., Canters, G.W., Stobbelaar, G., Duine, J.A., Frank, J. and Tsugita, A. (1985). Eur. J. Biochem. 153, 75-80.

Husain, M. and Davidson, V.L. (1985). J. Biol. Chem. 260, 14626-14629.

Husain, M. and Davidson, V.L. (1986). J. Biol. Chem. 261, 8577-8580.

Husain, M., Davidson, V.L. and Smith, A.J. (1986). Biochem. 25, 2431-2436.

Husain, M., Davidson, V.L., Gray, K.A. And Knaff, D.B. (1987). Biochemistry, 26, 4139-4143.

Ishii, Y., Hase, T., Fukumori, Y., Matsubara, H. and Tobari, J. (1983). J. Biochem. 93, 107-119. Kenney, W.C. and McIntire,W. (1983). Biochemistry, 22, 3858-3868.

Lawton, S.A. and Anthony, C. (1985a). Biochem. J. 228, 719-726.

Lawton, S.A. and Anthony, C. (1985b). J. Gen. Microbiol. 131, 2165-2171.

Lim, L.W., Mathews, F.S., Husain, M. and Davidson, V.L. (1986). J.Mol. Biol. 189, 257-258.

Matsumoto, T. and Tobari, J. (1978). J.Biochem. 84, 461-465.

Matsumoto, T., Hiraoka, B.Y. and Tobari, J. (1978). Biochi. Biophys. Acta 522, 303-310.

Matsumoto, T., Shirai, S., Ishii, Y. and Tobari, J. (1980). J.Biochem. 88, 1097-1102. McIntire, W.S. and Stults, J.T. (1986). Biochim. Biophys. Res. Comm. 141, 562-568.

McWhirter, R.P. and Klapper, M.H. (1987a). In Flavins and flavoproteins (D.E. Edmondson and D.B.McCormick, eds.), pp. 709-712, Walter de Gruyter & Co., Berlin.

McWhirter, R.P. and Klapper, M.H. (1987b). In Proceedings of Ist International Symposium on PQQ and Quinoproteins.

van der Meer, R., Jongejan, J.A. and Duine, J.A. (1987). FEBS. Lett. 221, 299-304.

Nunn, D.N. and Anthony, C. (1988). Nucleic Acids Research 16, 7722.

Nunn, D.N. and Lidstrom, M.E. (1986a). J. Bact. 166, 581-590.

Nunn, D.N. and Lidstrom, M.E. (1986b). J. Bact. 166, 591-597.

Ohta, S. and Tobari, J. J. Biochem. 90, 215-224.

O'Keeffe D.T. and Anthony, C. (1980a). Biochem. J. 192, 411-419.

O'Keeffe D.T. and Anthony, C. (1980b). Biochem. J. 192, 481-484.

Salisbury, S.A., Forrest, H.S., Cruse, W.B.T. and Kennard, O. (1979). Nature 280, 843-844.

Santos, H. and Turner, D.L. (1988). Biochim. Biophys. Acta. 954, 277-286.

Tobari, J. (1984). In "Microbial Growth on C_1-compounds". (Crawford, C.L. and Hanson, R.S., eds.).pp 106-112. American Society of Microbiology. Washington.

Tobari, J. and Harada, Y. (1981). Biochim. Biophys. Acta. 101, 502-508.

Vellieux, M.N., Frank, J., Swarte, M.B.A., Grodendik, H., Duine, J. and Drenth, J. (1986). Eur. J. Biochem. 154, 383-386.

Westerling, J., Frank, Jzn., J and Duine, J.A. (1979). Biochem. Biophys. Res. Comm. 87, 719-724.

Wood, P.M. (1988). In Bacterial Energy Transduction (C.Anthony, ed.), pp. 183-230. Academic Press, London.

METHANOL DEHYDROGENASE: MECHANISM OF ACTION

J. FRANK[1], M. DIJKSTRA[1], C. BALNY[2], P.E.J. VERWIEL[3] and J.A. DUINE[1]

[1]Dept. of Microbiol. & Enzymol., Delft University of Technology
Julianalaan 67, 2628 BC Delft, The Netherlands.
[2]INSERM, U 128, BP 5051, 34033 Montpellier Cedex, France.
[3]PML-TNO, Lange Kleiweg 137, 2288 GJ Rijswijk, The Netherlands.

INTRODUCTION

Properties of methanol dehydrogenase

Methanol dehydrogenase (MDH, EC 1.1.99.8) was first described in 1964 by Anthony and Zatman (Anthony and Zatman 1964, 1965, 1967a,b). Since that time a large number of MDH's have been found with similar properties (Anthony 1986) including the dimeric MDH from Hyphomicrobium X (Duine *et al.* 1978). Methanol dehydrogenases are dye-linked enzymes, efficient electron transfer occurring exclusively to cationic electron acceptors at relatively high pH values ($>$ 9). Enzymatic activity requires the presence of ammonia and sometimes higher amines as activators. Like many other quinoproteins, MDH's are characterized by a broad substrate specificity, besides methanol a wide range of primary alcohols is oxidized and some enzymes also oxidize secondary alcohols. Formaldehyde, and sometimes acetaldehyde, is also a substrate. Modulation of the substrate specificity by a modifier protein has been observed in several organisms (Bolbot and Anthony, 1980) and it has been suggested that this protein functions as a regulator of formaldehyde oxidation.

Most MDH's are dimeric proteins, containing two PQQ's, some are monomeric with one PQQ (Patel and Felix 1976, Dijkstra *et al.* 1985). The enzymes are usually isolated in a form containing the semiquinone of PQQ (MDHsem) which does not react with substrate. Oxidation with artificial electron acceptors leads to inactivation unless substrate or a carbonyl reagent is present (Duine and Frank 1980). This feature complicates the study of the catalytic cycle, as does the presence of endogenous substrate of unknown origin in the preparations (Anthony and Zatman 1964, Goldberg 1976, Bamforth and Quayle 1978, Duine *et al.* 1978).

The mechanism of action in vitro

The successful preparation of an oxidized form of MDH (Duine and Frank 1980) and the demonstration that this form reacted with alcohol to yet another

---

Abbreviations. PQQ, PQQH˙and PQQH$_2$ are the quinone, the semiquinone and the quinol forms of 2,7,9-tricarboxy-1H-pyrrolo-[2,3-*f*] quinoline-4,5-dione, respectively; Wurster's blue or TMPD˙,the free radical of N,N,N',N'-tetramethyl-p-phenylene-diamine; Ches, 2-(N-cyclohexylamino) ethanesulfonic acid; Mops, morpholino-propanesulphonic acid; MDH, methanol dehydrogenase; MDHred, reduced MDH; MDHsem, the semiquinone form of MDH; MDHox, the fully oxidized form of MDH; MDHox.HCN, the complex of MDH with cyanide; MDHox.S, the complex of MDH with substrate; MDHox.CP, cyclopropanol inactivated MDH; ESR, electron spin resonance; Hplc, high performance liquid chromatography.

J. A. Jongejan and J. A. Duine (eds.), PQQ and Quinoproteins, 13–22.

form of the enzyme, led to the formulation of a first reaction scheme (Duine *et al.* 1981). In this scheme an unstable oxidized form (MDH_{ox}) was postulated which reacts either with substrate to a reduced form of the enzyme (MDH_{red}) or with cyanide to a stable oxidized complex ($MDH_{ox}.HCN$), which can still react with alcohol. Since NH_4^+-salts were required to obtain ($MDH_{ox}.HCN$), the role of the activator was erroneously assigned to this oxidation step. A further refinement of the scheme was possible when it was found that MDH_{sem} could be reduced to MDH_{red} by a one-electron donor such as the methylviologen radical.

A different reaction sequence, based on the participation of a three-electron reduced form of PQQ, was proposed by Abeles and coworkers (Mincey et al 1981). ESR experiments performed in our laboratory, however, did not provide support for the existence of a three-electron reduced enzyme form (de Beer *et al.* 1983), while a revised reaction scheme formulated thereafter (Parkes and Abeles 1984) had to be rejected for other reasons, as will be also dicussed below.

The mechanism of action in vivo

In studies aimed to reveal the properties of MDH *in vivo*, significantly different behaviour was observed. Thus stimulation of MDH activity by ammonium salts did not occur in the case of an anaerobically prepared cell free extract of *Hyphomicrobium* X (Duine *et al.* 1979). Methanol-dependent reduction of cytochrome *c* could, however, be demonstrated in such extracts, a property that was lost upon admission of oxygen.

Interestingly, all methylotrophic bacteria studied sofar appear to contain at least two soluble cytochromes *c*, designated as cytochrome c_L and cytochrome c_H, according to the value of their isoelectric point (O'Keeffe and Anthony 1980a, b). Both cytochromes displayed fairly high rates of autoreduction which increased by raising the pH, and, in the case of cytochrome c_L was stimulated by the presence of MDH (O'Keeffe and Anthony 1980a). This feature was not seen with the cytochromes extracted from *Hyphomicrobium* X but instead a very rapid reaction was observed between the oxidized form of cytochrome c_L and MDH_{red} (Dijkstra et al. 1988b).

Methanol-dependent reduction of cytochrome c_L, insensitive to the presence of oxygen, was demonstrated by Anthony and coworkers for the components purified from several organisms (Beardmore-Gray *et al.* 1983). However, the rates were very low and it appears now that the efficiency of substrate oxidation depends on the presence of an unidentified, oxygen sensitive, low molecular weight factor (vide infra, Dijkstra *et al.* 1988a)

Different redox forms of MDH

As PQQ occurs in three different redox states (PQQ, PQQH˙ and $PQQH_2$, the existence of at least three different enzyme forms can be anticipated. As mentioned already, these forms have indeed been obtained and shown to have different absorption spectra (Fig. 1, Duine and Frank 1980a). Evidence has been provided that they are genuine redox forms of MDH and not complexes between enzyme and electron acceptor, as has been suggested by Abeles and coworkers (Parkes and Abeles 1984). Acid extraction of the cofactor, followed by reversed phase hplc on C_{18}-silica showed that MDH_{sem} contains equal amounts of PQQ and $PQQH_2$, probably arising from disproportionation of PQQH˙. Similarly, only

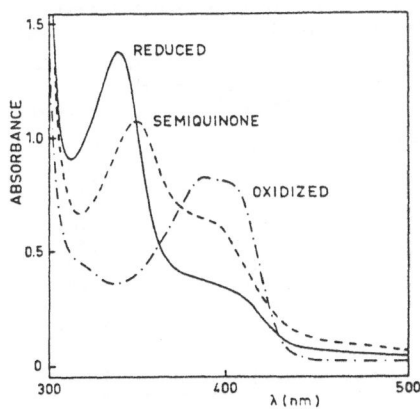

Fig. 1. *Absorption spectra of the different redox forms of MDH.*

$PQQH_2$ could be extracted from MDH_{red} (obtained by performing the isolation of MDH under anaerobic conditions). The redox state of PQQ in MDH_{ox} .HCN could only be deduced by comparison with an analogous enzyme form obtained upon irreversible inhibition of MDH with cyclopropanol (Frank *et al.* 1988).

MECHANISM OF ACTION

Kinetics of the oxidation of reduced MDH

Oxidation of MDH_{red} to MDH_{sem} by Wurster's blue occurred with the stoichiometry expected for the conversion of 2 molecules of $PQQH_2$ to $PQQH^{\cdot}$ in dimeric MDH (Frank *et al.* 1988). The product of oxidation of MDH_{sem} could only be observed in a dynamic experiment in which MDH_{red} was mixed with an excess of Wurster's blue. An intermediate with a spectrum similar to that of MDH_{ox}.HCN appeared within a few seconds after mixing, followed by a slow conversion to MDH_{red} (Frank *et al.* 1988).

Linear relationships between k_{obs} and the concentration of MDH were observed in all cases when cytochrome c_L or Wurster's blue were mixed with MDH, except for cytochrome c_L and MDH at pH 9.0. In the latter case saturation was observed, indicating that electron transfer in the complex of MDH and cytochrome c_L (k_2 in Eq. 1) becomes rate limiting (the kinetics were studied with an excess of MDH over electron acceptor in order to discriminate between the oxidation of MDH_{red} and MDH_{sem}).

$$A_{ox} + MDH_{red} \underset{k_{-1}}{\overset{k_1}{\rightleftharpoons}} [A_{ox}.MDH_{red}] \underset{k_{-2}}{\overset{k_2}{\rightleftharpoons}} [A_{red}.MDH] \underset{k_{-3}}{\overset{k_3}{\rightleftharpoons}} A_{red} + MDH_{sem}$$

Eq. 1

The bimolecular rate constants are summarized in Table 1.

At pH 9.0 Wurster's blue was rapidly reduced by MDH and an isotope effect of CD_3OH was not observed. The stimulation seen in the presence of NH_4Cl and cyanide is non-specific and reflects a general ionic strength effect, since NaCl had the same effect.

TABLE 1. The oxidation of MDH by Wurster's blue (TMPD$^{\cdot}$) and cytochrome c_L under different conditions.

Electron acceptor	Reaction conditions			k ($M^{-1}s^{-1}$)	
				MDH$_{red}$	MDH$_{sem}$
TMPD	10 mM CHES pH 9.0	+ 1 mM CH$_3$OH		220000	75000
"	"	+ 1 mM CD$_3$OH		–	78000
"	"	+ 1 mM cyanide		–	113000
"	"	+ 0.1 mM NH$_4$Cl		–	90000
"	"	+ 0.1 mM NaCl		–	92000
"	10 mM MOPS pH 7.0	+ 1 mM CH$_3$OH		3290	2280
Cyt. c_L	10 mM MOPS pH 7.0	+ 1 mM CH$_3$OH		190000	210000
"	"	+ 0.2 M NaCl		6000	3200
"	"	+ 100 mM phosphate		–	< 200
"	"	+ 0.1 mM EDTA		–	< 1000
"	10 mM CHES pH 9.0	(saturation)		0.33 s^{-1}	0.23 s^{-1}

At pH 7.0 the reaction between Wurster's blue and MDH is very slow, so that the high pH required in the dye-linked assay pertains to that step. In contrast, high oxidation rates are observed with cytochrome c_L at pH 7.0, but not at pH 9.0. Low concentrations of phosphate and EDTA are strongly inhibitory in this case, as has also been observed for oxidation of methanol *in vivo* (Higgins and Quayle 1970, Tonge *et al.* 1975, Mehta *et al.* 1987). It is known that several ions can bind to horse and tuna heart cytochrome c_L (Barlow and Margoliash 1966, Gopal *et al.* 1988), so that it is reasonable to suppose that inhibition is caused by binding of phosphate and EDTA to cytochrome c_L. This conclusion is supported by the observation that such effects are not observed when Wurster's blue is the oxidant. The effect of NaCl is most probably related to electrostatic interaction with complex formation between cytochrome c_L and MDH, as has been reported for other enzyme-enzyme interactions (Cheddar *et al.* 1985).

The nature of the oxidized form of MDH

The nature of the transient spectral form obtained upon oxidation of MDH$_{sem}$ to MDH$_{ox}$.S, (Fig. 2) has been deduced by comparison with the enzyme form

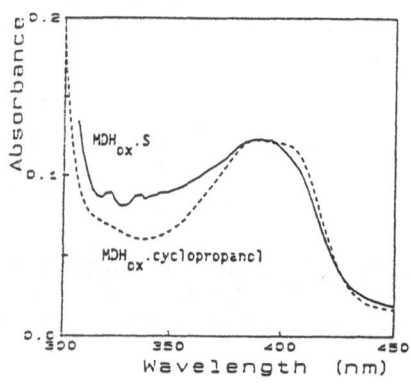

Fig. 2. *Absorption spectra of different oxidized forms of MDH.* (———) MDH oxidized with Wurster's blue in the presence of substrate. (-----) Irreversibly inactivated MDH, obtained after oxidation in the presence of cyclopropanol

resulting from irreversible inactivation of MDH by cyclopropanol (Dijkstra et al. 1984). As can been seen in Fig. 2, both enzyme forms have very similar spectra. The stable nature of the cyclopropanol inactivated enzyme form facilitated the investigation of the structure of its (modified) cofactor. Extraction yielded a product with an absorption spectrum different from that of PQQ, but closely ressembling that of the acetone adduct of PQQ (Duine and Frank 1980b). Upon reversed phase hplc, however, the retention time of the cofactor derivative clearly differed from that of both the acetone- and the 2-propanal adduct of PQQ (3 and 4 in Scheme 1). Elucidation of the structure of this cofactor derivative was achieved in two stages. First the derivative was extracted from cyclopropanol-treated MDH and characterized by hplc, coupled to photodiode array detection, and (partially) by ^{1}H-NMR. Larger amounts of modified cofactor were obtained by chemical synthesis. Free PQQ in solution does not react with cyclopropanol, but addition of a catalyst like Ag_2O, CuO or ZnO resulted in a slow formation of product. This product appeared to have the same retention time and spectrum as the product extracted from cyclopropanol-inactivated MDH. In addition, the same spectral features were found in the ^{1}H-NMR spectrum, consistent with the 3-propanal adduct of PQQ. Apparently the cyclopropanol ring opens prior to addition to PQQ at the C_4- or the C_5-carbonyl.

Since the exact site of addition might have an important bearing on the mechanism of action of MDH, this was further investigated by ^{13}C-NMR. Using 95 % ^{13}C-enriched PQQ, the signals of the ^{13}C-NMR spectrum could be assigned (Table 2). Comparison of this spectrum with that of the acetone adduct of PQQ, for which addition at C_5 has been unequivocally established by X-ray diffraction analysis (Salisbury et al. 1979), revealed that here too addition occurred at C_5. Further proof for this has been obtained by converting the product into a substance with a hydroxyquinoline moiety. This can be accomplished by reduction

TABLE 2. ^{13}C-NMR of PQQ and PQQ-adducts

Carbon atom [a]	Chemical shifts (ppm)		
	PQQ	acetone adduct	3-propanal adduct
CH_3		29.9	
CH_2			35.0
CH_2		51.1	38.6
CHOH			102.4
C_3	113.9 (d 66 [b] d 57)	112.1	111.9
C_{3a}	122.9 (q 58–62 very broad)	121.3	121.4
C_{9a}	126.1 (q 58)	121.6	122.2
C_2	127.7 (d 86, d 66)	127.2	128.0
C_8	130.7 (t 58)	125.6	125.7
C_{9b}	137.7 (d 61, d 59)	134.6	134.0
C_9	144.6 (q 58 sharp)	137.1	135.9
C_7	146.4 (d 82, d 57)	145.3	145.0
C_{5a}	147.9 (t 55–60 very broad)	162.0	162.7
C_2-COOH	161.4 (d 86, q 5 sharp)	161.2	161.0
C_7-COOH	165.7 (d 82, t 7, d 2 sharp)	165.3	164.9
C_9-COOH	166.6 (d 59)	169.2	169.1
C_4	173.4 (d 62, d 50, d 17)	190.3	191.0
C_5	179.9 (d 58, d 50, d 16)	74.9	88.8
$COCH_3$		207.0	

[a] The assignments for PQQ were deduced from HETCOR and selective proton-decoupling experiments.
[b] Coupling constants in Hz with an accuracy of ±3 Hz. All signals are broadened by further coupling, unless stated otherwise.

with NaBH$_4$ followed by dehydration as shown in Scheme 1. If addition occurs at C$_4$, the resulting phenolic compound (8) should be a strong chelator. Indeed, titration of 4-hydroxy-pyrroloquinoline with Cu^{2+}-ions led to a significant shift of the absorption spectrum. No such shifts were observed for the analogous products obtained from the acetone- and 3-propanal-adducts (4 and 5).

Conceivably, the enzyme forms with similar spectra have a common origin, arising from addition at C$_5$ of PQQ of different compounds as depicted in scheme 1. Spectral indications for the existence of a labile oxidized enzyme form, MDH$_{ox}$, obtained in the absence of cyanide and substrate, have been obtained in the past (Duine and Frank 1981)

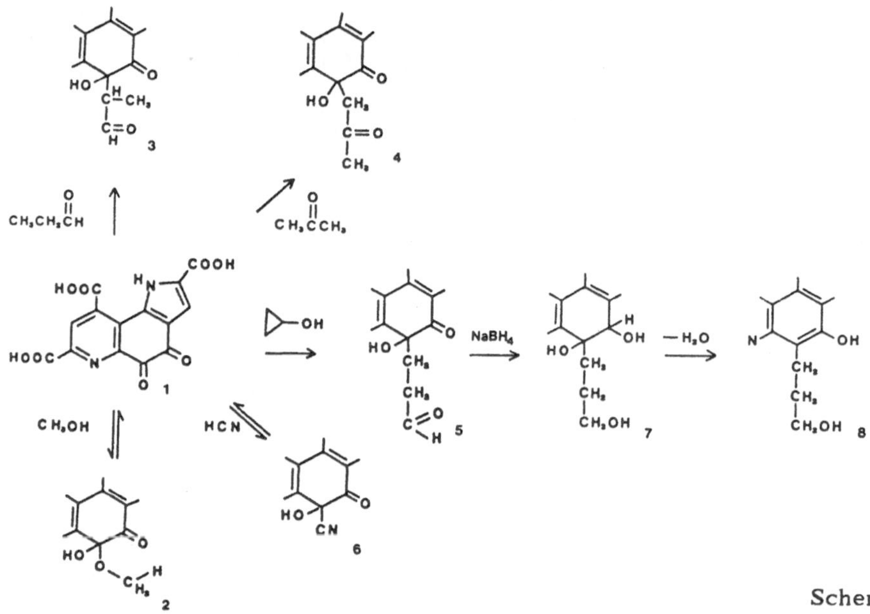

Scheme 1

Kinetics of the oxidation of the alcohol substrate

In Fig. 3 the time course of the reaction of MDH with an excess of Wurster's blue in the absence of activator is depicted. Both the absorbance at 612 nm, as a measure for the concentration of Wurster's blue, and at 410 minus 451 nm, representing the absorbance of the MDH forms corrected for the contribution of Wurster's blue, are shown. The 410 nm trace clearly shows that the Michaelis ES-complex is rapidly formed and the enzyme remains in a steady state as long as Wurster's blue is present. After that a slow conversion of MDH$_{ox}$.S to the MDH$_{red}$ form occurs. The reaction could be described by a single exponential and was found to be independent of the concentration of either Wurster's blue or MDH. Formation of aldehyde product occurred concomitantly with the spectral changes (Frank *et al.* 1988), while the nature of the electron acceptor had no effect. Obviously, the observed phenomenon represents the intramolecular oxidation of substrate in the ES-complex. The effects of various parameters on this reaction are summarized in Table 3. A large kinetic deuterium isotope effect is seen when CH$_3$OH is replaced by CD$_3$OH ($k_H/k_D = 7$), indicating a rate limiting

TABLE 3. The oxidation of the alcohol substrate in
the ES-complex under different conditions.

Electron acceptor	Condition		k (s^{-1})
TMPD˙	10 mM MOPS pH 7.0	+ 1 mM CH$_3$OH	0.064
"	10 mM CHES pH 9.0	"	0.06
"	"	+ 1 mM CD$_3$OH	0.008
"	" + 80 mM NH$_4$Cl	+ 1 mM CH$_3$OH	23.0
"	" + "	+ 1 mM CD$_3$OH	16.4
Cyt. c_L	50 mM MOPS pH 7.0	+ 1 mM CH$_3$OH	0.057
"	+ Factor "X"		0.155

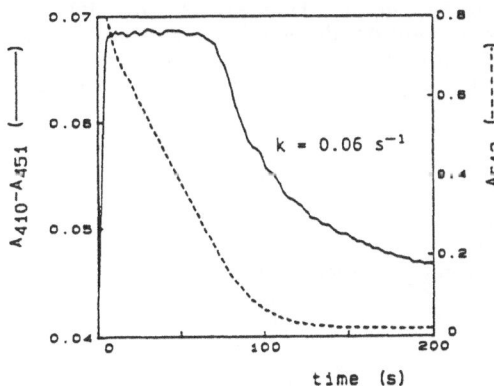

Fig. 3. *Time course of the reaction of MDH with an excess of Wurster's blue.* (——) Difference of absorbance at 410 and 451 nm, representative for the spectral changes occurring in MDH. (----) Absorbance at 612 nm, one of the absorption maxima of Wurster's blue.

hydrogen transfer step. It further indicates that externally added substrate displaces the endogenous substrate.

Activation in vitro and in vivo

At pH 9.0 the rate of alcohol oxidation is dramatically increased by NH$_4$Cl (Table 3). Clearly, the activator is required in this part of the redox cycle of MDH. Only a small deuterium isotope effect is seen at full activation, suggesting that substrate oxidation is only partly rate limiting (in an assay) at pH 9.0.

A different situation exists at pH 7.0 with cytochrome c_L as the electron acceptor. Without activator the rate of alcohol oxidation is essentially similar to that at pH 9.0. It appears, however, that activation under more physiological conditions cannot be achieved in the same way as at pH 9.0. Application of a concentration of NH$_4$Cl, equivalent to that required at pH 9.0, led to a decreased oxidation rate of· MDH by cytochrome c_L, obscuring a possible increase in the rate of substrate oxidation. However, stimulation is observed with factor X, which might be the natural activator.

Mechanism of substrate oxidation

MDH$_{ox}$.S is an intermediate in the reaction cycle. The spectral properties of the complex indicate addition of the alcohol substrate ˙at C$_5$ (**2** in Scheme 1).

Model studies performed on the addition of alcohols to carbonyl derivatives have revealed that addition is a general acid/base catalyzed process (Ogata and Kawasaki 1977). In view of the fact that the presence of MDH_{ox} could not be detected in the cycle, it is likely that in MDH the addition of substrate to PQQ is also a catalyzed reaction, for instance by a base in the active site (Scheme 2). The mechanism of subsequent oxidation of the substrate will be discussed below.

In our view inactivation of MDH by cyclopropanol is due to the aberrant behaviour of this substrate during the addition reaction. It has been suggested in the past that inactivation followed a radical mechanism (Dijkstra et al. 1984). For a number of reasons this seems less likely now. Model studies of the reaction between PQQ and cyclopropanol have revealed that a reaction only takes place in the presence of a metaloxide as catalyst. Since the reaction is equally well catalyzed by a redox-inactive oxide such as ZnO, the involvement of radicals does not seem likely. A more attractive hypothesis is that the metaloxide catalyzes proton abstraction leading to the cyclopropoxy-anion. This anion can either reversibly add to PQQ or its negative charge can be delocalized leading to ring-opening under formation of a reactive carbanion which subsequently adds irreversibly to C_5 of PQQ (Scheme 2). Formation of such a reactive carbanion in close

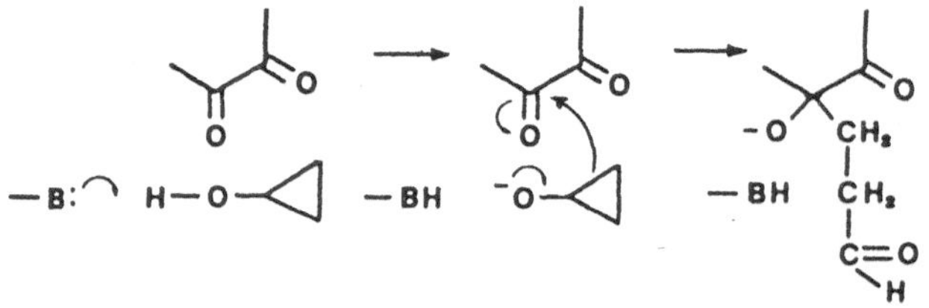

Scheme 2.

proximity to PQQ, for instance in a coordination complex between PQQ and the metaloxide, would favour the efficiency of the reaction. Irreversible addition of cyclopropanol to MDH_{ox} could proceed according to the same mechanism. The slow rate of the inactivation of MDH (1×10^{-3} s^{-1}, Dijkstra et al. 1984) supports the ionic character of the addition reaction, since model studies of base catalyzed ring-opening of cyclopropanol have shown a comparable rate (1×10^{-4} s^{-1}, Sherry and Abeles 1984) while ring-opening after generation of a free radical in cyclopropanol occurs much more rapidly (DePuy et al. 1972).

Cyclopropylmethanol appears to be oxidized by MDH without the formation of ring-opened products (results not shown), suggesting that the mechanism of oxidation of the alcohol does not involve free radicals either (MacInnes et al. 1982), although, in view of the experimental data obtained with methanol oxidase, this cannot be considered as conclusive evidence (Sherry and Abeles 1984).

Although it has been found now that activation by NH_4^+-salts or the natural activator takes place at the level of the oxidation of substrate, it is not clear yet how these agents participate in the mechanism of substrate oxidation. In the past, mechanisms involving p-amino-quinone type structures have been put forward to explain the rate enhancement (Forrest 1980, Duine et al. 1987). However, the recent finding that 1-methyl substituted PQQ is still active in alcohol dehydrogenase from Ps. testosteroni (Jongejan et al, this Proceedings) excludes the involvement of such structures. Possibly the activator has a similar role as that of a basic residue postulated to be present in the active site of the alcohol dehydrogenase from Ps. testosteroni (Jongejan et al., this Proceedings) and supposed to favour hydride transfer from the alcohol to PQQ. In that concept the activator could also play a role in scavenging potentially dangerous aldehyde product.

THE CATALYTIC CYCLE

The experimental data described here extend the catalytic reaction scheme proposed earlier (Duine *et al.* 1984) and is summarized in Scheme 3. Starting with

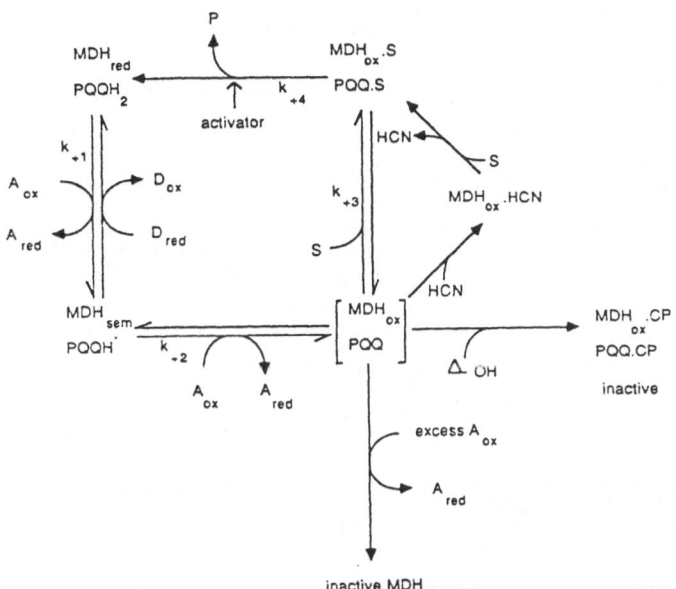

Scheme 3, MDH_{red}, reduced MDH; MDH_{sem}, the semiquinone form of MDH; MDH_{ox}, the fully oxidized form of MDH; MDH_{ox}.HCN, the complex of MDH with cyanide; MDH_{ox}.S, the complex of MDH with substrate; MDH_{ox}.CP, cyclopropanol inactivated MDH;

MDH_{red}, oxidation with cationic 1e-acceptors leads to MDH_{sem}. This enzym form does not react with alcohol substrate, but can be reduced by 1e-donors such as methylviologen or by lumiflavin mediated photoreduction (Massey and Hemmerich 1978). Further 1e-oxidation leads to MDH_{ox} which can react further in different ways. With substrate the Michaelis ES complex MDH_{ox}.S is formed, which decomposes into MDH_{red} and product. With cyanide a similar complex is formed, while no further reaction takes place, but alcohol is capable to displace cyanide. Irreversible modification of MDH_{ox} occurs when it reacts with cyclopropanol. In the absence of these substances MDH_{ox} probably reacts further with remaining electron acceptor and becomes inactivated. In the previous scheme the activator was thought to be involved in the oxidation of MDH_{red} and MDH_{sem} rather than in the oxidation of substrate. It has now been established that the opposite is true and the explication for the former, erroneous, conclusion is that addition of cyanide to MDH_{ox} only occurs in the absence of alcohol and activator is required to speed up the consumption of the endogenous substrate.

More insight into the catalytic cycle of MDH clearly requires more research, in particular to clarify the nature of the natural activator, the mechanism of the activation process and the associated mechanism of product channeling.

REFERENCES

Anthony C and Zatman LJ, 1964 Biochem. J. 92: 614–627
Anthony C and Zatman LJ, 1965 Biochem. J. 96: 808–812
Anthony C and Zatman LJ, 1967a Biochem. J. 104: 953–959
Anthony C and Zatman LJ, 1967b Biochem. J. 104: 960–969
Anthony C, 1986 Adv. Microbial Physiol. 27: 113–210
Bamforth CW and Quayle JR 1978 Biochem. J. 169: 677–686
Barlow GH and Margoliash E, 1966 J. Biol. Chem. 241: 1473–1477
Beardmore-Gray M, O'Keeffe DT and Anthoy C, 1983 J. Gen. Microbiol. 129: 923–933
Carver MA, Humphrey KM, Patchett RA and Jones CW, 1984 Eur. J. Biochem. 138: 611–615
Cheddar G, Meyer TE, Cusanovich MA, Stout CD and Tollin G, 1985 Biochemistry 25: 6502–6507
DePuy CH, Jones HL and Gibson DH, 1972 J. Am. Chem. Soc. 94: 3924–3929
De Beer R, Duine JA, Frank J and Westerling J, 1983 Eur. J. Biochem. 130: 105–109
Dudley Page M and Anthony, C, 1986 J. Gen. Microbiol. 132: 1553–1563
Duine JA, Frank J and Westerling J 1978 Biochim. Biophys. Acta 524: 277–287
Duine JA, Frank J and de Ruiter LG, 1979 J. Gen. Microbiol. 115: 523–526
Duine JA and Frank J, 1980a Biochem. J. 187: 213–219
Duine JA and Frank J, 1980b Biochem. J. 187: 221–226
Duine JA and Frank J, 1981 In: Microbial growth on C_1-compounds (H. Dalton, ed), pp 31–41, Heyden & Son Ltd, London.
Duine JA, Frank J, Jongejan JA and Dijkstra M, 1984 In: Microbial growth on C_1-compounds (RL Crawford and RS Hanson, eds), pp 91–96, American Society for Microbiology, Washington, D.C.
Duine JA, Frank J and Jongejan JA, 1987 Adv Enzymol. 59: 170–212
Dijkstra M, Frank J, Jongejan JA and Duine JA, 1984 Eur. J. Biochem. 140: 369–373
Dijkstra M, van den Tweel JJ, de Bont JAM, Frank J and Duine JA, 1985 J. Gen. Microbiol. 131: 3163–3169
Dijkstra M, Frank J and Duine JA, 1988a FEBS Letters 227: 198–202
Dijkstra M, Frank J, van Wielink JE and Duine JA, 1988b Biochem. J. 251: 467–474
Frank J, Dijkstra M, Duine JA and Balny C, 1988 Eur. J. Biochem. 174: 331–338
Forrest HS, Salisbury SA and Kilty CG, 1980 Bioch. Biophys. Res. Comm. 97: 248–251
Goldberg I, 1976 Eur. J. Biochem. 63: 233–240
Gopal D, Wilson GS, Earl RA and Cusanovich MA, J. Biol. Chem. 263: 11652–11656
Higgins IJ and Quayle JR, 1970 Biochem. J. 118: 201–208
MacInnes I, Nonhebel DC, Orsculik ST and Suckling CJ, 1982 J. Chem. Soc., Chem. Commun. 121–122
Massey V and Hemmerich P, 1978 Biochemistry 17: 9–16
Mehta PK, Mishra S and Ghose TK, 1987 J. Gen. Appl. Microbiol. 33: 221–229
Mincey T, Bell JA, Mildvan AS and Abeles RH, 1981 Biochemistry 20: 7502–7509
Ogata Y, Kawasaki A, 1970 In: The chemistry of the carbonyl group, (J Zabicky, ed.) pp 1–71, Interscience, London.
O'Keeffe DT and Anthony C, 1980a Biochem. J. 190: 481–484
O'Keeffe DT and Anthony C, 1980b Biochem. J. 192: 411–419
Parkes C and Abeles RH, 1984 Biochemistry 23: 6355–6363
Patel RN and Felix A, 1976 J. Bacteriology 128: 413–424
Salisbury SA, Forrest HS, Cruse WBT and Kennard O, 1979 Nature 280: 843–844
Sherry B and Abeles RH, Biochemistry 24: 2594–2605
Shinagawa E, Matsushita K, Nonobe M, Adachi O, Ameyama M, Oshiro Y Itoh S, Kitamura Y, 1986 Biochem. Biophys. Res. Comm. 139: 1279–1284
Tonge GM, Harrison DEF, Knowles CJ and Higgins IJ, 1975 FEBS Lett. 58: 293–299

QUINOPROTEIN ETHANOL DEHYDROGENASE FROM *PSEUDOMONAS*

Helmut GÖRISCH and Michael RUPP
Institut für Mikrobiologie, Universität Hohenheim, Garbenstraße 30,
D-7000 Stuttgart 70, Federal Republic of Germany

ABSTRACT

Dye-linked ethanol dehydrogenases from *Pseudomonas aeruginosa* ATCC
17 933 and *P. putida* ATCC 17 421 were purified to homogeneity and crystallized. The amino acid composition of the two enzymes is very similar and
the number of the aromatic amino acid residues found per subunit are almost
identical.
   With respect to their catalytic and molecular properties both ethanol
dehydrogenases are similar to the quinoprotein methanol dehydrogenases
known from methylotrophic bacteria. They show a high pH-optimum, need
ammonia or an amine as activator and are dimers of identical subunits of a
molecular mass of 60 000. The dimer is the catalytically active form. Each
subunit carries one prosthetic group pyrroloquinoline quinone, which can be
titrated by the suicide substrate cyclopropanone ethylhemiketal. In
contrast to the general methanol dehydrogenases the two ethanol
dehydrogenases have a low affinity for methanol and in addition to primary
alcohols they also oxidize secondary alcohols. With secondary alcohols
preferentially one of the two enantiomers is oxidized.
   The catalytic and spectral properties of the two enzymes are very
similar to the quinoprotein ethanol dehydrogenase isolated from *P. aeruginosa* LMD 80.53 (Groen et al., 1984. Biochem. J. 223: 921-924). However this
enzyme is reported to be a monomer of molecular mass 100 000.

INTRODUCTION

Methanol oxidizing bacteria possess dye-dependent alcohol
dehydrogenases which contain pyrroloquinoline quinone as prosthetic group
(Duine et al., 1980). The prosthetic group is also called methoxatin
(Salisbury et al., 1979). The usual function of this alcohol dehydrogenase
is to catalyze the oxidation of methanol. The isolated enzyme however
oxidizes a wide range of primary alcohols, while secondary alcohols are not
accepted as substrate. The pH-optima are around pH 9 and the K_m values for
methanol are in the range of 10 to 20 µM. The methanol dehydrogenases need
ammonia or methylamine as activator and phenazine methosulfate is used as
electron acceptor. In general the methanol dehydrogenases are dimers of
identical subunits with a molecular mass of 60 000.

J. A. Jongejan and J. A. Duine (eds.), PQQ and Quinoproteins, 23–33.
© *1989 by Kluwer Academic Publishers.*

From *P. aeruginosa* LMD 80.53 grown on ethanol recently a quinoprotein alcohol dehydrogenase was purified, which has a low affinity for methanol and oxidizes in addition to primary alcohols also secondary alcohols (Groen et al., 1984). With respect to the spectral and catalytic properties this ethanol dehydrogenase is very similar to the usual methanol dehydrogenases. Recently we found that *P. aeruginosa* ATCC 17 933 produces high levels of such a quinoprotein ethanol dehydrogenase. This enzyme was purified to homogeneity and crystallized (Rupp and Görisch, 1988). In addition we purified and crystallized a quinoprotein ethanol dehydrogenase from *P. putida* ATCC 17 421, grown on ethanol. Both enzymes show catalytic properties similar to the ethanol dehydrogenase described for *P. aeruginosa* LMD 80.53 (Groen et al., 1984).

MATERIALS AND METHODS

Chemicals. CM-Sepharose was obtained from Pharmacia (Uppsala, Sweden). All other chemicals were obtained as described recently (Rupp and Görisch, 1988).

Organisms and growth conditions. *P. putida* ATCC 17 421 and *P. aeruginosa* ATCC 17 933 were grown on a mineral medium supplemented with 0.5% ethanol as described (Rupp and Görisch, 1988).

Enzyme isolation. 10 g of frozen cells of *P. putida* 17 421 were suspended in 12 ml of 50 mM Tris/HCl buffer, pH 7.9, and cells were disrupted by ultrasonic treatment. Cell debris was removed by centrifugation at 60 000 x g for 60 min at 4°C. All subsequent steps were performed at 4°C. The crude extract was applied to a DEAE-Sephacel column (3 x 13 cm) equilibrated with 10 mM Tris/HCl buffer, pH 7.9, and quinoprotein ethanol dehydrogenase was eluted by the same buffer. The active fractions were pooled and applied to a CM-Sepharose column equilibrated with 10 mM ammonium acetate/acetate buffer, pH 6.5. The enzyme was eluted by a linear gradient from 0 to 1 M NaCl in 10 mM ammonium acetate/acetate buffer, pH 6.5. The active fractions were concentrated by an Amicon YM 10 membrane and applied to a Sephacryl S-200 column (2.6 x 100 cm). The column was equilibrated with 100 mM Tris/HCl buffer, pH 7, containing 100 mM NaCl, and quinoprotein ethanol dehydrogenase from *P. putida* ATCC 17 421 was eluted by the same buffer. Active fractions were pooled and stored at -80°C.

Quinoprotein ethanol dehydrogenase from *P. aeruginosa* ATCC 17 933 was purified as described previously (Rupp and Görisch, 1988).

Enzyme assay. The activity of quinoprotein ethanol dehydrogenase was measured spectrophotometrically by following the reduction of either phenazine methosulfate/2,6-dichlorophenolindophenol (PMS/DCPIP) or Wurster's blue as described (Rupp and Görisch, 1988). One unit of enzyme activity is defined as that amount, either reducing 1 μmol of DCPIP or 2 μmol of Wurster's blue per min. With Wurster's blue as electron acceptor the specific activity increases by a factor of 2.6 for the enzyme from *P. putida* ATCC 17 421 and by 2.8 for the enzyme from *P. aeruginosa* ATCC 17 933.

Crystallization. Crystallization was performed by the hanging drop method (McPherson, 1982). The reservoir contained 1 ml solution. 5-μl samples of an ethanol dehydrogenase solution (2 mg/ml) were mixed with 1 μl of reservoir solution.

Temperature inactivation. Solutions with 0.5 mg protein/ml of quinoprotein ethanol dehydrogenase from *P. putida* or *P.aeruginosa* were prepared in 100 mM Tris/HCl, 100 mM NaCl, pH 7, or the same buffer containing 10 mM $CaCl_2$, respectively. Samples were incubated in small test tubes at various temperatures and after rapid cooling residual activities were determined.

Isoelectric point. Isoelectric points were determined by isoelectric focussing experiments and by pH-dependent binding analysis (Yang and Langer, 1985).
All other methods used have been described recently (Rupp and Görisch, 1988).

RESULTS

Enzyme isolation. Quinoprotein ethanol dehydrogenase from *P. putida* ATCC 17 421 is purified 150-fold applying the purification steps described in Materials and Methods. Table 1 summarizes the procedure. Chromatography on CM-Sepharose is the most effective step. The purified enzyme shows a specific activity of 9 U/mg, corresponding to a catalytic centre activity of 9 with PMS/DCPIP as electron acceptors, while with Wurster's blue as electron acceptor a specific activity of 23 U/mg is found, which corresponds to a catalytic centre activity of 23.

Table 1. Purification of quinoprotein ethanol dehydrogenase from
     P. putida.

	Total prot. (mg)	Total act.[b] (U)	Spec. act.[b] (U/mg)	Yield (%)	Purification
Crude extract[a]	830	46.8	0.06	100	1
DEAE-Sephacel	200	46	0.23	98	3.8
CM-Sepharose	17	45.2	2.7	97	45
Sephacryl S-200	2.7	24.2	9	52	150

[a]Starting with 10 g of wet cell paste.
[b]With PMS/DCPIP as electron acceptors.

Analytical gel electrophoresis of the native enzyme reveals a single band of protein. By activity staining also one band of enzyme activity is found, which corresponds to the stained protein band. The enzyme sediments in a single sharp boundary in sedimentation velocity experiments and polyacrylamide gel electrophoresis in the presence of sodium dodecyl sulfate also shows only one band of protein.

Crystallization. Ethanol dehydrogenase from *P.putida* is readily crystallized in the presence of 2.5 M $(NH_4)_2SO_4$. Fig.1 depicts a typical bundle of thin plates. The enzyme from *P.aeruginosa* crystallizes in the presence of PEG 1550 and crystals up to 0.25 mm in the shortest dimension are formed within 4 to 5 days.

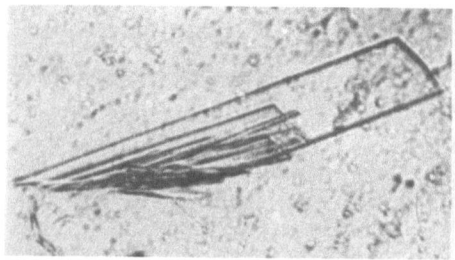 

Fig.1. Crystals of quinoprotein ethanol dehydrogenase from *Pseudomonas*. Left: *P. putida*; ethanol dehydrogenase (2 mg/ml) was crystallized from 10 mM glycin/NaOH, pH 8, in the presence of 2.5 M $(NH_4)_2SO_4$.
Right: *P. aeruginosa*; ethanol dehydrogenase (2 mg/ml) was crystallized from 10 mM glycin/NaOH, pH 8, containing 3 mM $CaCl_2$ in the presence of 22% polyethylene glycol 1550.

Relative molecular mass and subunit structure. The relative molecular mass M_r of ethanol dehydrogenase from *P. putida* was found to be 62 000 by gel electrophoresis in the presence of dodecyl sulfate. Two bands are detected after crosslinking with dimethyl suberimidate. The faster moving one corresponds to a M_r of 63 000 while the slower moving band shows a M_r of 120 000. Therefore the enzyme from *P. putida* is a dimer with subunits of identical size. Ethanol dehydrogenase from *P. aeruginosa* is also a dimeric enzyme with subunits of a M_r of 60 000. By dansylation only lysine was found as N-terminal amino acid and thus the two subunits are probably identical (Rupp and Görisch, 1988).

Absorption spectrum. The purified ethanol dehydrogenases from *P. putida* and *P. aeruginosa* show identical absorption spectra, which are similar to those of quinoprotein methanol dehydrogenases from methylotrophic bacteria. Both ethanol dehydrogenases show absorption maxima at 280 and 340 nm, with a shoulder at 290 nm. The absorption ratio A_{280}/A_{340} for both enzymes is 6.2.

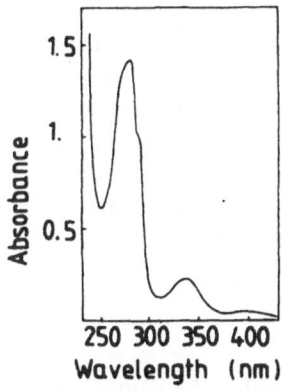

Fig. 2. Absorption spectrum of ethanol dehydrogenase from *P. aeruginosa*. The enzyme (1.1 mg/ml) was dissolved in 100 mM Tris/HCl buffer, pH 7, containing 100 mM NaCl and 10 mM CaCl$_2$.

Inactivation by suicide substrate. Cyclopropanone ethylhemiketal is a suicide substrate for quinoprotein alcohol dehydrogenases, which are able to oxidize secondary alcohols (Dijkstra et al., 1984). As shown in Fig.3, two molecules of suicide substrate will completely inactivate one molecule of dimeric ethanol dehydrogenase from *P. putida*. Therefore there are two independent catalytic centres per enzyme dimer with one molecule of PQQ per subunit. One molecule of ethanol dehydrogenase from *P. aeruginosa* is also inactivated by two molecules of cyclopropanone ethylhemiketal.

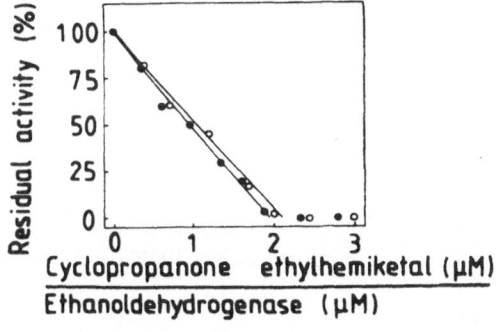

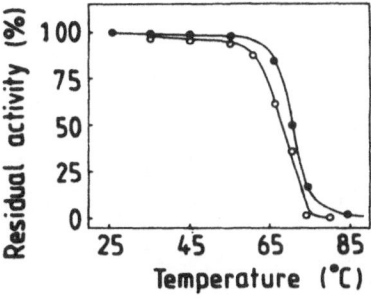

Fig.3. Left: Inactivation by suicide substrate. To solutions of ethanol dehydrogenase (6.5 μM) in 100 mM Na$_4$P$_2$O$_7$/HCl, 10 mM ethylamin, pH 9, were added solutions of various concentrations of cyclopropanone ethylhemiketal. After 15 min at room temperature, residual activity was determined: o-o-o ethanol dehydrogenase from *P. putida*; ●-●-● ethanol dehydrogenase from *P. aeruginosa*.
Right: Heat inactivation. Samples with 0.5 mg/ml of ethanol dehydrogenase in 100 mM Tris/HCl, 100 mM NaCl, pH 7, were treated for 10 min at the temperatures indicated. After cooling in ice, residual activities were determined: o-o-o ethanol dehydrogenase from *P. putida*; ●-●-● ethanol dehydrogenase from *P. aeruginosa*.

Amino-acid composition. The amino-acid composition of quinoprotein
ethanol dehydrogenase from *P. putida* is presented in Table 2. Based on a
molecular mass of 60 000 the subunit contains 575 to 590 residues. The
amino-acid composition of ethanol dehydrogenase from *P. aeruginosa* is also
shown. Both enzymes do not contain cysteine, both contain 21 tyrosine residues. With phenylalanine and tryptophan the numbers of residues determined differ by one. The number of aromatic amino acid residues found in
the enzyme from *P. aeruginosa* LMD 80.53 are quite different, Table 2.

Table 2. Amino-acid composition of ethanol dehydrogenases from
Pseudomonas.

Residue	Number of residues per subunit of ethanol dehydrogenases isolated from		
	P. aeruginosa[a] ATCC 17 933 (M_r 60 000)	*P. putida* ATCC 17 421 (M_r 60 000)	*P. aeruginosa*[b] LMD 80.53 (M_r 100 000)
asx	64	68	
thr	34	38	
ser	35	26	
glx	48	49	
pro	33	48	
gly	70	65	
ala	48	42	
val	38	40	
cys	0	0	
met	9	10	
ile	10	17	
leu	35	36	
tyr	21	21	27 (16)[c]
phe	27	28	33 (20)[c]
lys	40	46	
his	15	11	
arg	25	22	
trp	14	15	28 (17)[c]

[a]Data from Rupp and Görisch, 1988. Biol. Chem. Hoppe-Seyler 369: 431-439
[b]Data from Groen et al., 1984. Biochem. J. 223: 921-924
[c]Recalculated for a subunit molecular weight of 60 000.

Sedimentation coefficient. Ethanol dehydrogenase from *P. putida*
sediments in a single sharp boundary with a sedimentation coefficient of
$S_{20,W} = 6.7$ S. In active enzyme sedimentation experiments a value of
$S_{20,W} = 6.5$ S is found. The enzyme from *P. aeruginosa* shows the same
sedimentation properties (Rupp and Görisch, 1988).

Substrate specificity. The quinoprotein ethanol dehydrogenases from both *P. aeruginosa* and *P. putida* use a great number of primary and secondary alcohols and aldehydes as substrates, Table 3. With 2-butanol and 2-octanol one of the two enantiomers is preferentially oxidized. The enzyme from *P. aeruginosa* oxidizes also amino alcohols like ethanolamine and 1-amino-2-propanol. The K_m values for ethanol are 14 and 18 μM, while the K_m values for methanol are in the mM range, Table 3.

Table 3. Substrate specificity of ethanol dehydrogenases from *Pseudomonas*

	P. aeruginosa ATCC 17 933[a]	V', % (K_m, mM) P. putida ATCC 17 421[b]	P. aeruginosa LMD 80.53[c]
Primary alcohols:			
methanol	62 (94)	95 (2)	43 (8)
ethanol	100 (0.014)	100 (0.018)	100 (0.013)
2-chloro-ethanol	92	n.d.	
1-propanol	138 (0.021)	90	95 (0.01)
1-butanol	65	90	
1-hexanol	29	70	
1-octanol	72	70	80 (0.0035)
1-decanol	69	50	74 (0.002)
benzylalcohol	50	60	
Secondary alcohols:			
2-propanol	115 (0.68)	90	88 (0.64)
S(+)-2-butanol	87 (0.98)	70	
R(-)-2-butanol	41 (3.5)	40	
2-pentanol (racemate)	53	30	
S(+)-2-octanol	8	27	
R(-)-2-octanol	43	41	
Aldehydes:			
methanal	29	n.d.	90 (4)
ethanal	40 (4.5)	27	60 (0.12)
propanal	52	48	
heptanal	43	10	
octanal	n.d.	n.d.	60 (0.038)

[a]Data from Rupp and Görisch, 1988. Biol. Chem. Hoppe-Seyler 369: 431-439
[b]V'values have been determined at substrate concentrations of 5mM except for the higher alcohols hexanol and octanol (1 mM), in the presence of 0.1 mM PMS. V'values are reported with respect to the velocity found with ethanol.
[c]Data from Groen et al., 1984. Biochem. J. 223: 921-924.

Heat inactivation. When treated at various temperatures for 10 min the ethanol dehydrogenases from *P. putida* and *P. aeruginosa* show a quite similar behaviour, Fig.3. Both enzymes are stable up to 55°C. Above 55°C heat inactivation is observed and after 10 min at 70°C a residual activity of 30% is found with the enzyme from *P. putida*, while for the enzyme from *P. aeruginosa* a residual activity of 50% is found.

pH-Optimum. The ethanol dehydrogenases from *P. putida* and *P. aeruginosa* both show a sharp pH-optimum at pH 9, when the enzymatic activity is determined with PMS/DCPIP as electron acceptors. 40 mM H_3BO_3/40 mM sodium acetate/40 mM phosphoric acid/NaOH was used as test buffer between pH 6 and pH 10.5.

DISCUSSION

The quinoprotein alcohol dehydrogenases isolated from *P. aeruginosa* ATCC 17 933 and *P. putida* ATCC 17 421 are very similar with respect to their molecular and catalytic properties, Table 4. Both enzymes are dimers with one molecule of PQQ per subunit of M_r 60 000. The dehydrogenases show a high pH-optimum of pH 9, and ammonia or an amine is needed as activator. The enzymes oxidize a wide range of primary alcohols. The absorption spectra of both enzymes are identical with maxima at 280 and 340 nm. These spectra are similar to the spectrum of methanol dehydrogenase (EC 1.1.99.8) from methylotrophic bacteria. The dye-dependent alcohol dehydrogenases of the two *Pseudomonas* strains resemble in their molecular and catalytic properties the well characterized quinoprotein methanol dehydrogenases (Anthony, 1982). But in several respects they also differ from methanol dehydrogenase. Besides primary alcohols, also secondary alcohols are oxidized and higher aldehydes are also accepted as substrates, Table 3. Methanol, however is a poor substrate, and in addition to ammonia higher alkylamines are used as activators by the enzymes from *Pseudomonas*.

The quinoprotein alcohol dehydrogenases from *P. aeruginosa* ATCC 17 933 and *P. putida* ATCC 17 421 are induced when the organisms are grown on ethanol as sole source of carbon and energy. Therefore the enzymes will be regarded as ethanol dehydrogenases, since their usual function is the oxidation of ethanol, for which K_m values of 14 and 18 μM are found. The K_m values for methanol are in the mM range.

An enzyme with similar catalytic and spectral properties is induced by ethanol in *P. aeruginosa* LMD 80.53 (Groen et al., 1984), Table 4. This enzyme, however, is reported to be a monomer of M_r 100 000. Thus it appears, that two classes of quinoprotein ethanol dehydrogenases occur in *Pseudomonas*: The dimer of M_r 120 000 as found in *P. aeruginosa* ATCC 17 933 and *P. putida* ATCC 17 421 and the monomer as described for *P. aeruginosa* LMD 80.53 (Groen et al., 1984). *P. aeruginosa* and *P. putida* belong to the rRNA-homology group I of *Pseudomonas*. It will be interesting to learn, if members of other rRNA-homology groups of *Pseudomonas* also contain quinoprotein ethanol dehydrogenases.

A quinoprotein ethanol dehydrogenase with molecular and catalytic properties very similar to the two enzymes described in the present report was found in an unknown bacterium, which then was thought to be *Acinetobacter calcoaceticus* (Duine and Frank, 1981). The enzyme was induced, when the organism was grown on ethanol.

Table 4. Molecular properties of ethanol dehydrogenases from *Pseudomonas*.

	P. aeruginosa ATCC 17 933[a]	*P. putida* ATCC 17 421	*P. aeruginosa* LMD 80.53[b]
Molecular weight			
native enzyme	120 000	120 000	101 000
subunit	60 000	60 000	100 000
Sedimentation coefficient			
protein sedimentation	6.7 S	6.7 S	
active enzyme	6.5 S	6.5 S	
Number of (identical) subunits per native enzyme	2	2	1
Number of PQQ per native enzyme	2	2	2
Isoelectric point	8.2	9.1	>9
Absorption spectrum			
maxima (shoulder)	280 (290) 340	280 (290) 340	280 (290) 340
ratio A_{280}/A_{340}	6.2	6.2	6
Spec. activity (U/mg)[c]	19	23	17.5
Catalytic centre activity[c]	19	23	29
pH-Optimum	9	9	9.5
K_m (mM)			
ethanol	0.014	0.018	0.013
methanol	94	2	8

[a]Data from Rupp and Görisch 1988. Biol. Chem. Hoppe-Seyler 369: 431-439.
[b]Data from Groen et al., 1984. Biochem. J. 223: 921-924.
[c]With Wurster's blue as electron acceptor.

A dimeric quinoprotein ethanol dehydrogenase of M_r 120 000 with catalytic properties very similar to the *Pseudomonas* enzymes was also purified from *Rhodopseudomonas acidophila* grown aerobically on ethanol (Bamforth and Quayle, 1979). *R. acidophila* synthesizes the same enzyme when grown autotrophically in the light, using methanol as source of reducing power (Bamforth and Quayle, 1978).

Recently a dimeric alcohol dehydrogenase was isolated from an ethanol grown facultative methylotroph, *Pseudomonas BB1*. This enzyme shows a high affinity for ethanol but low affinity for methanol like the ethanol de-hydrogenases described above. However, when grown on methanol, *Pseudomonas BB1* synthesizes the same enzyme. It occurs as a monomer with M_r 60 000 and as a dimer, the ratio of both depending on the growth conditions (Dijkstra et al., 1985).

ACKNOWLEDGEMENTS

This work was supported in part by the Fonds der Chemischen Industrie.

REFERENCES

Anthony C, 1982. The Biochemistry of Methylotrophs. Academic Press, London.

Bamforth CW and Quayle JR, 1978. The Dye-Linked Alcohol Dehydrogenase of *Rhodopseudomonas acidophila*. Biochemical Journal 169: 677-686.

Bamforth CW and Quayle JR, 1979. Structural Aspects of the Dye-Linked Alcohol Dehydrogenase of *Rhodopseudomonas acidophila*. Biochemical Journal 181: 517-524.

Duine JA, Frank Jzn J and Verwiel PEJ, 1980. Structure and Activity of the Prosthetic Group of Methanol Dehydrogenase. European Journal of Bio-chemistry 108: 187-192.

Duine JA and Frank Jzn J, 1981. Quinoprotein Alcohol Dehydrogenase from a Non-methylotroph, *Acinetobacter calcoaceticus*. Journal of General Micro-biology 122: 201-209.

Dijkstra M, Frank Jzn J, Jongejan JA and Duine JA, 1984. Inactivation of quinoprotein alcohol dehydrogenases with cyclopropane-derived suicide substrates. European Journal of Biochemistry 140: 369-373.

Dijkstra M, van den Tweel WJJ, deBont JAM, Frank Jzn J and Duine JA, 1985. Monomeric and Dimeric Quinoprotein Alcohol Dehydrogenase from Alcohol-grown *Pseudomonas BB1*. Journal of General Microbiology 131: 3163-3169.

Groen B, Frank Jzn J and Duine JA, 1984. Quinoprotein alcohol dehydrogenase from ethanol-grown *Pseudomonas aeruginosa*. Biochemical Journal 223: 921-924.

McPherson A, 1982. Preparation and Analysis of Protein Crystals. Wiley, New York.

Rupp M and Görisch H, 1988. Purification, Crystallization and Characterization of Quinoprotein Ethanol Dehydrogenase from *Pseudomonas aeruginosa*. Biological Chemistry Hoppe-Seyler 369: 431-439.

Salisbury SA, Forrest HS, Cruse WBT and Kennard O, 1979. A novel coenzyme from bacterial primary alcohol dehydrogenases. Nature, London 280: 843-844.

Yang VC and Langer R, 1985. pH-Dependent Binding Analysis, A New and Rapid Method for Isoelectric Point Estimation. Analytical Biochemistry 147: 148-155.

PQQ-DEPENDENT METHANOL DEHYDROGENASE FROM CLOSTRIDIUM THERMOAUTOTROPHICUM.

Debra K. Winters and Lars G. Ljungdahl
Department of Biochemistry and Center for Biological Resource Recovery, University of Georgia, Athens, GA 30602, U.S.A.

KEY WORDS: Clostridium thermoautotrophicum, acetogens, methanol dehydrogenase, PQQ.

ABSTRACT: The incorporation of methanol into acetate by acetogenic bacteria has been investigated. A PQQ-dependent methanol dehydrogenase, that oxidizes methanol and formaldehyde to formate, has been purified from C. thermoautotrophicum.

INTRODUCTION

Anaerobic acetogenic bacteria, that grow autotrophically on H_2/CO_2, use the recently described Wood's acetyl-CoA pathway for the fixation of CO_2 [Wood et al. 1986; Ljungdahl, 1986]. They grow also on CO and on methanol plus CO_2. Not much is known about the metabolism of methanol in the acetogens, however, the following reactions may summarize it.

$$CH_3OH + H_2O \longrightarrow CO_2 + 6H^+ + 6e \qquad (1)$$

$$3CO_2 + 6H^+ + 6e \longrightarrow 3CO + 3H_2O \qquad (2)$$

$$3CH_3OH + 3CO \longrightarrow 3CH_3COOH \qquad (3)$$

$$\text{Sum: } 4CH_3OH + 2CO_2 \longrightarrow 3CH_3COOH + H_2O \qquad (4)$$

Here we report on results obtained with Clostridium thermoautotrophicum grown on ^{14}C or deuterium labeled methanol and on the purification and properties of a PQQ-dependent methanol dehydrogenase from this acetogen.

MATERIALS AND METHODS

C. thermoautotrophicum strain JW701/3 [Wiegel et al.,1981] was grown at $58^O C$ under 100% CO_2 in a medium previously described [Ljungdahl et al., 1985] but with 0.5% of methanol instead of glucose as substrate. Pseudomonas aeruginosa (ATCC 10145) was grown as described by Duine and Frank [1981]. From it was prepared the apoenzyme of glucose dehydrogenase, which was used for determination of PQQ by reconstitution [Duine et al., 1983; Ameyama et al., 1985].
    PQQ was obtained as gifts from O. Ghisalba (Ciba-Geigy Ltd) and J.A. Duine (Delft University). It was also prepared from C. thermoautotrophicum using methods described by Duine and Frank 1980]. The isolated compound had the same spectra and chromatographic behavior as PQQ. It reconstituted the glucose dehydrogenase apoenzyme of P aeruginosa. About 190 ug PQQ was

35

J. A. Jongejan and J. A. Duine (eds.), PQQ and Quinoproteins, 35–39.
© 1989 by Kluwer Academic Publishers.

obtained from 100 g of wet cells of C. thermoautotrophicum.

Methanol dehydrogenase of C. thermoautotrophicum was determined anaerobically using a dye-linked assay [Duine et al., 1987] at 55°C and pH 8.5 in a mixture consisting of 75 mM pyrophosphate, 65 uM DCPIP, 0.5 mM TMPD (Wurster's blue), and 1.5 mM methanol. Formaldehyde dehydrogenase activity was assayed similarly but with phosphate replacing pyrophosphate and with 1.5 mM formaldehyde. A unit of enzyme activity was defined as one nmol of DCPIP reduced/min. Protein was determined according to Elliott and Brewer [1978]. Formate and formaldehyde were determined with formyl-H_4folate synthetase [Ljungdahl et al., 1970] and formaldehyde dehydrogenase [Ogushi et al., 1986], respectively.

The purification of the methanol dehydrogenase from C thermoautotrophicum was under strictly anaerobic conditions [Yamamoto et al., 1983]. It involved the preparation of an extract using a French pressure cell, passage of the extract through a DEAE Sepharose column, ammonium sulfate fractionation, and chromatography on Phenyl Sepharose, Bio-Gel HTP, Ultrogel ACA 34 and an HPLC sizing column TSK-G4000SW from LKB Instrument, Inc.

RESULTS AND DISCUSSION

Acetate produced by C. thermoautotrophicum when grown on [^{14}C]methanol had a ^{14}C-labeling ratio (CH$_3$/COOH) of 3.3. With ^{14}CO$_2$ instead of methanol this ratio was 0.7. In these experiments methanol was oxidized to CO$_2$ and CO$_2$ was converted to methanol. With CD$_3$OH as substrate, unlabeled, mono, di, and tri deuterated acetates were formed in the proportions 38, 24, 15, and 23, respectively. Methanol is clearly a more direct precursor of the methyl group of acetate than of the carboxyl group. It is incorporated directly into the methyl group without loss of hydrogen, but it apparently is also oxidized to formaldehyde and formate. The results suggest a metabolism of methanol in acetogens as outlined in Fig.1. Here the synthesis

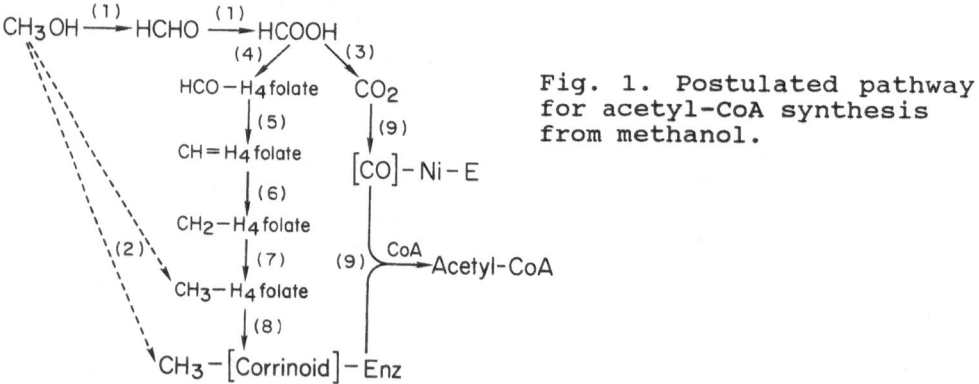

Fig. 1. Postulated pathway for acetyl-CoA synthesis from methanol.

of acetyl-CoA from CO$_2$ is shown. [CO]-Ni signifies CO bound to the nickel of carbon monoxide dehydrogenase, that catalyzes the condensation of the methyl group on the corrinoid-enzyme, CO, and CoA to form acetyl-CoA (9). The methyl group of methanol may be incorporated into acetate via methyl-H_4folate or more directly via the corrinoid-enzyme (2). The latter alternative is the most likely [van der Meijden et al., 1984]. It would involve a methyl-corrinoid transferase similar to the methanol: 5-hydroxybenzimidazolylcobamide methyltransferase found in methanogens. A PQQ-dependent methanol dehydrogenase is present in C. thermoautotrophicum [Duine et al., 1984]. It oxidizes methanol to formaldehyde and formate (1). Formaldehyde may then react chemically with H_4folate to form methylene-H_4folate and formate yields formyl-H_4folate (4), both of which are precursors to the methyl group of acetate. To form the carboxyl group of acetate, formate must first be oxidized to CO_2 and reduced to CO (3 and 9).

The methanol dehydrogenase of C.thermoautotrophicum was purified from an extract of 90 g of wet cells about 100-fold to a spec. activity of 0.83. During the purification PQQ slowly dissociated from the enzyme, which when pure had a spec. activity of 0.1 when assayed without preincubation with PQQ. The enzyme has an M_r of 110,000. It is composed of 2 apparently identical subunits and binds 2 mol each of PQQ and zinc per mol. Methanol and formaldehyde are about equally good substrates with K_ms of 50 uM and 40 uM, respectively. They are almost completely oxidized to formate in the regular assay mixture. A small amount of formaldehyde is formed from methanol which indicates that it is an intermediate. Primary alcohols from ethanol to n-hexanol, 2-methoxy ethanol, ethylene glycol, acetaldehyde, and propionaldehyde are good substrates, whereas secondary alcohols are slowly oxidized if at all. The enzyme is active between pH 7 and 10 with an optimum at about 8.2. It is rather oxygen sensitive; most of its activity is lost within a day if it is stored in air. In contrast to many other methanol dehydrogenases the C. thermoautotrophicum enzyme is not activated by NH_4^+ or amines. They are in fact inhibitory.

Duine et al. (1984) suggested that clostridial methanol dehydrogenase could be NAD(P)-dependent, however, neither NAD, NADP, FAD, nor FMN influenced the enzyme assay or replaced PQQ in restoring activity. Attempts to localize the enzyme in the cell indicated that it is loosely connected to the membrane. The activity was about equally distributed between the membrane and the cytoplasmic fractions. ATPase, tightly bound to the membrane, and formyl-H_4folate synthetase, a cytoplasmic enzyme were used as controls (Hugenholtz et al., 1987).

Clostridium thermoaceticum like many other acetogens grows on methanol (Wiegel and Garrison, 1985) and it has a methanol dehydrogenase similar to that described here (Winters Ivey, 1987). Clearly, these bacteria have much larger metabolic diversity than what was considered based on original observations with C. thermoaceticum. The findings reported here leads to speculations that the acetogens may have 2 routes for the synthesis of the methyl group of acetate, the well described H_4folate pathway and the reduction of formate to formaldehyde or methanol by methanol dehydrogenase.Furthermore, the fact that ethanol and acetaldehyde are oxidized to acetate by the methanol dehydrogenase suggest that these compounds and other alcohols may serve as substrates for acetogenic bacteria.

38

REFERENCES

Ameyama M, Nonobe M, Shinagawa E, Matsushita K and Adachi O, 1985. Method of enzymatic determination of pyrroloquinoline quinone. Anal. Biochem. 151:263-267.

Duine JA and Frank J, 1980. The prosthetic group of methanol dehydrogenase: purification and some of its properties. Biochem. J. 187:221-226.

Duine JA and Frank J, 1981. Quinoprotein alcohol dehydrogenase from a non-methylotroph, Acinetobacter calcoaceticus. J. Gen. Microbiol. 122:201-209.

Duine JA, Frank Jsn J and Dijkstra M, 1987. Quinoproteins in the dissimilation of C_1 compounds. In: van Verseveld HW and Duine JA. Microbial growth on C_1 compounds, pp. 103-112. Martinus Nijhoff Publ. Dordrecht.

Duine JA, Frank J and Jongejan JA, 1983. Detection and determination of pyrroloquinoline quinone, the coenzyme of quinoproteins. Anal. Biochem. 133:239-243.

Duine JA, Frank Jzn J, Jongejan JA and Dijkstra M, 1984. Enzymology of the bacterial methanol oxidation step. In: Crawford RL and Hanson RS. Microbial growth on C_1 compounds, pp. 91-96.Am. Soc. Microbiol. Washington.

Elliott JI and Brewer JM, 1978. The inactivation of yeast enolase by 2,3-butane dione. Arch. Biochem. Biophys. 190:351-357.

Hugenholtz J, Ivey DM and Ljungdahl LG, 1987. Carbon monoxide-driven electron transport in Clostridium thermoautotrophicum membranes. J. Bacteriol. 169:5845-5847.

Ljungdahl LG, 1986. The autotrophic pathway of acetate synthesis in acetogenic bacteria. Ann. Rev. Microbiol. 40:415-450.

Ljungdahl LG, Brewer JM, Neece SH and Fairwell T, 1970. Purification, stability, and composition of formyltetrahydrofolate synthetase from Clostridium thermoaceticum. J. Biol. Chem. 245:4791-4797.

Ljungdahl LG, Carreira LH, Garrison RJ, Rabek NE and Wiegel J, 1985. Comparison of three thermophilic acetogenic bacteria for production of calcium-magnesium acetate. Biotech. Bioeng. Symp. 15:207-223.

Ogushi S, Ando M and Tsuru D, 1986. Formaldehyde dehydrogenase from Pseudomonas putida: the role of a cysteinyl residue in the enzyme activity. Agric. Biol. Chem. 50:2503-2507.

van der Meijden P, van der Drift C and Vogels GD, 1984. Methanol conversion in Eubacterium limosum. Arch. Microbiol. 138:360-364.

Wiegel J, Braun M and Gottschalk G, 1981. Clostridium thermoautotrophicum species novum, a thermophile producing acetate from molecular hydrogen and carbon dioxide. Curr. Microbiol. 5:255-260.

Wiegel J and Garrison R, 1985. Utilization of methanol by Clostridium thermoaceticum. Abstr. Annu. Meet. Am. Soc. Microbiol. I115, p. 165.

Winters Ivey DK, 1987.Metabolism of methanol in acetogenic bacteria. Dissertation, Univ. of Georgia. Athens. Georgia.

Wood HG, Ragsdale SW and Pezacka E, 1986. The acetyl-CoA pathway: a newly discovered pathway of autotrophic growth. Trends Biochem. Sci. 11:14-18.

Yamamoto I, Saiki T, Liu SM and Ljungdahl LG,1983. Purification and properties of NADP-dependent formate dehydrogenase from Clostridium thermoaceticum, a tungsten-selenium-iron protein. J.Biol.Chem. 258:1826-1832.

NEW PQQ-ENZYME: AROMATIC ALCOHOL AND ALDEHYDE DEHYDROGENASES IN _RHODOPSEUDOMONAS ACIDOPHILA_ M402.

Kei Yamanaka
Inistitute of Applied Biochemistry, University of Tsukuba, Tsukuba, Ibaraki, 305 Japan.

Key words: Phototrophic bacteria, Aromatic alcohol dehydrogenase, Aromatic aldehyde dehydrogenase, PQQ, _Rhodopseudomonas acidophila_

Abstract: Two NAD^+-independent, dye-linked aromatic alcohol and aldehyde dehydrogenases of _Rhodopseudomonas acidophila_ M402 were purified. PQQ was demonstrated in both enzymes: the characteristic absorption spectra and activation by PQQ on apo-dehydrogenase preparations. Anaerobically-grown cells do not show the dehydrogenation activity on any alcohols, but the activities were appeared when the bacterium grew on the medium supplemented PQQ under anaerobic conditions.
Absorption spectrum of the purified aromatic alcohol dehydrogenase showed absorption maxima at 277, 416, 523 and 552 nm with a shoulder at 290 nm. This indicates the presence of PQQ and also cytochrome(s). Therefore, this enzyme should be classified as quinohemoprotein aromatic alcohol dehydrogenase.

INTRODUCTION
        Primary alcohol dehydrogenases were found from a variety of methylotrophic bacteria and also from non-methylotrophic bacteria(Groen et al., 1984). PQQ was demonstrated as cofactor for these enymes. Bamforth and Quayle(1978) found another type of dye-linked alcohol dehydrogenase from a phototrophic bacterium, _Rhodopseudomonas acidophila_ strain 10050. We also found a new dye-linked alcohol dehydrogenase which was active on aromatic and aliphatic alcohols, but was inert on methanol, from an isolated strain of _Rhodopseudomonas acidophila_ strain M402 (Yamanaka et al. 1983). Molecular properties including substrate specificities were compared. The latter enzyme has the high affinity on aromatic alcohols. In this paper, we describes the participation of PQQ in this aromatic alcohol dehydrogenase.

J. A. Jongejan and J. A. Duine (eds.), PQQ and Quinoproteins, 40–42.
© _1989 by Kluwer Academic Publishers._

MATERIALS AND METHODS

Organism and growth conditions: _R. acidophila_ M402 was used throuout in this experiment(Yamanaka et al.,1983).Culture was carried out with malate-mineral medium supplemented 5 mM vanillyl alcohol under aerobic-dark conditions in a 10-liter jar fermantor(MBS MF-114) at 30° C. Ultracentrifugation was carried out with Beckman Model L5-50B at 149,000xg for 90 min to remove interferring pigments.

Enzyme assay: The dehydrogenase activities were measured with DCPIP reduction at 30°C, pH 7.0 with vanillyl alcohol for alcohol dehydrogenase, and _m_-anisaldehyde for aldehyde dehydrogenase as substrates(Yamanaka and Tuyuki, 1984).

Detection of PQQ: Methanol extracts were prepared from and purified on a Sephadex G-25 column and detected at 249 nm. Content of PQQ was assayed by HPLC , and with apo-glucose dehydrogenase of membrane of _E. coli_.

Preparation of apo-dehydrogenase: Purified alcohol and aldehyde dehydrogenase preparations were treated with 3M guanidine.These preparations had no activity without PQQ.

RESULTS AND DISCUSSION

PMS-dependent two dehydrogenases were purified with the similar procedures to a single protein band on disc electrophoresis.Successful separation of two enzymes was achieved with the chromatography on DEAE-celluose at pH 8.0 based on the difference of their isoelectric points(Tables 1 and 2).

Detection of PQQ: Crude cell-free extracts and cultural filtrate (did not contain yeast extract) were chromatographed on Sephadex G-25. One peaak of A $_{249}$ was obtained at the same elution location of that PQQ. These fractions were active for the activation of apo-glucose dehydrogenase of _E. coli_.

Activation of apo-dehydrogenases by PQQ.Activities of two apo-dehydrogenase preparations were recovered by addition of PQQ. 100 nM of PQQ was almost saturated for recovery.

TABLE 1   Purification   of   Aromatic Alcohol   Dehydrogenase

Step	Protein (mg)	Dehydrogenase activity Total units	Sp.act(u/mg)	Fold	Yield (%)
Crude extract	9,380	778.5	0.096	1.0	100
Ammonium sulfate ppt.	9,080	454.2	0.05	0.5	58.3
DEAE-cellulose(pH 7.5)	4,580	644.7	0.14	1.5	82.8
DEAE-cellulose(pH 8.0)	276	466.1	1.68	17.6	59.9
BIOGEL HTP	46	422.1	9.10	95.7	54.2
Sepharose CL-6B	22	325.5	17.49	182.2	41.8

TABLE 2   Purification of Aromatic Aldehyde Dehydrogenase

Step	Protein (mg)	Dehydrogenase activity Total units	Sp.act(u/mg)	Fold	Yield (%)
Crude extract	2810	1724	0.61	1	100
Ultracentrifugation	2280	1872	0.82	1.3	109
DEAE-cellulose(pH 7.5)	376	1383	3.68	6.	80
DEAE-cellulose(pH 8.0)	38	702	18.47	30.3	41
BIOGEL HTP		8.6309	35.93	58.9	18
Sepharose CL-6B		1.1 84	76.36	125.2	5

Induction of dehydrogenase by PQQ under anaerobic-light
conditions.

TABLE 3. Effect of PQQ on the Induction of Dehydrogenases.

Carbon source in medium	Growth (A_{650})	Protein (mg)	Dehydrogenase activity (units) Vanillyl alcohol		Anisaldehyde	
			-PQQ	+PQQ	-PQQ	+PQQ
Malate	0.07	–	n.d.	n.d.	n.d.	n.d.
M + V	0.01	0.62	0.06	0.04	0.46	0.6
M + Yeast ext.	1.95	96.8	n.d.	n.d.	1.05	0.9
Vanillyl alcohol	0.01	–	n.d.	n.d.	n.d.	n.d.
V + Y	0.47	53.4	n.d.	n.d.	n.d.	n.d.
M + V + Y	1.35	144.4	0.99	1.14	3.94	52.5
M + PQQ	0.01	0.49	0.20	n.d.	0.14	n.d.
M + Y + PQQ	1.56	103.0	3.6	2.3	43.0	42.8
V + PQQ	0.01	0.28	n.d.	n.d.	n.d.	n.d.
V + Y + PQQ	0.73	101.0	10.4	7.6	21.9	25.0
M + V + Y + PQQ	1.38	102.7	3.4	3.0	31.8	41.5

*PQQ was added 910 ng/100 ml of medium.

REFERENCES
Bamforth C W and Quayle J R, 1978. Biochem. J. 169: 677.
Bamforth C W and Quayle J R, 1979. Biochem. J 181: 517-524.
Groen B, Frank J Jzn and Duine J A, 1984. Biochem.J. 223: 921-924.
Groen B W, van Kleef M A G and Duine J A, 1986. Biochem J. 234: 611-615.
Yamanaka K, Moriyama M, Minoshima R and Tsuyuki Y, 1983. Agric. Biol.Chem. 47:1257-1267.
Yamanaka K and Tsuyuki Y,1983. Agric.Biol.Chem 47: 2173-2183.

STRUCTURAL STUDIES ON THE PQQ-LIKE COFACTOR
OF NITROALKANE OXIDASE FROM FUSARIUM OXYSPORUM

Katsuyuki Tanizawa, Takeo Moriya, Toshiko Kido, Hidehiko Tanaka, and
Kenji Soda, Laboratory of Microbial Biochemistry, Institute for Chemical
Research, Kyoto University, Uji, Kyoto 611, Japan

Key words: nitroalkane oxidase, PQQ-like compound, FAD

ABSTRACT

Nitroalkane oxidase from Fusarium oxysporum exhibits absorption maxima
at 274, 340 and 450 nm, and shoulders at 380 and 470 nm. The absorption
and fluorescence spectra of the enzyme suggest the presence of both FAD and
a PQQ-like compound as cofactors. The bound FAD is easily resolved from
the enzyme by dialysis against 1 M KBr. The KBr-treated enzyme has no
absorption peak in the 380-500 nm region, but still exhibits the absorption
peak at 340 nm. The 340-nm chromophore can be isolated from the enzyme
protein only by treatment with protein denaturants. Although the isolated
compound shows spectrophotometric properties similar to PQQ, it does not
activate apoglucose dehydrogenase. The compound also differs from PQQ in
chromatographic properties. However, mass spectroscopic analysis indicates
that it is composed of mass fragments very close to PQQ. These results
suggest that nitroalkane oxidase contains a new cofactor, which is not
identical with PQQ but structurally similar to it.

INTRODUCTION

In studies of the microbial metabolism of nitro compounds, we have
found two new FAD enzymes catalyzing the decompsition of various nitroalkanes; 2-nitropropane dioxygenase from a yeast, Hansenula mrakii (Kido et
al., 1976) and nitroalkane oxidase from a fungus, Fusarium oxysporum (Kido
et al., 1978). The latter enzyme was shown to catalyze the oxidative
denitrification of several nitroalkanes producing alkyl aldehydes, nitrite,
and hydrogen peroxide (Kido et al., 1978):

$$RCH_2-NO_2 + O_2 + H_2O \longrightarrow RCHO + HNO_2 + H_2O$$

The absorption spectrum of the purified nitroalkane oxidase indicated the
presence of an unknown chromophore, which resembled PQQ, in addition to FAD
in the enzyme protein (Tanizawa and Soda, 1987). In this report, we describe structural studies on the PQQ-like compound firmly bound with nitroalkane oxidase from F. oxysporum.

MATERIALS AND METHODS

Chemicals. Authentic PQQ was purchased from Mitsubishi Gas Chemical
Co., Tokyo, Japan. Synthetic derivatives of PQQ were kindly provided by

J. A. Jongejan and J. A. Duine (eds.), PQQ and Quinoproteins, 43–45.
© 1989 by Kluwer Academic Publishers.

Dr. O. Adachi. The other chemicals were of highest purity commercially
available.

   Enzyme preparation and assay. Nitroalkane oxidase was purified to
homogeneity from a cell-free extract of F. oxysporum as described previously (Tanizawa and Soda, 1987). The enzyme was assayed by measurement of the
nitrite formed according to the method of Tanizawa et al. (1980).

   Isolation of the PQQ-like compound. The purified enzyme (100 mg) was
incubated at 37°C with 6 M guanidine hydrochloride for 6 h. The denatured
enzyme solution was applied on a Sephadex G-25 column (1 x 50 cm). Fractions containing compounds with low molecular weight (Mr 200-1000) were
pooled and lyophilized. The residue was dissolved in a small volume of
0.1% trifluoroacetic acid and chromatographed on an Ultron N-C$_{18}$ reversed
phase column (0.46 x 15 cm, Shinwa Kako, Kyoto) with a Jasco HPLC system.
A 40-min linear gradient of 0-80% acetonitrile in 0.1 % trifluoroacetic
acid was employed for elution at a flow rate of 0.8 ml/min. The elution
was monitored with a Waters 990J photodiode array detector.

   Mass spectroscopy. Mass spectra of the authentic PQQ and the isolated
compound were obtained with a Shimadzu QP1000 GC-MS equipped with a wide
bore capillary column (0.58 mm x 15 m). Ion sources were either electron
impact (E.I.) or chemical ionization (C.I.). Electron energy and ion
source temperature were 70eV and 250°C, respectively.

RESULTS AND DISCUSSION

   The purified nitroalkane oxidase exhibits absorption maxima at 274,
340 and 450 nm, and shoulders at 380 and 470 nm. The absorption spectrum
in the 380-500 nm region indicates the presence of FAD as a coenzyme.
However, when the enzyme was dialyzed against 1 M KBr at pH 7.0 to resolve
the bound FAD, the resulting inactive enzyme still showed an absorption at
around 340 nm. The chromophore was removed from the enzyme only by denaturation of the enzyme protein with 6 M guanidine hydrochloride; the denatured protein isolated by gel filtration with Sephadex G-25 had no absorption at 340 nm and the separated chromophore showed a spectrum with an
absorption maximum at about 350 nm. These results suggest that the enzyme
contains not only FAD but also a PQQ-like compound, which is non-covalently
but strongly bound with the enzyme.
   In the reversed phase HPLC, the PQQ-like compound was eluted at 28.6
min under the conditions employed, whereas the authentic PQQ appeared at
23.5 min under the same conditions, suggesting that it is more hydrophobic
than PQQ. In addition, the isolated compound did not activate apoglucose
dehydrogenase both from E. coli and Acinetobacter calcoaceticus. Various
synthetic derivatives of PQQ available were also examined by HPLC, but none
of them was co-chromatographed with the compound. These results clearly
show that the chromophore contained in nitroalkane oxidase is neither PQQ
nor the known PQQ derivatives. However, mass spectroscopic analysis of the
isolated compound indicated that it is composed of mass fragments very
close to PQQ.
   The native enzyme was irreversibly inactivated by incubation with
carbonyl reagents such as phenylhydrazine and 2,4-dinitrophenylhydrazine.
The phenylhydrazone of the PQQ-like compound was isolated by HPLC with a
40-min linear gradient of 0-100% methanol in 10 mM sodium phosphate buffer
(pH 7.0) containing 10 mM ammonium chloride. It showed a spectrum similar

to that of the phenylhydrazine-adduct of the authentic PQQ (van der Meer et al., 1987). Thus, it is concluded that the 340-nm compound in nitroalkane oxidase has an o-quinone structure like PQQ. Based on the results described above and those reported by Shinagawa et al. (1986), the compound may be a PQQ-derivative with a hydrophobic modification at the C-9 carboxyl group.

REFERENCES

Kido T, Soda K, Suzuki T and Asada K, 1976. A new oxygenase, 2-nitropropane dioxygenase of Hansenula mrakii. Enzymologic and spectrophotometric properties. J. Biol. Chem. 251: 6994-7000.

Kido T, Hashizume K and Soda K, 1978. Purification and properties of nitroalkane oxidase from Fusarium oxysporum. J. Bacteriol. 133: 53-58.

Shinagawa E, Matsushita K, Nonobe M, Adachi O, Ameyama M and Ohshiro Y, 1986. The 9-carboxyl group of pyrroloquinoline quinone, a novel prosthetic group, is essential in the formation of holoenzyme of D-glucose dehydrogenase. Biochem. Biophys. Res. Commun. 139: 1279-1284.

Tanizawa K, Hirasawa T and Soda K, 1980. Enzymatic microdetermination of nitroalkanes. Anal. Lett. 13: 645-654.

Tanizawa K and Soda K, 1987. Nitroalkane oxidase: function and coenzyme. In: Biochemistry of PQQ and quinoproteins. Yabuta Seminar (Abstracts), Kyoto, pp. 29-32.

van der Meer RA, Jongejan JA and Duine JA, 1987. Phenylhydrazine as probe for cofactor identification in amine oxidoreductases. Evidence for PQQ as the cofactor in methylamine dehydrogenase. FEBS Lett. 221: 299-304.

REGULATION OF METHANOL DEHYDROGENASE SYNTHESIS IN *PARACOCCUS DENITRIFICANS*

N. Harms, R.J.M. van Spanning, L.F. Oltmann, A.H. Stouthamer, Department of Microbiology, Biological Laboratory, Free University, P.O. Box 7161, 1007 MC Amsterdam, The Netherlands.

KEY WORDS
Methanol dehydrogenase, *Paracoccus denitrificans*, MDH operon

ABSTRACT

The region downstream from the methanol dehydrogase (MDH) structural gene has been cloned and sequenced. MDH promoter activity have been studied by using a broad-host-range promoter probe vector.

INTRODUCTION

We recently reported the isolation and the nucleotide sequence of the methanol dehydrogenase (MDH) structural gene from *P. denitrificans* (Harms et al., 1987). Downstream from this gene the start of a second open reading frame was found. In the serine-type methylotroph *Methylobacterium* sp. strain AM1 the MDH structural gene, *moxF*, is located in one operon with three other genes, *moxJ*, *moxG* and *moxI* (Anderson and Lidstöm, 1988). These genes encode proteins with molecular weights of 30kD, 20kD and 12kD, respectively.

In a physiological study on the regulation of MDH synthesis in *P. denitrificans* we have found that two separate mechanisms are involved in this process (de Vries et al., 1988).

In order to study the regulation of MDH oxidation in *P. denitrificans* we started to clone and sequence the full MDH operon of this organism and to study the regulation of MDH synthesis on a molecular level.

MATERIALS AND METHODS

The bacteria used in this study were: *E. coli* SM10: (Simon et al., 1983); *P. denitrificans* PD1222, a restriction-deficient derivative of the wild type strain PD1103 (Harms, 1988). Media and growth conditions for *P. denitrificans* and *E. coli* have been described previously (Harms et al., 1987).

DNA manipulations were performed as described previously (Harms et al., 1987).

Conjugations were performed as follows: A mixture of 100µl of donor and 100µl of recipient cells from exponentially growing bacterial cultures was spread on a LB agar plate. After incubation for 48 h at 35°C the cells were scraped from the plates and plated in appropiate dilutions on selective plates.

ß-Galactosidase activity was determined as described by Miller (Miller, 1972).

J. A. Jongejan and J. A. Duine (eds.), PQQ and Quinoproteins, 46–49.

RESULTS AND DISCUSSION

   <u>Cloning and nucleotide sequence of the region downstream from the</u>
<u>MDH structural gene.</u> The full MDH structural gene has been reported to
be located on the clone pNH3 (Harms et al., 1987). The insert of this
clone was used as a hybridization probe to find additional clones linked to this region. One clone, pNH33, was found that appeared to contain
a stretch of DNA, 3.4 kbp in length, downstream from the MDH gene (Fig
1). Nucleotide sequence analysis of this clone revealed that downstream
from the MDH gene an open reading frame (ORF-2) is located which
encodes a protein of 29 kD (Fig 2). At the N-terminal part of this protein 24 amino acids were found that could form a signal sequence. Downstream from ORF-2 a start codon and a putative Shine and Dalgarno sequence could be found. This indicates that a third open reading frame
might be present. The MDH operon of *P. denitrificans*, therefore, shows
resemblance with the *moxFJGI* operon of *Methylobacterium* sp. strain AM1
(Anderson and Lidström, 1988).

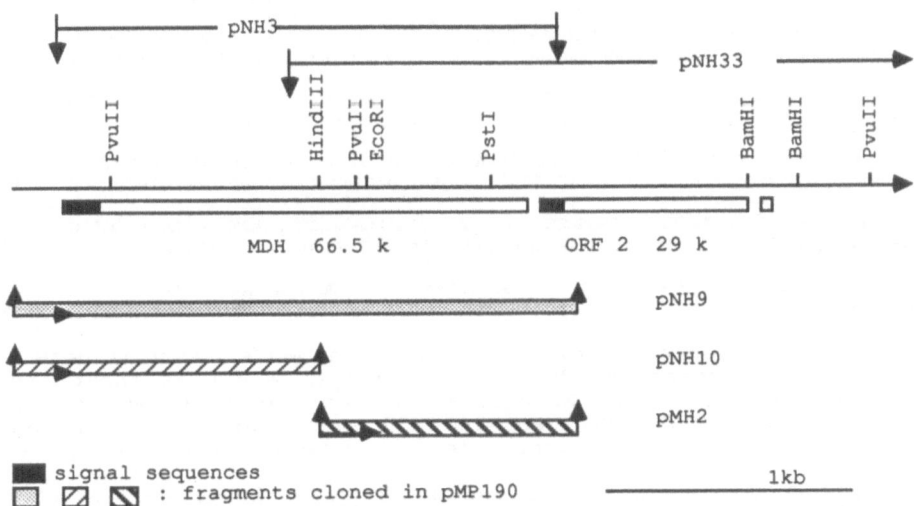

Fig 1. The MDH operon of *P. denitrificans*.

   Between the MDH gene and ORF-2 a stretch of 75 nucleotides was
found that contained a long inverted repeat, indicating the possibility
of transcription termination at that point (Harms et al., 1987). When
ORF-2 forms one operon with the MDH gene this inverted repeat might be
a regulatory signal. To shed more light on the regulation of methanol
oxidation we have started transcriptional studies on the MDH gene and
following ORF's.
   <u>Regulation of MDH synthesis.</u> We have used the broad host range
promoter probe vector pMP190 (Zaat et al., 1987) to clone and
characterize promoters in *P. denitrificans*. This vector is a derivative
of the IncQ plasmid RSF1010 and may be transferred to restriction-
deficient *P. denitrificans* strains with high frequency. Promoter
activity is coupled to ß-galactosidase production, so that regulation
can be studied in the parent strain without disruption of the parent
gene.

```
ACCTGACGCGGCGGGCCGGCCCGCCCTTCCTTCCCGGCGGGCCGGCCTGCGTGACCCTATTCAGTCTGGAGAAACCGAA2080
Tyr                                                                            2158

ATGCTCATTGATTTCCGCCAGGTTTGCGGCGCGGGGGCGGCGGCCCTGGCCCTTGCCTCGCCCGCTCTGGCA*GACACG 2159
MetLeuIleAspPheArgGlnValCysGlyAlaGlyAlaAlaAlaLeuAlaLeuAlaSerProAlaLeuAla AspThr 2236

ACGAACCTGCGCGTCTGCGCCTCGACCAAGGACGCGCCCTTCTCGGACGCGCAGGGCGCCGGCTTTGAAAACAAGATC  2237
ThrAsnLeuArgValCysAlaSerThrLysAspAlaProPheSerAspAlaGlnGlyAlaGlyPheGluAsnLysIle  2314

GCCCAAGTGCTGGCCGACGAGATGGGCGCCACGCTTGATCTGGTGTGCTGGAAAAAGACGCGATCTATCTGGTCCGCA  2315
AlaGlnValLeuAlaAspGluMetGlyAlaThrLeuAspLeuValMetLeuGluLysAspAlaIleTyrLeuValArg  2392

GACGGCATCGAAAAGGATCTGTGCGACGTTCTGGTCGGCGTGGATGCGGGCGACGAGCGGCTGCTGACCACCCGGCCC  2393
AspGlyIleGluLysAspLeuCysAspValLeuValGlyValAspAlaGlyAspGluArgLeuLeuThrThrArgPro  2470

TATTACCGCTCGGGCTATGCCTTCGTCACGCGGCAGGACCGCAACTTCGAGGGCGACAAGTGGCAGGACGTGGATCAG  2471
TyrTyrArgSerGlyTyrAlaPheValThrArgGlnAspArgAsnPheGluGlyAspLysTrpGlnAspValAspGln  2548

GAGGGGTTCGACACCTTCTCGTATCGCCTGCATTCCCCGGCCGAGACGATCCTGAAATATACCGGCCGCTACGAATAC  2549
GluGlyPheAspThrPheSerTyrArgLeuHisSerProAlaGluThrIleLeuLysTyrThrGlyArgTyrGluTyr  2626

AACCTGATCTACCAGGCCTCGCTGACCAATTTCGAGGACCGGCGAAACAAATACACCCAGGTCGAGGCCAGCCGCGTC  2627
AsnLeuIleTyrGlnAlaSerLeuThrAsnPheGluAspArgArgAsnLysTyrThrGlnValGluAlaSerArgVal  2704

ATCACGGAGGTCGCGGATGGCGGGGCCGATCTGGCCATCGTCTTCGCCCCCGAGGCGGCGCGCTATGTCCGCGACTCG  2705
IleThrGluValAlaAspGlyGlyAlaAspLeuAlaIleValPheAlaProGlaAlaAlaArgTyrValArgAspSer  2782

CGCGAGCCCTTGCGTATGACCCTGATTACCAATGAAATCGAACGCTCGGACGGCGTCATCATCCCGCTGCAATATTCG  2783
ArgGluProLeuArgMetThrLeuIleThrAsnGluIleGluArgSerAspGlyValIleIleProLeuGlnTyrSer  2860

CAGTCCGTGGGCGTGTCCAAGACCCATCCCGAGCTTCTGGGGCCGATCGAACAGGCCCTGCAATCCGGCAAAGCCAGG  2861
GlnSerValGlyValSerLysThrHisProGluLeuLeuGlyProIleGluGlnAlaLeuGlnSerGlyLysAlaArg  2938

ATTGATGCCATCCTGACCGAGGAAGGCATCCCCCTGCTGCCGTCAAGCTGAAGGAAATGGAATGACGAAG           2839
IleAspAlaIleLeuThrGluGluGlyIleProLeuLeuProSerSer         MetThrLys              3009
```

Fig.2. Nucleotide sequence downstream from the MDH structural gene. The MDH gene ends at nucleotide 2082. The inverted repeat is indicated by underlining. The start of ORF-2 is at nucleotide 2159. The start of ORF-3 is at nucleotide 3001. Possible SD sequences are indicated in bold. The putative signal sequence cleavage site is indicated by an *.

Three DNA fragments of clone pNH3 were inserted in pMP190 resulting in the clones pNH9, pNH10 and pMH2 (Fig 1). These clones were transferred to PD1222. ß-Galactosidase production was determined after growth on a variety of carbon sources. In Table 1 it is shown that promoter activity by clone pNH9 was present under conditions that MDH can be found (de Vries et al., 1988), but absent under conditions of MDH repression. Some promoter activity could be found in PD1222/pNH9 cells incubated during 16 hours in a medium without carbon source (results not shown). The catabolite repression-like mechanism might be responsible for this low level of ß-galactosidase expression. In PD1222/pNH10 cells less ß-galactosidase activity could be found compared with PD1222 cells harboring clone pNH9. This may be caused by differences in copy number or by effects of the C-terminal part of the MDH gene on the MDH promoter. The long inverted repeat downstream from the MDH gene seems not to function as transcription terminator under MDH inducing conditions. Wild type cells containing pMH2 show no promoter activity, which might indicate that no promoter is present in this region, and that the MDH structural gene and ORF-2 are located in one operon, just as was found in *Methylobacterium* sp. strain AM1.

Experiments are in progress to characterize the MDH promoter activity further. These studies together with RNA studies might give more insight in the regulation of methanol oxidation in *P. denitrificans*.

Table 1. Activity of ß-galactosidase in *P. denitrificans* containing different plasmids and grown on different carbon sources.

strain/plasmid	growth			
	MeOH	MeNH$_2$	Choline	Glucose
PD1222/pNH9	++	++++	+++	-
PD1222/pNH10	nd	+	+	-
PD1222/pMH2	nd	-	-	-

nd : not determined;
-:1-10; +:20-40; ++:40-80; +++:80-170; ++++:170-200 Miller units

REFERENCES

Anderson DJ, and Lidström ME, 1988. The *MoxFG* region encodes four polypeptides in the methanol-oxidizing bacterium *Methylobacterium* sp. strain AM1. J Bacteriol 170:2254-2262.

Harms N. 1988. Genetic and physiological studies on methanol metabolism in *Paracoccus denitrificans*. Thesis Vrije Universiteit Amsterdam, The Netherlands.

Harms N, de Vries GE, Maurer K, Veltkamp E, and Stouthamer AH. 1985. Isolation and characterization of *Paracoccus denitrificans* mutants with defects in the metabolism of one-carbon compounds. J Bacteriol 164:1064-1070.

Harms N, de Vries GE, Maurer K, Hoogendijk J and Stouthamer AH, 1987. Isolation and nucleotide sequence of the methanol dehydrogenase structural gene from *Paracoccus denitrificans*. J Bacteriol 169:3969-3975.

Miller JH. 1972. Assay of ß-galactosidase. p352-359 In: Experiments in molecular genetics. Cold Spring Harbor Laboratory, Cold Spring Harbor, NY.

Simon R, Priefer U, and Pühler A. 1983. Vector plasmids for in vivo and in vitro manipulations of gram-negative bacteria. p 98-106. In: Puhler A (ed). Molecular genetics of bacteria-plant interactions. Springer Verlag Berlin Heidelberg.

De Vries GE, Harms N, Maurer K, Papendrecht A, and Stouthamer AH. 1988. Physiological regulation of *Paracoccus denitrificans* methanol dehydrogenase synthesis and activity.J Bacteriol 170:3731-3737

Zaat SAJ, Wijffelman CA, Spaink HP, Van Brussel AAN, Okker RJH, and Lugtenberg BJJ. 1987. Induction of the *nodA* promoter of *Rhizobium leguminosarum* Sym plasmid pRL1JI by plant flavanones and flavones. J Bacteriol 169:198-204.

METHANOL DEHYDROGENASE FROM NOCARDIA SP. 239

P.W. VAN OPHEM AND J.A. DUINE.

Delft University of Technology, Department of Microbiology and Enzymology,
Julianalaan 67, 2628 BC Delft, The Netherlands.

INTRODUCTION

Gram-negative, methylotrophic bacteria convert methanol into
formaldehyde via NAD(P)-independent methanol dehydrogenase (MDH) with
pyrroloquinoline quinone (PQQ) as the cofactor. The activity of MDH can
be measured in vitro at pH 9 in the presence of ammonium salts as
activator and with artificial dyes as electron acceptor (Anthony, 1982).
In the Gram-positive bacterium Nocardia sp. 239 such an enzyme is
absent, although it excretes substantial amounts of PQQ into the medium
during growth on methanol (Hazeu et al., 1983).
   Cell free extracts of Nocardia sp. 239 show methanol dehydrogenase
activity with DCPIP, while the presence of NAD is required (Duine et al.,
1984). It was proposed that this "n-MDH" forms part of a multi enzyme
complex, together with NAD-dependent formaldehyde dehydrogenase and NADH
dehydrogenase, while NAD and PQQ function as cofactor in the complex.
   Continuing work on this topic was directed to characterize this
complex and its constituents.

METHODS

Cultivation of the organism. Nocardia sp. 239 was grown batch-wise on
1 % methanol as described by Duine et al. (1984) or fed-batch-wise by
supplying medium with 1 % methanol at a rate of 21 ml/h.
   Preparation of cell free extract. Cell free extract was prepared as
described (Duine et al., 1984).
   PQQ-determination. PQQ was determined with a biological assay, using
quinoprotein alcohol dehydrogenase apo-enzyme (Groen et al., 1986). To
detect free as well as protein-bound PQQ, measurements occurred of the
sample as such and after denaturation (3 min at 100 $^\circ$C).
   Enzyme assays. n-MDH (anaerobically), formaldehyde dehydrogenase and
NADH dehydrogenase activities were measured as described (Duine et al.,
1984). NAD-dependent alcohol dehydrogenase was determined under the
conditions according to Dijkhuizen et al. (1988). Protein was determined
with the Bradford assay (Bradford, 1976).

RESULTS AND DISCUSSION

Methanol dehydrogenase activity. Methanol dehydrogenase activity
could only be demonstrated under anaerobic conditions in cell free
extracts of cells grown under methanol limitation. In contrast to the

50

J. A. Jongejan and J. A. Duine (eds.), PQQ and Quinoproteins, 50–53.
© 1989 by Kluwer Academic Publishers.

TABLE 1: Stability of n-MDH activity in cell free extracts.

The cell free extract was prepared anaerobically and was kept under nitrogen atmosphere at room temperature.

Time (h)	n-MDH activity (nmol DCPIP/min/mg)
0	45
1.5	5.4
3.5	0.6

earlier report, n-MDH activity was very unstable, disappearing within a few hours (Table 1). NAD-dependent alcohol dehydrogenase activity, assayed as described for thermophilic methanol-using Bacillus strains (Dijkhuizen et al., 1988), was absent. The conclusion is, therefore, that Nocardia sp. 239 uses n-MDH for methanol conversion. Since the lability of this enzyme excluded its purification, it was attempted to obtain information on this enzyme by purifying the cell free extract on DEAE-Sepharose and looking for PQQ-containing proteins in the fractions. As shown in fig. 1, two chromatographically different PQQ-containing proteins are present (indicated with I and II).

PQQ-protein I. PQQ-protein I was further purified by FPLC Mono-Q-Sepharose and Superose-12 gelfiltration (Table 2). The final preparation was almost homogeneous, as judged by polyacrylamide gelelectrophoresis. It has the following characteristics: a M_r of 81.000 Da; maxima in the absorption spectrum at 280 and 345 nm (Fig. 2); neither methanol, formaldehyde nor NADH dehydrogenase activity could be detected.

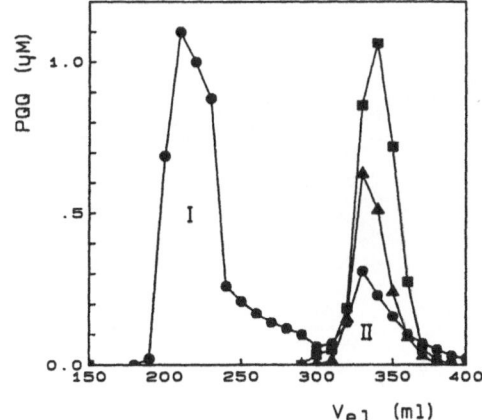

● = PQQ after denaturation
▲ = Formaldehyde dehydro-genase activity (arbitrary units)
■ = NADH dehydrogenase activity (arbitrary units)
I = PQQ-protein I
II = PQQ-protein II

FIGURE 1: Chromatography of cell free extract on DEAE-Sepharose.

TABLE 2: Purification scheme of PQQ-protein I.

	Total protein (mg)	Total PQQ (nmol)	Yield (%)	Purific. factor
Cell free extract	1700	185	100	1
DEAE-Sepharose	191	129	70	6.2
Mono-Q	5.9	24	13	37.1
Superose	2.4	12	6.4	45.2

PQQ-protein II. PQQ-protein II eluted together with an NAD-/ factor-dependent formaldehyde dehydrogenase and NADH dehydrogenase activity. It was attempted to separate the activities from each other by hydroxyl apatite, Mono-Q-Sepharose and gelfiltration chromatography, all without success. Also polyacrylamide gelelectrophoresis was unable to achieve separation, since all activities were found in the same protein band (M_r 178.000). Therefore, the tentative conclusion is that PQQ-protein II forms part of a multi enzyme complex, probably similar to the one previously proposed (Duine et al., 1984). However, in contrast to the previous report, no methanol dehydrogenase activity was detected.

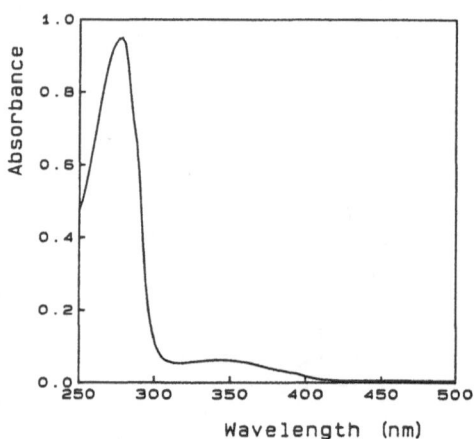

FIGURE 2: Absorption spectrum of PQQ-protein I.

REFERENCES

Anthony C, 1982. In: The Biochemistry of Methylotrophs. Academic Press, London.

Bradford MM, 1976. A rapid and sensitive method for the quantitation of microgram quantities of protein utilizing the principle of protein-dye binding. Anal. Biochem. 72: 248 - 254.

Dijkhuizen L, Arfman N, Attwood MM, Brooke AG, Harder W and Watling EM, 1988. Isolation and initial characterization of thermotolerant methylotrophic Bacillus strains. FEMS Microbiol. Lett. 52: 209 - 214.

Duine JA, Frank J and Berkhout MPJ, 1984. NAD-dependent, PQQ-containing methanol dehydrogenase: a bacterial dehydrogenase in a multienzyme complex. FEBS Lett. 168: 217 - 221.

Groen BW, van Kleef MAG and Duine JA, 1986. Quinohaemoprotein alcohol dehydrogenase aopenzyme from Pseudomonas testosteroni. Biochem. J. 234: 611 - 615.

Hazeu W, de Bruyn JC and van Dijken JP, 1983. Nocardia sp. 239, a facultative methanol utilizer with the ribulose monophosphate pathway of formaldehyde fixation. Arch. Mocrobiol. 135: 205 - 210.

PROPERTIES OF DYE-LINKED FORMALDEHYDE DEHYDROGENASE FROM HYPHOMICROBIUM ZV 580 GROWN ON METHYLAMINE

F.P.Kesseler, I. Baduns and A.C. Schwartz
Botanisches Institut der Universität Bonn, Kirschallee 1,
D-5300 Bonn 1, F.R. Germany

Dye-linked formaldehyde dehydrogenase (EC 1.2.99.3) is induced in Hyphomicrobium ZV 580 during growth on methylamine. Properties distinguishing it from the dye-linked general aldehyde dehydrogenase of methanol-grown ZV 580 include its specificity for formaldehyde (Km 0,076 - 0,1 mM), a tetrameric structure with Mr 210 000 (subunits 54 000), the pH optimum of 8.0, and the cofactor or coenzyme. The enzyme does not reduce NAD and NADP, with or without glutathione present. The irreversible inhibition by phenylhydrazine and features of the absorption spectra suggest that the enzyme might be a quinoprotein.

Key words: Formaldehyde dehydrogenase, dye-linked; formaldehyde oxidation; methylamine oxidation; induction; properties; Hyphomicrobium.

INTRODUCTION

Hyphomicrobium is endowed with two ways of formaldehyde oxidation, (a) via the methenyl-THF pathway (1), which is constitutive in ZV 580, and (b) via dye-linked aldehyde dehydrogenases (EC 1.2.99.3) (2,3,4), which are induced and repressed in this strain. The dehydrogenase which is induced in methanol-grown cells of ZV 580, is a dimer of Mr 76 000, and contains an iron-sulfur centre (4, and unpublished results). A different enzyme is induced in ZV 580 during growth on methylamine. Its purification and some properties will be described here.

MATERIALS AND METHODS

Maintenance and culture of ZV 580, as well as preparation of cell-free extracts have been described (3). The enzyme was assayed with a Clark oxygen electrode, 2.8 mM phenazine methosulfate (PMS) and 1.1 mM formaldehyde in 50 mM K$^+$ phosphate buffer pH 8.0 at 30° C. - The determination of metals by emission spectral analysis was carried out by Mikroanalytisches Labor Pascher, D-5480 Remagen.

RESULTS AND DISCUSSION

Purification of the enzyme. The fraction obtained at 40-90 % saturation of the cell-free extract with ammonium sulfate was chromatographed on DEAE cellulose in 50 mM Tris-HCl buffer pH 8.5. The enzyme did not bind to this

54

J. A. Jongejan and J. A. Duine (eds.), PQQ and Quinoproteins, 54–56.
© 1989 by Kluwer Academic Publishers.

column, and was concentrated by filtration with an Amicon PM-30 membrane. Subsequent separation on Sephacryl S-300 superfine in 50 mM K$^+$ phosphate buffer pH 7.2 was followed by binding to Phenyl-Sepharose in 10 mM K$^+$ phosphate buffer pH 6.8 and elution with a gradient of 30-0 % ammonium sulfate in this buffer. In the last step the enzyme bound to DEAE-Sephacel in 10 mM Tris-HCl pH 8.5, and was eluted with a gradient of 0-0.5 M NaCl. In a good

preparation 6 mg (UV method) of almost pure enzyme with a specific activity of 4.5 µmol/min/mg protein of formaldehyde oxidized were obtained from 185 g of wet cells after 130-fold purification with 6 % yield. Polyacrylamide-gel electrophoresis (PAGE) showed only 1-2 minor impurities.

<u>Molecular mass.</u> Mr 210 000 of the native enzyme was determined by chromatography on Sephacryl S-300 superfine with a set of protein standards. Mr 54 000 of the subunits was obtained from SDS-PAGE with protein standards and with and without mercapto-ethanol (Fig.1). The enzyme is a tetramer with 4 non-covalently linked subunits.

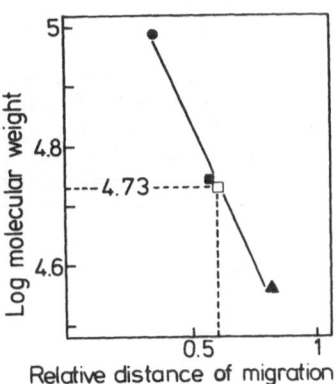

Fig. 1. Mr 54 000 of the subunits. Standards:
● Phosphorylase b 97 400
■ glutamate dehydrogenase 55 400
▲ lactate dehydrogenase 36 500

<u>Specificity for formaldehyde.</u> Apparent Michaelis constants at the fixed concentration of 2.8 mM PMS amount to 0.076 mM for formaldehyde, 0.35 mM for acetaldehyde, 0.22 mM for propionaldehyde and 0.313 mM for n-butyraldehyde. They show that formaldehyde has the highest affinity to the enzyme. In contrast, the dehydrogenase from methanol-grown cells of ZV 580 displayed a rather high Km for formaldehyde (12 mM), and the highest affinity to n-butyraldehyde (Km 0.033 mM) among the aliphatic aldehydes (cf. also 2).

Parallel lines in the Lineweaver-Burk diagrams for both formaldehyde and PMS in a preliminary kinetic analysis indicated that the enzyme reacts in a ping-pong mechanism. It was also evident that the oxygen concentration of the saturated buffer limits the rates at the highest substrate concentrations to a minor degree. Vmax values were about 35 µmol/min/mg protein of formaldehyde oxidized. Kms in these series of determinations were 0.1 mM for formaldehyde and 0.21 mM for PMS.

<u>Coenzyme.</u> The observation of a ping-pong mechanism suggested that the enzyme contains a cofactor participating in the redox reaction. However, it was free of metals, except for impurities in the range of 0.8 g atoms/mol of the tetramer (Fe), and lower. The native enzyme is slightly yellowish, and is partially reduced. The absorption spectrum of the enzyme oxidized with ferricyanide (Fig. 2) shows

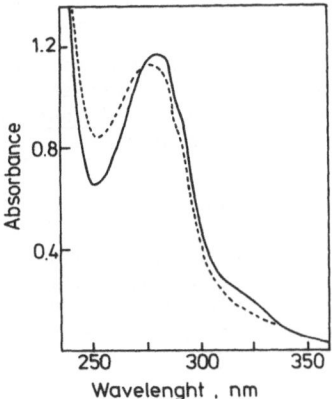

Fig. 2. Absorption spectra of the enzyme. Protein 1,4 mg/ml
(——) reduced
(– – –) oxidized

little structure and resembles spectra of quinoproteins such as glucose dehydrogenase (5, 6) and alcohol dehydrogenase (7). Only slight changes occur upon reduction with borohydride, whereas a maximum at 350 nm is observed in noted quinoproteins upon reduction. The difference spectrum of our enzyme (reduced minus oxidized) displays a maximum at 316 nm.

Further information on the nature of the coenzyme was obtained with specific inhibitors. The enzyme is completely and irreversibly inhibited by phenylhydrazine, which was applied as described for dinitro-phenylhydrazine (8). 10 μl of 2 mM phenylhydrazine in 0.1 N HCl were added to 1.4 mg enzyme in 1 ml of 50 mM K^+ phosphate buffer pH 8. The mixture was incubated for 16 h at 37 ° C, and dialyzed for the removal of unreacted inhibitor. The control was subjected to the same treatment, except that the inhibitor was omitted. The amount of enzyme available was too low to permit a photometric analysis of the adduct formed by coenzyme and inhibitor.

Cyclopropanol did not inhibit the enzyme. It was applied in equimolar concentration, and in 10- and 50-fold excess (9).

The results reported here would support, and not exclude, that the enzyme might be a quinoproteine. Exact proof of this hypothesis must be left to further experiments.

Other properties. The enzyme reduces PMS, and via PMS, oxygen and dichlorophenol indophenol. DCIP may also be reduced directly at about 0.2 % of the rate observed in the presence of PMS. The enzyme does not reduce NAD, NADP (both with and without glutathione present), horse heart cytochrome c and methylene blue. The dehydrogenase from methanol-grown ZV 580 reduces the last two acceptors.

The pH optimum of the enzyme is 8.0 (50 mM K^+ phosphate buffer). The pI is 5.67 (isoelectric focusing according to Pharmacia Informations).

The enzyme is inhibited by Tris in concentrations, which are used in buffers. 50 % inhibition was obtained with 66 mM Tris-HCl in the assay buffer. Cl^- is without effect. Partial reversal by increased concentrations of formaldehyde suggests that the inhibition may be competitive.

The enzyme is equally active in 50 mM phosphate, Hepes and Hepps buffer. In comparison, the enzyme from methanol-grown ZV 580, and the enzyme from Pseudomonas AM1 (10) are inhibited by higher concentrations of phosphate anions.

References

1. Marison IW, Attwood MM, 1982. J. Gen. Microbiol. 128, 1441-1446.
2. Marison IW, Attwood MM, 1980. J. Gen. Microbiol. 117, 305-313.
3. Köhler J, Schwartz AC, 1982. Can. J. Microbiol. 28, 65-72.
4. Kesseler FP, Schwartz AC, 1987. Biol. Chem. Hoppe-Seyler 368, 566-567.
5. Ameyama M, Matsushita K, Ohno Y, Shinagawa E, Adachi O, 1981. FEBS Lett. 190, 179-183.
6. Dokter P, Frank Jzn J, Duine JA, 1986. Biochem. J. 239, 163-167.
7. Groen B, Frank Jzn J, Duine JA, 1984. Biochem. J. 223, 291-297.
8. Lobenstein-Verbeek CL, Jongejan JA, Frank J, Duine JA, 1984. FEBS Lett. 170, 305-309.
9. Groeneveld A, Dijkstra M, Duine JA, 1984. FEMS Microbiol. Lett. 25, 311-314.
10. Anthony C, 1975. Biochem. J. 146, 289-298.

PHYSIOLOGICAL SIGNIFICANCE AND BIOENERGETIC ASPECTS OF GLUCOSE DEHYDROGENASE

Oense M. Neijssel[1], Ronald W.J. Hommes[1], Pieter W. Postma[2] and David W. Tempest[3], Departments of [1]Microbiology and [2]Biochemistry, Biotechnology Centre, University of Amsterdam, Amsterdam, The Netherlands, and of [3]Microbiology, University of Sheffield, Sheffield S10 2TN, England.

Key words: glucose dehydrogenase, bioenergetics, growth yield, enzyme regulation, chemostat culture, *Acinetobacter calcoaceticus*, *Klebsiella aerogenes*, *Escherichia coli*, *Pseudomonas* species.

ABSTRACT

The regulation of the PQQ-linked glucose dehydrogenase in different organisms is reviewed. It is concluded that this enzyme functions as an auxiliary energy-generating mechanism, because it is maximally synthesized under conditions of energy stress. It is now definitively established that the oxidation of glucose to gluconate generates metabolically useful energy. The magnitude of the contribution of the oxidation of glucose to gluconate via this enzyme to the growth yield of organisms such as *Acinetobacter calcoaceticus* is not yet clear.

INTRODUCTION

The production of gluconate from glucose by bacterial cultures has been discovered already in the previous century (Boutroux, 1880). It is therefore surprising that it has lasted just less than 100 years before the detailed enzymic mechanism of this oxidation was elucidated, particularly when one realizes the extent of progress that has been made in the understanding of the mammalian biochemistry of glucose metabolism during the same period of time. And even today the study of the properties, regulation and functional significance of the PQQ-linked glucose dehydrogenase in bacteria is limited to a relatively small group of scientists. More detailed investigations into the mechanism of gluconate formation were first published by Dalby and Blackwood (1955), who showed that cell-free extracts of *Aerobacter aerogenes* contained a glucose dehydrogenase that was stimulated by Mg^{2+} and did not show a requirement for the cofactors that were known at that time. Hauge (1964) showed that glucose dehydrogenase from *Bacterium anitratum* (= *A. calcoaceticus*) contained an unknown cofactor and Niederpruem and Doudoroff (1965) demonstrated that this unknown compound could bind to the apoenzyme of glucose dehydrogenase present in cells of *Rhodopseudomonas sphaeroides*, forming an active glucose dehydrogenase. Finally, Duine *et al.* (1979) were able to establish that this cofactor was PQQ.

It was soon discovered that the glucose dehydrogenase that was present in a wide variety of organisms, such as *Klebsiella aerogenes*, *Pseudomonas*, *Acetobacter*, and *Gluconobacter* species (Neijssel *et al.*, 1983; van Schie *et al.*, 1984) was PQQ dependent. Surprisingly, however, it was also shown that many other organisms contain the apo-enzyme of glucose dehydrogenase and are seemingly unable to synthesize PQQ, for instance *Acinetobacter lwoffi* (van Schie *et al.*, 1984), *Agrobacterium radiobacter* (Ameyama *et al.*, 1985), *Agrobacterium tumefaciens*, *Azotobacter vinelandii* (van Schie *et al.*, 1987a), *Escherichia coli* (Hommes *et al.*, 1984), *Rhizobium leguminosarum* (van Schie *et al.*, 1987a), and *Salmonella typhimurium* (Hommes *et al.*, 1986). This is

57

J. A. Jongejan and J. A. Duine (eds.), PQQ and Quinoproteins, 57–67.

truly remarkable, since in the absence of PQQ one could consider this apo-enzyme as a nonsense protein and therefore a genetic burden, against which selection supposedly would take place.

It is interesting to establish what the functional significance of this (apo-)enzyme is in the physiology of the organism. This contribution tries to analyze the environmental factors that influence the synthesis and the activity of the enzyme, and the bioenergetic consequences of the oxidation of aldoses via the PQQ-linked glucose dehydrogenase.

REGULATION OF GLUCOSE DEHYDROGENASE SYNTHESIS

An analysis of the literature shows that there are two groups of organisms which clearly differ with respect to their regulation of the synthesis of the PQQ-linked glucose dehydrogenase. In organisms, such as *A. calcoaceticus* and *Pseudomonas* species, substantial levels of glucose dehydrogenase are present under all aerobic growth conditions that have been tested, though the enzyme levels may vary with the growth conditions. It has, for instance, been observed that the activity of the enzyme in *A. calcoaceticus* decreases with increasing growth rates (van Schie, 1987), whereas in *Pseudomonas aeruginosa* the activity decreased at low growth rates (Ng and Dawes, 1973). In oxygen-limited cultures of *A. calcoaceticus*, growing on acetate in the presence of glucose, holo-glucose dehydrogenase was synthesized, but was not functional due to the low oxygen tension extant in the culture (van Schie, 1987). When *Pseudomonas aeruginosa* was grown under fully anaerobic conditions and carrying out a nitrate respiration, apo-glucose dehydrogenase was synthesized (van Schie *et al.*, 1984). This could indicate that in this organism synthesis of PQQ is a stricly aerobic process.

In contrast, the levels of glucose dehydrogenase in *K. aerogenes* were much more dependent on the culture conditions (Table 1). Carbon-, ammonia-

Table 1

Limitation	Dilution rate h^{-1}	Glucose dehydrogenase activity nmol WB reduced.min^{-1}.mg protein^{-1}
Carbon	0.31	2
Ammonia	0.30	7
Ammonia + DNP	0.29	36
Sulphate	0.31	5
Sulphate + DNP	0.30	27
Phosphate	0.23	32
Potassium	0.34	33
Potassium*	0.25	38

Glucose dehydrogenase activities in cell-free extracts of *Klebsiella aerogenes* (without extra PQQ added), grown in aerobic chemostat culture on glucose or on *glycerol in the presence and absence of 2,4-dinitrophenol (1 mM) at 35 °C and pH 6.9. WB = Wurster's Blue. Data from Hommes *et al.* (1985) and Neijssel *et al.* (1983).

or sulphate-limited cultures showed low enzyme activities and did not produce gluconate or 2-ketogluconate, which shows that the *in vivo* activities
were virtually zero. On the other hand, potassium- or phosphate-limited
cultures possessed considerable quantities of the enzyme and were prolific
gluconate and 2-ketogluconate producers (Neijssel and Tempest, 1976). Anaerobically grown *K. aerogenes* did not contain any glucose dehydrogenase activity and these cells possessed only small quantities of apo-enzyme, irres-
pective of whether the cells were fermenting glucose or were carrying out a
fumarate or nitrate respiration (Hommes, 1988).

It was shown previously that addition of the uncoupler 2,4-dinitrophenol
(DNP, 1 mM final concentration) to the medium of an ammonia-limited chemostat culture of *K. aerogenes* had dramatic effects: the production of extracellular polysaccharides was suppressed and instead great quantities of
gluconate and 2-ketogluconate were produced (Neijssel, 1977). This led us to
investigate the effect of this uncoupler on the synthesis of glucose dehydrogenase in ammonia- or sulphate-limited chemostat cultures of *K. aerogenes*. The results, shown in Table 1, clearly indicate that the presence of
DNP in the medium led invariably to the synthesis of high levels of the
enzyme. From these observation one is forced to conclude that the trigger
for glucose dehydrogenase synthesis in *K. aerogenes* must be the energetic
status of the cell, since it is extremely unlikely that DNP itself can act
as an inducer or derepressor. It is also obvious that glucose does not play
this role, since a potassium-limited culture growing on glycerol as the
carbon source also showed a high enzyme activity. In addition, it has been
shown that addition of DNP to chemostat cultures of *Pseudomonas fluorescens*
also stimulates synthesis of the PQQ-linked glucose dehydrogenase (G.P.M.A.
Hardy, personal communication).

The question is then, why potassium- and phosphate-limited cultures also
show high activities of glucose dehydrogenase and what these cultures have
in common with cultures growing in the presence of DNP. An answer could be
provided by the experiments carried out by Mulder *et al.* (1986). On the
basis of their results they proposed that, since in potassium-limited cells
of *E. coli* two potassium uptake systems are present, a low affinity system
called Trk and a high affinity system (Kdp), these two systems would
generate a futile cycle of potassium ions across the cell membrane leading
to a severe spilling of energy. Thus, there is analogy between cells growing
in the presence of DNP and those growing in potassium-limited environments:
under both circumstances the cells suffer from a futile cycle, in the first
case of protons and in the second of potassium ions. We propose that a
similar explanation is valid for phosphate-limited cultures, since Willsky
and Malamy (1974, 1976) have shown that *E. coli* possesses two uptake systems
for phosphate with different affinities.

Nelson and Kennedy (1972) demonstrated that *E. coli* contains two uptake
systems for magnesium ions. This would indicate that magnesium-limited cultures could possess high glucose dehydrogenase activities. The results of
Hardy *et al.* (1988) and Boiardi *et al.* (1988) show that in this type of
culture the situation is more complex, since magnesium-limited cultures of
K. aerogenes did not produce much gluconate and or 2-ketogluconate. However,
when extra calcium was added to the medium (up to 1 mM final concentration)
holo-enzyme activity was restored and reached levels similar to those of
potassium-limited cultures. Boiardi *et al.* (1988) propose that, although a
magnesium-limited growth environment promotes the synthesis of the holoenzyme, a functional holo-enzyme is not formed because the binding of PQQ to
the apo-enzyme does not take place due to the low magnesium concentration in
the culture. Addition of calcium ions to the culture enables binding of PQQ
and the result is that the culture is still growing magnesium-limited but
now with a functional holo-enzyme.

As mentioned above, many organisms contain only the apo-glucose dehydrogenase. This prompted us to investigate the amounts of apo-enzyme and holoenzyme in cell of K. aerogenes grown in aerobic chemostat cultures. The results, shown in Table 2, are surprising. Carbon-limited cells contained

Table 2

Limitation	Glucose dehydrogenase activity nmol WB reduced.min^{-1}.mg protein^{-1}		
	-PQQ	+PQQ	Increase (%)
Carbon	25	66	160
Ammonia	13	170	1200
Sulphate	26	240	820
Phosphate	120	200	67
Potassium	150	200	33

Glucose dehydrogenase activities in cell-free extracts of *Klebsiella aerogenes* grown in aerobic chemostat cultures (D = 0.3 h^{-1}; pH 6.0; 35 °C), and the increase (%) in activity after preincubation with 100 μM PQQ for 15 minutes at 35 °C prior to the addition of glucose. WB = Wurster's Blue. Data from Hommes (1988).

relatively low levels of holo- and apo-enzyme, but all other cultures in which glucose was present in excess of the growth requirements, possessed roughly similar amounts of the glucose dehydrogenase protein and these data suggest that the low activities present in ammonia- or sulphate-limited cells are caused by a diminished synthesis of PQQ. One may argue that during the preparation of the cell-free extract some PQQ may have been dissociated from the holo-enzyme causing wide variations in the observed activities. This was checked by adding PQQ to the culture vessel of a chemostat. If the glucose dehydrogenase activity, as expressed in the culture, is directly proportional to the amount of PQQ available, an increased concentration of this cofactor should be followed by an increase in enzyme activity. Addition of extra PQQ to a potassium-limited culture had no effect on the rates of oxygen consumption and carbon dioxide production, whereas the oxygen consumption of an ammonia-limited culture of K. aerogenes increased instantaneously with almost 40%; the production of carbon dioxide showed a small but significant decrease; Similarly, addition of extra PQQ to a sulphate-limited culture of this organism led to an increase of the *in vivo* activity of glucose dehydrogenase from 2.5 to 12 mmol.h^{-1}.g cells (Hommes, 1988). On the basis of these results it can be concluded that during the preparation of the cell-free extracts some loss of PQQ took place, but that ammonia-and sulphate-limited cells contained less holo-enzyme than apo-enzyme.

This observation also points to another interesting phenomenon: the immediate increase of the rate of oxygen consumption after a pulse of PQQ indicates that the segment of the respiratory chain from cytochrome *b*, commonly assumed to be the acceptor of the electrons from PQQH$_2$ (Beardmore-Gray and Anthony, 1986; Matsushita *et al*. 1987), and the terminal cytochrome must have had extra capacity to allow this transport of electrons to take place. The fact that the rate of carbon dioxide production showed a small decrease indicates that the rate of NADH oxidation by the respiratory chain was diminished, but not to the same extent as the increase in the rate of electron transport from $PQQH_2$ to oxygen.

When one compares the activities shown in Table 1 and 2 it is apparent that there is a difference between the two sets of data. Further analysis shows that the culture pH value has a great influence on both the activity and the synthesis of glucose dehydrogenase. In this respect there is a clear difference between *A. calcoaceticus* and *K. aerogenes*. Experiments performed by van Schie *et al.* (1987c) indicate that the glucose dehydrogenase *A. calcoaceticus* is synthesized and active in between pH values of 5.0 and 8.2. In *K. aerogenes* the enzyme has an acidic pH optimum ranging from 5.0 to 6.0 (Matsushita *et al.*, 1982; Hommes,1988). The optimal pH at which cell-free extracts from *Aerobacter aerogenes* oxidized glucose to gluconate was shown to be 5.5 to 6.5 (Dalby and Blackwood, 1955). The effect of the culture pH value on the *in vivo* activity of glucose dehydrogenase in *K. aerogenes* is shown in Table 3. One can calculate from these data that at pH 5.0, 77% of

Table 3

pH	$q_{glucose}$	$q_{gluconate}$	$q_{2-ketogluconate}$	GDH
5.0	570	210	230	440
5.5	620	220	270	490
6.0	510	200	180	380
6.5	440	100	70	170
7.0	250	10	20	30
7.5	200	0	0	0
8.0	220	0	0	0

The influence of the culture pH value on the steady state rates of gluconate and 2-ketogluconate production by K^+-limited cultures of *Klebsiella aerogenes* (D = 0.3 h^{-1}; 35 °C). Rates (q) are expressed in $nmol.min^{-1}$.(mg dry weight of cells)$^{-1}$. GDH = total *in vivo* activity of glucose dehydrogenase. Data from Hommes (1988).

the glucose was metabolized via glucose dehydrogenase and at pH 5.5 this value was even higher: 79%! Detailed enzymatic analyses showed that maximal synthesis of the enzyme occurred between culture pH values of 5.5 and 6.0. At a culture pH value of 8.0 no gluconate was produced, but when the culture pH value was suddenly lowered to a value of 5.2, immediate production of gluconate took place, indicating that some glucose dehydrogenase was synthesized at this high pH value. It could be shown that the lack of glucose dehydrogenase activity at pH 8.0 was not due to the dissociation of the cofactor PQQ from the enzyme (Hommes, 1988).

Although *E. coli* is only able to synthesize apo-glucose dehydrogenase (Hommes *et al.*, 1984), at least under the culture conditions that have been tested until now, the synthesis of this protein is regulated similar way as in *K. aerogenes*. Thus, glucose-limited cultures contained the lowest amounts of apo-enzyme and glucose-excess cultures contained considerable quantities

62

Table 4

Limitation	Glucose dehydrogenase activity nmol WB reduced.min^{-1}.mg protein^{-1}
Glucose	15
Glucose + PQQ	12
Sulphate	310
Sulphate + PQQ	270
Phosphate	350
Phosphate + PQQ	290

Glucose dehydrogenase activities in cell-free extracts of *Escherichia coli* B/r, grown in chemostat culture (pH 5.5; 35 °C; in the presence and absence of 0.2 µM PQQ), after preincubation with 0.1 mM PQQ (15 minutes at 35 °C). WB = Wurster's Blue. Data from Hommes (1988).

of this protein. Addition of 1 mM DNP to the medium of batch cultures and chemostat cultures led again to increased synthesis of the apo-enzyme (Hommes *et al.* 1984; Hommes, 1988). This organism offered us also the unique opportunity to study the influence of PQQ on the synthesis of the apo-enzyme, but the data in Table 4 show that there was no effect. On the other hand the type of glucose metabolism carried out by *E. coli* growing in the

Table 5

Limitation	Dilution rate	$q_{glucose}$	$q_{gluconate}$	GA/Glc
Glucose	0.15	2.0	0	0
Glucose + PQQ	0.15	1.9	0	0
Ammonia	0.14	2.9	0	0
Ammonia + PQQ	0.15	3.6	0.3	8
Sulphate	0.15	2.7	0	0
Sulphate + PQQ	0.16	7.3	4.4	60
Phosphate	0.15	3.2	0	0
Phosphate + PQQ	0.15	10.0	6.3	63
Potassium	0.16	4.6	0	0
Potassium + PQQ	0.19	6.0	1.1	18

Rates of glucose utilization and of gluconate production in variously-limited chemostat culture of *Escherichia coli* B/r growing in the absence and presence of 0.2 µM PQQ (pH 5.5; 35 °C). Rates (q) are expressed in mmol.h^{-1}. (g dry weight)$^{-1}$. GA/Glc = % of glucose converted into gluconate. Data from Hommes (1988).

presence of PQQ was altered dramatically (Table 5). For instance under phosphate-limited growth conditions 63% of the glucose in the medium was converted into gluconate when PQQ was added to the medium.

In conclusion, the data discussed in this contribution and of others (de Bont *et al.* 1984, van Schie, 1987) provide overwhelming evidence that the PQQ-linked glucose dehydrogenase is regulated by the energy status of the cell. In *K. aerogenes* this regulation is mainly effected via the synthesis of the cofactor PQQ.

GROWTH YIELDS AND ENERGETICS AT THE MOLECULAR LEVEL

Earlier investigations seemed to indicate that the oxidation of glucose to gluconate via glucose dehydrogenase did not generate biologically useful energy (Campbell *et al.*, 1956; Mackechnie and Dawes, 1969; Uspenskaya and Loitsyanskaya, 1979). Thus, although it was shown in the previous section that an energetic stress in organisms led to an increased synthesis of the enzyme, it still remained to be shown whether active involvement of the enzyme in glucose metabolism really contributed to the generation of energy in the cell. Experiments by van Schie *et al.* (1985, 1987b) have established that in membrane vesicles from different organisms generation of a proton motive force and subsequent energization of solute transport could be demonstrated when these vesicles were provided with glucose and/or xylose.

Table 6

electron donor	concentration mM	initial uptake rate nmol.min-1.mg membrane protein-1
glucose	20	1.3
xylose	20	1.2
PMS/ascorbate	0.2/20	1.35
NADH	10	0
gluconate	20	0
glucono-δ-lactone	20	0
malate	10	0.4

Effect of various electron donors on alanine transport in membrane vesicles obtained from acetate-limited cells of *Acinetobacter calcoaceticus*. PMS = phenazine methosulphate. Uptake studies were carried out at pH 6.6. Data from van Schie *et al.* (1987b).

The data in Table 6 give an example of such measurements and they demonstrate that oxidation of $PQQH_2$ via the respiratory chain is just as effective as oxidation of PMS+ascorbate. Although one could conclude that this matter is now settled, other observations suggest that the situation is not so clear-cut. When *A. calcoaceticus* was grown in an acetate-limited chemostat culture at pH 7.0, addition of glucose to the growth medium led to increased growth yields. In contrast, when xylose was added as the co-substrate to such acetate-limited cultures no increase in growth yield could be observed (van Schie *et al.*, 1987c). This problem was further analyzed and it was found that addition of xylose to the medium could lead to an increased growth yield provided that the culture pH value was 8.2. The explanation for these intriguing results was found to be the stability of the lactone produced via the glucose dehydrogenase: glucono-δ-lactone is spontaneously hydrolyzed at pH values of 6 and higher, whereas xylono-γ-lactone is stable at pH 7.0 and is hydrolyzed to xylonate at pH 8.2. Glucono-δ-lactone is stable at pH 5.0 and indeed, when *A. calcoaceticus* was grown in an acetate-limited chemostat culture at this pH value (with glucose present as the co-substrate) no increase of the growth yield could be observed. This would suggest that the hydrolysis of the lactone would contribute to the energetic efficiency of the oxidation of glucose and xylose. This is hard to understand, particularly in view of the data presented in Table 6, where it is shown that glucono-δ-lactone or gluconate cannot drive alanine transport. But the problem becomes even more complex when the energetic efficiency of the oxidation of $PQQH_2$ via the respiratory chain is calculated on the basis of these growth yield data: 1 xylose reduction equivalent would equal 3 reduction equivalents from acetate (NADH) with respect to energetic efficiency (van Schie *et al.*, 1987c). However, studies with labeled glucose or xylose indicated that some metabolism of the aldonic acid took place (9.2 % of the glucose and 2.6 % of the xylose that was consumed was recovered as carbon dioxide) and this could have a great effect on the calculation. This, and the fact that it was shown in later studies that this strain of *A. calcoaceticus* (LMD 79.41) acquired the capacity to grow on glucose as the sole carbon and energy source after prolonged cultivation (8-10 volume changes) on mixtures of acetate and glucose (van Schie, 1987), indicates that one has to interpret the calculations of the energetic efficiency of the oxidation of $PQQH_2$ with extreme caution.

CONCLUDING REMARKS

When one reviews the functional significance of the PQQ-linked glucose dehydrogenase in the physiology of different organisms, one is forced to conclude that the enzyme plays a role in the energy metabolism of the cell. Nevertheless, it is clear that the situation today is rather incomprehensible from a microbial physiologist's or ecologist's point of view:
1) The enzyme has a relatively high K_m for glucose (around 1 mM) and the active site is facing the periplasm. This means that the enzyme will function only in environments that contain relatively high levels of glucose (more precisely: aldose).
2) In many organisms the synthesis of PQQ and the glucose dehydrogenase protein are not coordinately regulated; one can observe frequently that only the apo-enzyme is synthesized.
3) Many organisms seemingly are able only to synthesize apo-glucose dehydrogenase, but have retained the capacity to synthesize this protein even after having been grown for decades in many different laboratory media without PQQ.
4) Some organisms can oxidize glucose via this enzyme, but cannot grow on gluconate and need another carbon source for growth.

Whereas one can argue that PQQ is present in many environments as a vitamin and that therefore reconstitution in Nature is very well possible, one wonders whether many natural environments exist that contain sufficient PQQ, substrate, and oxygen to allow the enzyme to function. Moreover, if such an environment would exist, some organisms would need even an *extra* substrate to be able to grow. It is this variety of problems that is such an interesting subject of further study!

ACKNOWLEDGEMENT

Part of the research reported in this contribution was supported by the Foundation for Fundamental Biological Research (BION), which is subsidized by the Netherlands Organization for Scientific Research (NWO).

REFERENCES

Ameyama M, Shinagawa E, Matsushita K, and Adachi O, 1985. Growth stimulating activity for microorganisms in naturally occurring substances and partial characterization of the substance for the activity as PQQ. Agricultural and Biological Chemistry 49: 699-709.

Beardmore-Gray M, and Anthony C, 1986. The oxidation of glucose by *Acinetobacter calcoaceticus*: the interaction of the quinoprotein glucose dehydrogenase with the electron transport chain. Journal of General Microbiology 132: 1257-1268.

Boiardi JL, Buurman ET, Hardy GPMA, Teixeira de Mattos MJ, and Neijssel OM, 1988. The effect of magnesium and calcium on the synthesis of PQQ in *Klebsiella aerogenes* and *Pseudomonas* species. Poster abstract, First International Symposium on PQQ and Quinoproteins, Delft.

Bont JAM de, Dokter P, Schie BJ van, Dijken JP van, Frank Jzn J, Duine JA, and Kuenen JG, 1984. Role of quinoprotein glucose dehydrogenase in gluconic acid production by *Acinetobacter calcoaceticus*. Antonie van Leeuwenhoek 50: 76-77.

Boutroux L, 1880. Sur une fermentation nouvelle du glucose. Comptes Rendus Hebdomadaires des Seances de l'Academie des Sciences 91: 236-238.

Campbell JJR, Ramakrishna T, Linnes AG, and Eagles BA, 1956. Evaluation of the energy gained by *Pseudomonas aeruginosa* during the oxidation of glucose to 2-ketogluconate. Canadian Journal of Microbiology 2: 304-310.

Dalby A, and Blackwood AC, 1955. Oxidation of sugars by an enzyme preparation from *Aerobacter aerogenes*. Canandian Journal of Microbiology 1: 733-742.

Duine JA, Frank Jzn J, and Zeeland JK van, 1979. Glucose dehydrogenase from *Acinetobacter calcoaceticus*: a quinoprotein. FEBS Letters 108: 443-446.

Hardy GPMA, Teixeira de Mattos MJ, and Neijssel OM, 1988. The regulation of the PQQ-linked glucose dehydrogenase in chemostat cultures of *Pseudomonas species*. Poster abstract, First International Symposium on PQQ and Quinoproteins, Delft.

Hauge JG, 1964. Glucose dehydrogenase of *Bacterium anitratum*: an enzyme with a novel prosthetic group. Journal of Biological Chemistry 239: 3630-3639.

Hommes RWJ, 1988. The role of the PQQ-linked glucose dehydrogenase in the physiology of *Klebsiella aerogenes* and *Escherichia coli*. PhD thesis, University of Amsterdam.

Hommes RWJ, Hell B van, Postma PW, Neijssel OM, and Tempest DW, 1985. The functional significance of glucose dehydrogenase in *Klebsiella aerogenes*.

Archives of Microbiology 143: 163-168.

Hommes RWJ, Loenen WAM, Neijssel OM, and Postma PW, 1986. Galactose metabolism in *gal* mutants of *Salmonella typhimurium* and *Escherichia coli*. FEMS Microbiology Letters 36: 187-190.

Hommes RWJ, Postma PW, Neijssel OM, Tempest DW, Dokter P, and Duine JA, 1984. Evidence of a quinoprotein glucose dehydrogenase apoenzyme in several strains of *Escherichia coli*. FEMS Microbiology Letters 24: 329-333.

Mackechnie I, and Dawes EA, 1969. An evaluation of the pathways of metabolism of glucose, gluconate and 2-oxogluconate by *Pseudomonas aeruginosa* by measurement of molar growth yields. Journal of General Microbiology 55: 341-349.

Matsushita K, and Ameyama M, 1982. D-Glucose dehydrogenase from *Pseudomonas fluorescens*, membrane-bound. Methods in Enzymology 89: 149-154.

Matsushita K, Nonobe M, Shinagawa E, Adachi O and Ameyama M, 1987. Reconstitution of pyrroloquinoline quinone-dependent D-glucose oxidase respiratory chain of *Escherichia coli* with cytochrome *o* oxidase. Journal of Bacteriology 169: 205-209.

Mulder MM, Teixeira de Mattos MJ, Postma PW, and Dam K van, 1986. Energetic consequences of multiple K$^+$ uptake systems in *Escherichia coli*. Biochimica et Biophysica Acta 851: 223-228.

Neijssel OM, 1977. The effect of 2,4-dinitrophenol on the growth of *Klebsiella aerogenes* NCTC 418 in aerobic chemostat cultures. FEMS Microbiology Letters 1: 47-50.

Neijssel OM, and Tempest DW, 1975. The regulation of carbohydrate metabolism in *Klebsiella aerogenes* NCTC 418 organisms, growing in chemostat culture. Archives of Microbiology 106: 251-258.

Neijssel OM, Tempest DW, Postma PW, Duine JA, and Frank Jzn J, 1983. Glucose metabolism by K$^+$-limited *Klebsiella aerogenes*: evidence for the involvement of a quinoprotein glucose dehydrogenase. FEMS Microbiology Letters 20: 35-39.

Nelson DL, and Kennedy EP, 1972. Transport of magnesium by a repressible and a nonrepressible system in *Escherichia coli*. Proceedings of the National Academy of Sciences of the U.S.A. 69: 1091-1093.

Ng FMW, and Dawes EA, 1973. Chemostat studies on the regulation of glucose metabolism in *Pseudomonas aeruginosa* by citrate. Biochemical Journal 132: 129-140.

Niederpruem DJ, and Doudoroff M, 1965. Cofactor-dependent aldose dehydrogenase from *Rhodopseudomonas sphaeroides*. Journal of Bacteriology 89: 697-705.

Schie BJ van, 1987. The physiological function of gluconic acid production in *Acinetobacter* species and other gram-negative bacteria. Implications for energy conservation. PhD thesis, University of Technology, Delft.

Schie BJ van, Dijken JP van, and Kuenen JG, 1984. Non-coordinated synthesis of glucose dehydrogenase and its prosthetic group PQQ in *Acinetobacter* and *Pseudomonas* species. FEMS Microbiology Letters 24: 133-138.

Schie BJ van, Hellingwerf KJ, Dijken JP van, Elferink MGL, Dijl JM van, Kuenen JG, and Konings WN, 1985. Energy transduction by electron transfer via a pyrrolo-quinoline quinone-dependent glucose dehydrogenase in *Escherichia coli*, *Pseudomonas aeruginosa*, and *Acinetobacter calcoaceticus* (var. *lwoffi*). Journal of Bacteriology 163: 493-499.

Schie BJ van, Mooy OH de, Linton JD, Dijken JP van, and Kuenen JG, 1987a. PQQ-dependent production of gluconic acid by *Acinetobacter*, *Agrobacterium*, and *Rhizobium* species. Journal of General Microbiology 133: 867-875.

Schie BJ van, Pronk JT, Hellingwerf KJ, Dijken JP van, and Kuenen JG, 1987b. Glucose-dehydrogenase-mediated solute transport and ATP synthesis in *Acinetobacter calcoaceticus*. Journal of General Microbiology 133: 3427-3435.

Schie BJ van, Rouwenhorst RJ, Bont JAM de, Dijken JP van, and Kuenen JG, 1987c. An in vivo analysis of the energetics of aldose oxidation by *Acinetobacter calcoaceticus*. Applied Microbiology and Biotechnology 26: 560-567.

Uspenskaya SN, and Loitsyanskaya MS, 1979. Effectiveness of the utilization of glucose by *Gluconobacter oxydans*. Microbiology 48: 306-310.

Willsky GR, and Malamy MH, 1974. The loss of the phoS periplasmic protein leads to a change in the specificity of a constitutive inorganic phosphate transport system in *Escherichia coli*. Biochemical and Biophysical Research Communications 60: 226-233.

Willsky GR, and Malamy MH, 1976. Control of the synthesis of alkaline phosphatase and the phosphate binding protein in *Escherichia coli*. Journal of Bacteriology 127: 595-609.

Quinoprotein D-glucose dehydrogenases in Acinetobacter calcoaceticus LMD 79.41: Purification and characterization of the membrane-bound enzyme distinct from the soluble enzyme

Kazunobu Matsushita, Emiko Shinagawa, Osao Adachi and Minoru Ameyama

Laboratory of Applied Microbiology, Department of Agricultural Chemistry, Faculty of Agriculture, Yamaguchi University, Yamaguchi 753, Japan

Key words: D-Glucose dehydrogenase, membrane-bound, quinoprotein, ubiquinone, Acinetobacter calcoaceticus

ABSTRACT

Acinetobacter calcoaceticus is known to contain soluble and membrane-bound quinoprotein D-glucose dehydrogenases while other oxidative bacteria such as Pseudomonas or Gluconobacter contain only membrane-bound enzyme. The two different forms were believed to be the same enzyme or interconvertible. Present results show that the two different forms of glucose dehydrogenase are distinct from each other in their enzymatic and immunological properties as well as in their molecular size.
The soluble and membrane-bound glucose dehydrogenases were separated after French press-disruption by repeated ultracentrifugation, and then purified to nearly homogeneous state. The soluble enzyme was a polypeptide of 55 Kdaltons, while the membrane-bound enzyme was a polypeptide of 83 Kdaltons which is mainly monomeric in detergent solution. Both enzymes showed different enzymatic properties including substrate specificity, optimum pH, kinetics for glucose, and reactivity for ubiquinone-homologues. Furthermore, the two enzymes could be distinguished immunochemically; the membrane-bound enzyme is cross-reactive with an antibody raised against membrane-bound enzyme purified from Pseudomonas but not with antibody elicited against the soluble enzyme, while the soluble enzyme is not cross-reactive with the antibody of membrane-bound enzyme.
Data also suggest that the membrane-bound enzyme functions by linking to the respiratory chain via ubiquinone though the function of the soluble enzyme remains unclear.

INTRODUCTION

Bacterial D-glucose dehydrogenase (GDH) is a quinoprotein, having pyrroloquinoline quinone as the prosthetic group, and is linked to the respiratory chain. The GDH has been found in a wide variety of bacteria, in which the enzyme is usually tightly bound to the outer surface of the cytoplasmic membrane (Matsushita et al., 1986). Thus, GDHs of Pseudomonas fluorescens, Gluconobacter suboxydans and Escherichia coli have been purified in the presence of detergent only after solubilization from the membranes (Matsushita et al., 1980; Ameyama et al., 1981; Ameyama et al., 1986). Unlike other bacteria, however, Acinetobacter calcoaceticus contains a soluble form of GDH in addition to the membrane-bound GDH (Hauge, 1960a; Duine et al., 1982). The soluble GDH was originally purified by Hauge (1960a, 1964) and recently has been purified by two

69

groups (Dokter et al., 1986; Geiger & Görisch, 1986). Thus it has been shown that the soluble enzyme is a dimer of identical subunits, which are 48-54 Kdaltons, and is able to oxidize disaccharides as well as monosaccharides. On the other hand, attempted purification of membrane-bound GDH by solubilization and fractionation (Hauge & Hallberg, 1964) revealed the enzyme to be the same as the soluble GDH. Recently, Dokter et al. (1987) made a similar observation in which the enzyme in the membrane vesicles turned out to be able to oxidize disaccharides as well as monosaccharides after solubilization with Triton X-100.

However, our recent observation with immunoblotting using antibody prepared against Pseudomonas GDH showed that the membrane of A. calcoaceticus contained GDH peptide of higher molecular weight (83 Kdaltons) than the soluble form, which seemed to be closely related to other membrane-bound GDHs (Matsushita et al., 1986). Only recently, the notion described above has been supported by a genetic experiment in which isolated gdh gene produces a protein of 83 Kdaltons (Cleton-Jansen et al., 1988).

We have, therefore, examined the relationship between the soluble and membrane-bound GDHs in A. calcoaceticus LMD 79.41. The results show that the enzymes are completely different in terms of molecular size, immunogenecity, kinetics and reaction with ubiquinone (Q). Data also suggest that only the membrane-bound GDH is functional in the glucose oxidase respiratory chain.

MATERIALS AND METHODS

Bacterial strain and growth conditions.
A. calcoaceticus LMD 79.41 was kindly provided by Dr. J. Duine. The organism was grown at 30°C aerobically to the late exponential phase. The growth medium prepared with tap water included 0.54% sodium succinate, 0.4% ammonium sulfate, 0.14% K2HPO4, 0.08% KH2PO4 and 0.02% MgSO4/7H2O. Usually final klett unit and medium pH were 250-270 and 7.7-7.9, respectively, after 18 h growth.

Preparation of soluble and membrane fractions.
The cells were collected by centrifuging, washed with 50 mM K phosphate, pH 7.5, and suspended in the same buffer. The suspension was passed twice through a French pressure cell press at 16000 psi, and then centrifuged at 12000 x g for 20 min to remove intact cells and cell debris. The supernatant was separated into crude soluble and membrane fractions by centrifugation at 120000 x g for 90 min. Centrifugation was repeated to obtain soluble and membrane fractions from the crude counterparts.

Purification of soluble and membrane-bound enzymes.
Soluble GDH was purified from the soluble fraction prepared by ultracentrifugation as described above, instead of treating cells with 1% Triton X-100 which was used by the previous workers (Dokter et al., 1986; Geiger & Görisch, 1986). The soluble fraction was dialysed extensively against 20 mM K phosphate, pH 7.0, and then ultra-centrifuged again to remove the contaminated membrane fraction. The supernatant was applied to a DEAE-Toyopearl column equilibrated with 20 mM K phosphate, pH 7.0. After washing the column with the same buffer, the enzyme was eluted by a linear gradient upto 100 mM K phosphate, pH 7.0. The enzyme was further purified by a CM-Toyopearl, hydroxyapatite and Phenyl-Sepharose chromatography essentially as described previously (Geiger & Görisch, 1986).

The membrane-bound enzyme was purified from the membrane fraction prepared as described above. The enzyme was solubilized with 1% Triton X-114 in the presence of 0.3 M KCl after washing the membrane fraction with 0.2% Triton X-100. The solubilized supernatant was further separated to upper and lower phases by phase separation. The lower phase was dialysed against 0.1% Triton X-114, and then ultra-centrifuged to remove membranous aggregates. Furthermore, the enzyme was purified from the supernatant by DEAE- and CM-Toyopearl column chromatography in the presence of 0.1% Triton X-114.

Enzyme assays.
    GDH activity was measured spectrophotometrically using phenazine methosulfate (PMS) and 2,6-dichlorophenol indophenol (DCIP) as electron acceptors as described (Matsushita et al., 1980). Quinone reductase activity of GDH was also measured spectrophotometrically with several Q homologues (Matsushita et al., 1980 & 1982).

Other analytical procedures.
    Sodium dodecyl sulfate-polyacrylamide gel electrophoresis (SDS-PAGE), preparation of antibodies and immunoblotting analysis were performed as described (Matsushita et al., 1986 & 1988). Sucrose density gradient centrifugation and protein determination were carried out as described (Matsushita et al., 1980).

RESULTS

Separation and differentiation of soluble and membrane-bound GDHs.
    Soluble and membrane-bound GDHs were separated by repeated ultra-centrifugation after French press-disruption. Crude soluble and membrane fractions were heavily cross-contaminated each other, and thus the ratio of PMS/DCIP reductase to Q-2 reductase activities in both fractions was largely changed by repeating ultracentrifugation (Table 1). Even after repeated ultracentrifugation, the soluble fraction was contaminated partially with the membrane fraction, which can be shown by immunoblotting analysis as shown later. The soluble and membrane fractions thus obtained

Table 1. GDH activity in soluble and membrane fractions

Fractions	GDH activity				Q-2/PMS
	PMS/DCIP reductase		Q-2 reductase		
	units	units/mg	units	units/mg	%
Crude soluble	1101	10.8	46.8	0.50	4.6
Soluble	865	11.3	32.9	0.31	2.7
Crude membrane	229	5.2	82.3	1.87	36
Membrane	133	3.8	73.4	2.10	55
Triton supernatant	90	9.4	46.8	4.90	52

were used to compare the property of both soluble and membrane-bound GDHs. As shown in Table 2, D-fucose, D-xylose, D-galactose and D-ribose as well as D-glucose were good electron donors for GDH of the membrane fraction, while GDH of the soluble fraction showed a low activity for these sugars but not for disaccharides such as maltose and lactose compared with D-glucose. These results are consistent with the data produced by previous investigators (Hauge, 1960b; Hauge & Hallberg, 1964; Duine et al., 1982; Dokter et al., 1987). Most importantly, however, the substrate specificity of the membrane-bound enzyme was shown to be retained after solubilization from the membrane fraction with 1% Triton X-100, which is contrary to the previous results (Hauge & Hallberg, 1964; Dokter et al., 1987). And also, as shown in Table 1, Q-2 reductase activity, relative to PMS/DCIP reductase activity, was considerably higher in the membrane fraction than in the soluble fraction. Such a character of the membrane-bound enzyme was also retained even after solubilization with Triton X-100.

Furthermore, kinetics of GDH was different in the soluble and membrane fractions, which was shown in Lineweaver-Burk plots for glucose. GDH of the soluble fraction exhibited a curve with negative cooperativity,

Table 2. Substrate specificity of GDHs in several fractions of A. calcoaceticus

| Substrate | Relative activity(%) | | |
	Soluble fraction	Membrane fraction	Triton supernatant
D-Glucose	100	100	100
D-Fucose	28	110	119
D-Xylose	15	81	81
D-Galactose	30	79	73
D-Ribose	8	46	54
Maltose	93	12	13
Lactose	72	4	5

while such a negative cooperativity was not observed in the membrane fraction as well as the solubilized fraction (data not shown).

These results suggest that the membrane-bound enzyme is distinct from the soluble enzyme which has been purified and characterized so far.

Purification of soluble and membrane-bound GDHs.

The soluble GDH of A. calcoaceticus has already been purified to homogeneity by several groups (Hauge, 1960a, 1960b & 1964; Dokter et al., 1968; Geiger & Görisch, 1986). In our case, the soluble fraction was used as the starting material and ultracentrifugation was repeated after dialysis of the soluble fraction such that cross-contamination with membrane-bound enzyme was excluded as much as possible. After the dialysis and ultracentrifugation, the enzyme was adsorbed on a DEAE column, which is contrary to the notion obtained by the previous investigators (Hauge, 1960a; Geiger & Görisch, 1986). After eluted from the DEAE column, the enzyme was purified by further several column chromatography, which is summarized in Table 3. As shown, the enzyme was purified about 350-fold from the soluble fraction, and the purified enzyme showed a specific activity of 2200 μmol of glucose oxidized/min/mg protein at pH 6.5 with PMS/DCIP as electron acceptors.

On the other hand, membrane-bound GDH was purified from the membrane fraction in the presence of 0.1% Triton X-114. Since the enzyme is tightly bound to membrane, 0.2% Triton X-100 scarcely solubilizes the enzyme so that it can be used to delete some impure proteins lightly attached to the membranes. By the same reason, solubilization of the enzyme with 1% Triton X-114

Table 3. Purification of soluble GDH

| Fractions | Protein | GDH activity | |
	mg	units	units/mg
Soluble fraction	2132	12726	6
Dialysis and ultra-centrifugation	1061	8490	8
DEAE-Toyopearl	56.2	7609	135
CM-Toyopearl	5.6	6682	1193
Hydroxyapatite	2.3	5011	2178
Phenyl-Sepharose	1.6	3534	2209

was enhanced by the presence of 0.3 M KCl. Thus, the enzyme was purified after solublization by the procedure described in Materials and Methods, the summary of which is shown in Table 4. As shown, the enzyme was purified about 50-fold from the membrane fraction and the purified enzyme showed a specific activity of 570 μmol of glucose oxidized/min/mg protein at pH 8.5 with PMS/DCIP as electron acceptors.

Table 4. Purification of membrane-bound GDH

Fractions	Protein	GDH activity	
	mg	units	units/mg
Membrane fraction	458	4895	10.7
Washed membrane with Triton X-100	384	4693	12.2
Triton X-114 extract	74.1	3981	53.8
Phase separation	40.2	3200	80.0
Dialysis and ultracentrifugation	20.0	1829	91.6
DEAE-Toyopearl	3.5	1337	382
CM-Toyopearl	1.3	742	571

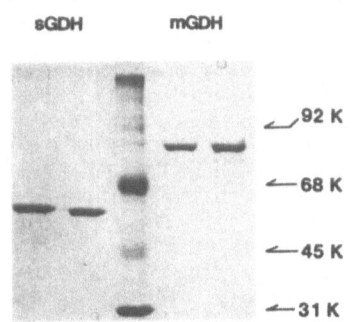

Fig. 1. SDS-PAGE of purified soluble (sGDH) and membrane-bound (mGDH) glucose dehydrogenases from A. calcoaceticus.

The soluble and membrane-bound GDHs thus purified showed a single polypeptide band in SDS-PAGE and their molecular weights were estimated to be 55,000 and 83,000, respectively (Fig. 1). Molecular size in the native form of both enzymes was examined by sucrose density gradient centrifugation (Fig.2). As shown, soluble GDH showed almost the same molecular size in the absence or presence of 0.2% Triton X-100, which is estimated to be about 110,000 - 120,000, relative to yeast alcohol dehydrogenase (Mr 150,000) as a maker. On the other hand, membrane-bound GDH showed several aggregates even in the presence of Triton X-100. Major components might be monomer with 1% Triton, and monomer and dimer both with 0.2 % Triton. Molecular weight of the monomer form was estimated to be 75,000-85,000.

Kinetic properties of soluble and membrane-bound GDHs.
        Possible difference of properties between soluble and membrane-bound GDHs was already shown from the experiments that were performed using soluble and membrane fractions, as described above. In order to elucidate more definitively the relationship between soluble and membrane-bound GDHs, kinetic properties of the purified enzymes were examined. artificial electron acceptors, both

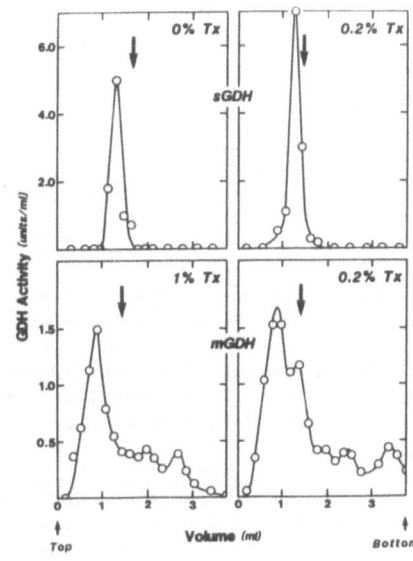

Fig. 2. Sucrose density gradient centrifugation of purified soluble and membrane-bound GDHs.

When PMS and DCIP were used as purified enzymes exhibited

considerably different optimum pHs; pH 8.5 for membrane-bound GDH and pH 6.5 for soluble GDH (Fig. 3). As noticed already in the experiment using the soluble and membrane fractions, purified soluble GDH exhibited a biphasic curve in reciprocal plots for glucose, suggesting the enzyme having a negative cooperativity for the reaction with glucose (Fig. 4). The same phenomenon was observed for lactose with the soluble enzyme (data not shown), but not for glucose with the membrane-bound GDH (Fig. 4). The Km values of the soluble GDH were 24.5 and 26.7mM for glucose and lactose, respectively, which was comparable to the value (30 mM) with the soluble fraction (Table 5). The purified membrane-bound GDH showed a Km value of 4.2 mM for glucose, which is also comparable with Km value(2mM) obtained

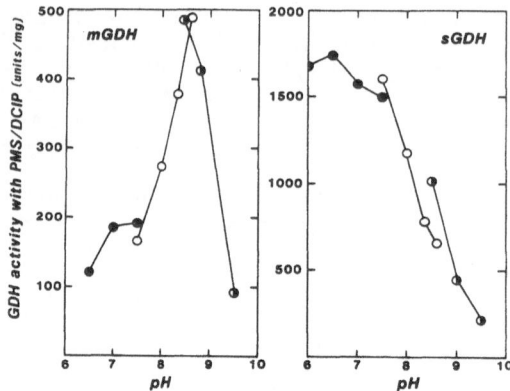

Fig. 3. pH profiles of GDH activity measured with PMS/DCIP as electron acceptors in purified soluble and membrane-bound GDHs.

Table 5. Kinetics for sugar of purified soluble and membrane-bound GDHs

GDHs	Substrate	Km	Vmax
		mM	units/mg
sGDH	Glucose	24.5	3205
	Lactose	26.7	1789
mGDH	Glucose	4.2	614

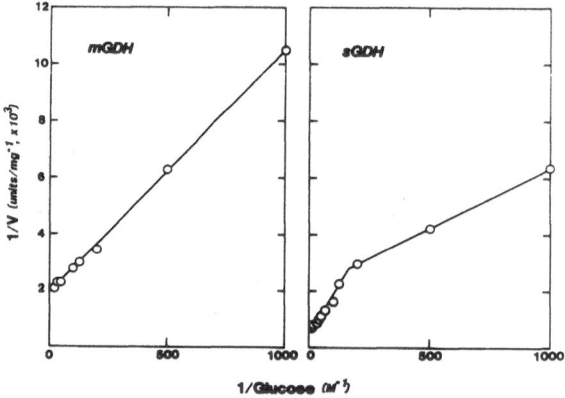

Fig. 4. Lineweaver-Burk plots for glucose of purified soluble and membrane-bound GDHs.

in the membrane fraction. Unlike soluble GDH, purified membrane-bound GDH had no activity for lactose, thus indicating that substrate specificity of the membrane-bound GDH never be changed even by purification.

Reactivity of purified soluble and membrane-bound GDHs with Q homologues.
As shown in Table 1, GDH in the soluble and membrane fractions has a considerably different reactivity for Q-2, suggesting that there is a large difference in the reactivity for Q homologues between soluble and membrane-bound GDHs. With the purified soluble and membrane-bound enzymes, therefore, the reactivity for Q-1 or Q-2 were first examined. Membrane-bound enzyme showed higher reactivity to Q-1 than soluble GDH when the activity was compared relative to PMS reductase, although both enzymes were able to react with Q-1 at almost the same velocity (Table 6). Q-2 reductase activity was much higher with the membrane-bound GDH than the soluble enzyme, and the difference between both enzymes was much

bigger in Q-2 than in Q-1 reductase activities. The distinctiveness of soluble and membrane-bound GDHs became more obvious when optimum pH of Q-2 reductase activity was compared. As shown in Fig. 5, membrane-bound GDH showed two pH optima at acidic region below pH 5 and also at pH 8, while soluble GDH had a single pH optimum at pH 8. The difference between soluble and membrane-bound GDHs was much more critical in the reactivity for Q homologues having longer side-chain (Q-6 or Q-9),which had to be solubilized with octyl-glucoside because of their insolubility in aqueous solution. Membrane-bound GDH was able to react with Q-6 or Q-9 at fairly high rate when adequate concentration of octylglucoside was used (Table 6 and Fig. 6). On the other hand, soluble GDH had appreciably no activity for such Q homologues having longer side-chain at both acidic and alkaline pHs. Interestingly, Q-6 reductase activity of membrane-bound GDH exhibited optimum pH only at acidic region (Fig. 5).

Table 6. Kinetics for PMS and Q homologues of soluble and membrane-bound GDHs.

	sGDH		mGDH	
	Km	Vmax	Km	Vmax
	µM	units/mg	µM	units/mg
PMS	1900	4098	74	1098
Q-1	178	276	148	348
Q-2	17.7	27.0	12.0	208
Q-6	–	0.0	4.4	45.5
Q-9	–	0.0	3.4	40.7

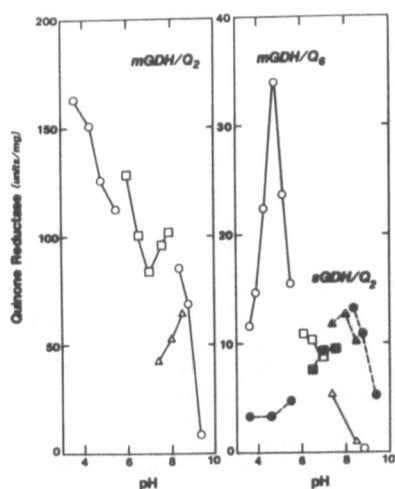

Fig. 5. pH profiles of Q-2 or Q-6 reductase activity of soluble and membrane-bound GDHs.

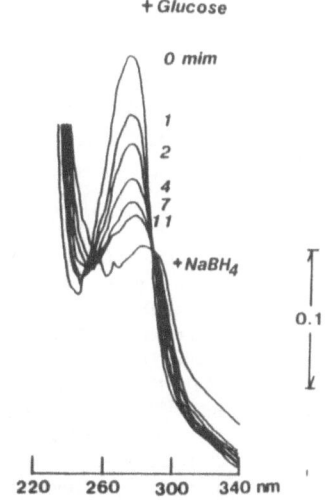

Fig. 6. Reduction of Q-9 solubilized in octylglucoside with membrane-bound GDH.

Immunological relationship between soluble and membrane-bound GDHs.

It has been shown that the membranes of A. calcoaceticus contain a polypeptide of 83 Kdaltons cross-reactive with an antibody raised against GDH purified from the membranes of P. fluorescens (Matsushita et al., 1986). Using the same antibody and an antibody raised against the purified soluble GDH, cross-reactivity of soluble and membrane-bound GDHs was examined by an immunoblotting analysis. As shown in Fig. 7A, the antibody specific for Pseudomonas GDH did not cross-react with soluble

GDH, but with membrane-bound GDH. Furthermore, the antibody prepared against soluble GDH was also shown not to cross-react with membrane-bound GDH even at a relatively high concentration (Fig. 7B). These results clearly indicate that soluble and membrane-bound GDHs are different structurally from each other. Data also showed that the soluble fraction contained a polypeptide of 83 Kdaltons cross-reacted with the antibody for membrane-bound GDH as well as 55 Kdaltons-polypeptide cross-reactive with the antibody specific for soluble GDH. This finding is considered for the soluble fraction to be contaminated partially with the membrane fraction.

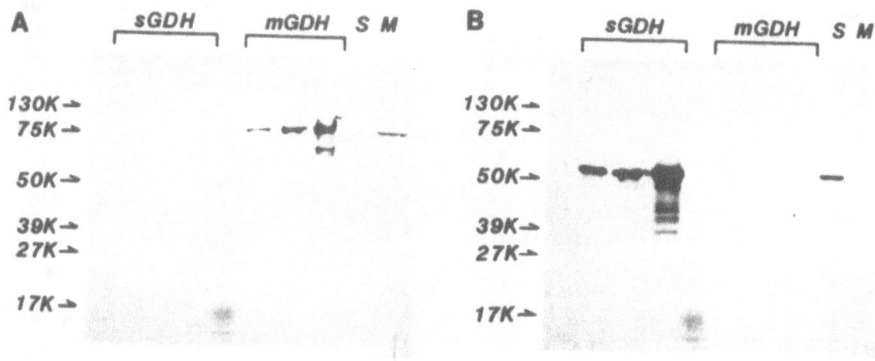

Fig. 7. Immunoblotting analysis of soluble and membrane-bound GDHs, and soluble (S) and membrane (M) fractions with an antibody directed for GDH purified from Pseudomonas (A) and an antibody for the soluble GDH (B).

DISCUSSION

The soluble GDH of A. calcoaceticus has been purified to homogeneity and characterized well by several groups (Hauge, 1960a, 1960b, & 1964; Dokter et al., 1986; Geiger & Gorisch, 1986). On the other hand, the membrane-bound form of the enzyme has not been purified and characterized, and thus seemed to be confused with the soluble enzyme. Hauge and Hallberg (1964) tried to purify it after solubilization from the the membranes with 1.5% deoxycholate; the enzyme was reported to resemble the soluble one as judged by its substrate specificity and chromatographic behavior. Recently, Dokter et al. (1987) have reported that solubilization of the membrane vesicles with 0.02% Triton X-100 changes the substrate specificity of membrane-bound GDH and disaccharides can then be oxidized.

We have solubilized and purified membrane-bound GDHs from P. fluorescens (Matsushita et al., 1980), G. suboxydans (Ameyama et al., 1981) and E. coli (Ameyama et al., 1986). These GDHs are all tightly bound to the membrane, and thus solubilized only with relatively high concentrations of both detergent and salt. Furthermore, these membrane-bound GDHs have a similar molecular weight ranging 83 to 88 Kdaltons and are closely related each other immunologically (Matsushita et al., 1986). These GDHs seem to function by constituting the primary part of the glucose oxidase respiratory chain, which was demonstrated by a reconstitution experiment of the whole system in E. coli (Matsushita et al., 1987). Thus, quinoprotein GDH would be generalized as working bound

to the membrane and linking to the respiratory chain. In this respect,
GDH of A. calcoaceticus would be considered to be exceptional.
    The results described here indicate that the membranes of A.
calcoaceticus contain a distinct enzyme from the soluble GDH purified and
characterized so far. The membrane-bound GDH was shown to have a
completely different character from the soluble GDH in all aspects,
including optimum pH, kinetics for glucose, substrate specificity, Q
reactivity, molecular size and immunogenecity. Especially, the
immunochemical evidence may deny the notion that the soluble form is the
degradation product or the precursor of the membrane-bound enzyme and thus
suggest that the membrane-bound enzyme is a distinct entity from the
soluble one.
    Unlike the soluble GDH, the membrane-bound enzyme was able to react
with Q-9, intrinsic Q in A. calcoaceticus, in octylglucoside solution. In
order to conclusively demonstrate the reactivity of Q-9 with the enzyme in
the native environment, however, the reaction has to be reproduced in an
artificial phospholipid bilayer. Such a reconstitution has been achieved
in the glucose oxidase respiratory chain of E. coli or G. suboxydans
(Matsushita et al., 1987 & unpublished), although it has not been done in
the case of A. calcoaceticus. By analogy to E. coli or G. suboxydans,
therefore, it seems reasonable to conclude that the membrane-bound GDH of
A. calcoaceticus is linking to the respiratory chain via Q. On the other
hand, it is not evident how does the soluble GDH link to the respiraotry
chain, since the enzyme is not bound to the membrane and does not react
with Q. Thus the biological role of soluble GDH remains unclear at this
moment.

ACKNOWLEDGEMENT

    We thank S. Sakurai and S. Kawanaka for their assistance with the
experiments. This work was supported in part by a grant from the Naito
Fundation.

REFERENCES

Ameyama M, Shinagawa E, Matsushita K and Adachi O, 1981. D-Glucose
dehydrogenase of Gluconobacter suboxydans. Agric. Biol. Chem. 45: 851-861.

Ameyama M, Nonobe M, Shinagawa E, Matsushita K, Takimoto K and Adachi O,
1986. Purification and characterization of quinoprotein apo-D-glucose
dehydrogenase from Escherichia coli. Agric. Biol. Chem. 50: 49-57.

Cleton-Jansen AM, Goosen N, Wenzel TJ and Van de Putte P, 1988. Cloning of
the gene encoding quinoprotein glucose dehydrogenase from Acinetobacter
calcoaceticus: Evidence for the presence of a second enzyme. J. Bacteriol.
170: 2121-2125.

Dokter P, Frank Jzn J and Duine JA, 1986. Purification and
characterization of quinoprotein glucose dehydrogenase from Acinetobacter
calcoaceticus L.M.D.79.41. Biochem. J. 239: 163-167.

Dokter P, Pronk JT, Van Schie BJ, Van Dijken JP and Duine JA, 1987. The in
vivo and in vitro substrate specificity of quinoprotein glucose
dehydrogenase of Acinetobacter calcoaceticus LMD79.41. FEMS Microbiol.
Lett. 43: 195-200.

Duine JA, Frank Jzn J and Van der Meer R, 1982. Different forms of quinoprotein aldose- (glucose-) dehydrogenase in Acinetobacter calcoaceticus. Arch. Microbiol. 131: 27-31.

Hauge JG, 1960a. Purification and properties of glucose dehydrogenase and cytochrome b from Bacterium anitratum. Biochim. Biophys. Acta 45: 250-262.

Hauge JG, 1960b. Kinetics and specificity of glucose dehydrogenase from Bacterium anitratum. Biochim. Biophys. Acta 45: 263-269.

Hauge JG, 1964. Glucose dehydrogenase of Bacterium anitratum: an enzyme with a novel prosthetic group. J. Biol. Chem. 239: 3630-3639.

Hauge JG and Hallberg PA, 1964. Solubilization and properties of the structurally-bound glucose dehydrogenase of Bacterium anitratum. Biochim. Biophys. Acta 81: 251-256.

Geiger O and Görisch H, 1986. Crystalline quinoprotein glucose dehydrogenase from Acinetobacter calcoaceticus. Biochemistry 25: 6043-6048.

Matsushita K, Ohno Y, Shinagawa E, Adachi O and Ameyama M, 1980. Membrane-bound D-glucose dehydrogenase from Pseudomonas sp.: solubilization, purification and characterization. Agric. Biol. Chem. 44: 1505-1512.

Matsushita K, Ohno Y, Shinagawa E, Adachi O and Ameyama M, 1982. Membrane-bound, electron transport-linked, D-glucose dehydrogenase of Pseudomonas fluorescens. Interaction of the purified enzyme with ubiquinone or phospholipid. Agric. Biol. Chem. 46: 1007-1011.

Matsushita K, Shinagawa E, Inoue T, Adachi O and Ameyama M, 1986. Immunological evidence for two types of PQQ-dependent D-glucose dehydrogenase in bacterial membranes and the location of the enzyme in Escherichia coli. FEMS Microbiol. Lett. 37: 141-144.

Matsushita K, Nonobe M, Shinagawa E, Adachi O and Ameyama M, 1987. Reconstitution of pyrroloquinoline quinone-dependent D-glucose oxidase respiratory chain of Escherichia coli with cytochrome o oxidase. J. Bacteriol. 169: 205-209.

Matsushita K, Shinagawa E, Adachi O and Ameyama M, 1988. Quinoprotein D-glucose dehydrogenase in Acinetobacter calcoaceticus LMD 79.41: the membrane-bound enzyme is distinct from the soluble enzyme. FEMS Microbiol. Lett. in press.

CLONING OF THE GENES ENCODING THE TWO DIFFERENT GLUCOSE
DEHYDROGENASES FROM *ACINETOBACTER CALCOACETICUS*

Anne-Marie Cleton-Jansen, Nora Goosen, Kees Vink, Pieter van de Putte

Laboratory of Molecular Genetics, University of Leiden, P.O. Box 9505,
2300 RA Leiden, The Netherlands

Keywords: glucose dehydrogenase, *Acinetobacter calcoaceticus*,
substrate specificity, DNA sequence, signal peptide.

ABSTRACT
Glucose dehydrogenase (GDH) is a PQQ dependent bacterial enzyme which
converts aldoses to their corresponding acids. *A. calcoaceticus* contains
two different PQQ dependent glucose dehydrogenases designated GDH-A which
is active *in vivo* and GDH-B of which only *in vitro* activity can be shown.
We cloned the genes coding for the two GDH enzymes. The DNA sequences of
both *gdh* genes were determined. There is no obvious homology between *gdh*A
and *gdh*B. Both GDH enzymes oxidize D-glucose *in vitro* but disaccharides
are specific GDH-B substrates and 2-deoxyglucose is specifically oxidized
by GDH-A.

INTRODUCTION

The quinoprotein glucose dehydrogenase (GDH) (E.C.1.1.99.17) is a
membrane bound enzyme which converts glucose and other aldoses to their
corresponding acids (Hauge, 1966). In this reaction pyrrolo
quinoline-quinone (PQQ) is used as a coenzyme (Duine et al., 1979). PQQ
dependent glucose dehydrogenase has been detected in a wide variety of
bacterial species, some of which synthesize the cofactor PQQ themselves
(e.g. *Acinetobacter calcoaceticus*) (Duine et al., 1979) while others are
dependent on the presence of PQQ in the culture medium (e.g. *Escherichia
coli*) (Hommes et al., 1984).
PQQ dependent GDH has been purified from several bacterial species:
Pseudomonas fluorescens (Matsushita et al., 1980), *Gluconobacter
suboxydans* (Ameyama et al., 1981) and *E. coli* (Ameyama et al. 1986). The
enzymes from these organisms have molecular weights, ranging from 83,000
- 88,000 under denaturing conditions. Antibodies raised against
purified GDH from *P. fluorescens* showed cross-reactivity with crude
membranes of *G. suboxydans*, *Klebsiella pneumoniae*, *Acetobacter aceti*, *P.
aeruginosa*, *E. coli* and *A. calcoaceticus* (Matsushita et al., 1986),
suggesting that the GDH enzymes from these bacterial species are closely
related.
Dokter et al. (1986) and Geiger and Görisch (1986) reported
independently the isolation of a PQQ dependent GDH from *A. calcoaceticus*
consisting of two identical subunits . The molecular weight of the subunit
was approximately 50,000 which is inconsistent with the results from the
immuno-cross-reactivity between anti-GDH antibodies and a M_r 85,000

J. A. Jongejan and J. A. Duine (eds.), PQQ and Quinoproteins, 79–85.
© *1989 by Kluwer Academic Publishers.*

protein in *A. calcoaceticus* (Matsushita et al., 1986). Moreover the purified Mr 50,000 GDH has a substrate specificity which differs from the *in vivo* substrate specificity of the *A. calcoaceticus* GDH. Cell free extracts and purified M_r 50,000 GDH oxidize disaccharides like lactose and maltose efficiently in the presence of an artificial electronacceptor. *A. calcoaceticus* whole cells, however, are not able to oxidize disaccharides. These results suggest that *A. calcoaceticus* contains two different PQQ dependent enzymes with GDH activity: one of M_r 85,000 which will be designated GDH-A and a dimer with two identical subunits of approximately M_r 50,000 GDH referred to as GDH-B.

In this paper we describe the cloning of the genes encoding GDH-A and GDH-B from *A. calcoaceticus*. The DNA sequences of *gdh*A (Cleton-Jansen et al., 1988b) and *gdh*B (Cleton-Jansen et al., submitted) show that GDH-A and GDH-B are indeed different enzymes and that GDH-B is not a degradation product of GDH-A. Mutants from both *gdh* genes are isolated and tested for their *in vitro* substrate specificity.

MATERIAL & METHODS

Bacterial strains, plasmids and culture conditions. *Escherichia coli, Acinetobacter calcoaceticus* and *Pseudomonas* strains were cultured in L-broth, on L-agar plates or in defined minimal medium as indicated (Miller, 1972). Plasmids were transformed to *E. coli* as described previously (Maniatis et al., 1978). Tetracycline was used at a concentration of 20 μg/ml, kanamycin at 50 μg/ml and ampicillin at 40 μg/ml for all strains.
Bacterial matings to transfer plasmids were performed as described (Goosen et al., 1987).

Chemicals and reagents. All restriction enzymes, T4 DNA ligase, Klenow fragment of DNA polymerase I and calf intestine phosphatase were obtained commercially. Purified PQQ and Würsters Blue, an electron acceptor-dye, made by one electron oxidation of N,N,N',N'-tetra methyl-p-phenylenedia-mine, were gifts of M. van Kleef. PQQ was used at a concentration of 2 μM for *A. calcoaceticus* and 12 μM for *E. coli* and *P. stutzeri*.

Analysis of plasmid DNA. Plasmid DNA from *E. coli* was isolated by the method of Birnboim and Doly as described (Maniatis et al., 1978). Electrophoresis of DNA fragments was carried out on 1% agarose gels in 40 mM Tris acetate, 2 mM EDTA (pH8).

Screening of the *A. calcoaceticus* genomic library with a synthetic oligonucleotide probe. We screened a genomic library of *A. calcoaceticus* genomic DNA with a mixture of 20 mer oligonucleotides derived from a part of the N-terminal protein sequence of the purified GDH-B protein. The oligonucleotide mixture consists of a pool of 64 different DNA fragments, in which all combinations of codons necessary for translation of this aminoacid sequence are present. The *A. calcoaceticus* genomic library in vector pLV21 was described by Goosen et al. (1987). The library was transformed to *E. coli* MH1 and transferred to Hybond-N (Amersham, UK) nylon membranes as described by the manufacturer. 50 pMol oligonucleotide mixture (5' GA^A_GAA^T_CTT^T_CGA^T_CAA^A_GAA^A_GGT 3') was labeled with [γ-^{32}P] ATP and hybridised with these membranes. After specific washing of the membranes positive clones were visualised by autoradiography.

DNA sequencing was performed with the dideoxymethod (Sanger et al., 1977). Sequences were compiled and analysed using the Sequence Analysis Software Package of the University of Wisconsin Genetics Computergroup (Deveraux et al., 1984).

Isolation of insertion mutants in both *gdh* genes. Insertion mutations in both *gdh* genes were constructed by gene disruption. The *gdh*A or *gdh*B gene was cloned in pGP173 (Goosen et al., 1987), a pBR322 derivative containing a mobilisation site, which enables the vector to be transferred to *A. calcoaceticus*. However, this plasmid cannot be propagated in this strain. Within the coding sequences of the *gdh* genes a DNA fragment encoding kanamycine resistance (*kan*) was cloned. The plasmids were transferred to *A. calcoaceticus* and transconjugants were selected for *kan* resistance. In these strains the chromosomal *gdh* gene and the disrupted *gdh-kan* gene have recombined. We selected for double recombinants by excluding transconjugants which are also ampicillin resistant, which is the antibiotic marker of pGP173.

RESULTS

Cloning of the *gdh*A gene. For cloning of the *gdh*A gene mutants were isolated that are affected in the *in vivo* GDH activity (Cleton-Jansen et al., 1988). A genomic library of *A. calcoaceticus* described by Goosen et al. (1987) was transferred to one of the GDH⁻ mutants and transconjugants were tested for complementation of their GDH activities. In this way a plasmid (pGP426) was found that complemented all 15 GDH⁻ mutants. We studied the expression of plasmid pGP426 in *Pseudomonas auruginosa* strain 2F32-106B, which has a mutation in *gdh* (Midgley & Dawes, 1973), in *P. stutzeri*, which naturally lacks the GDH enzyme (M. van Kleef, pers. comm.) and in *E. coli* PP1795, which is also mutated in the *gdh* gene (Cleton-Jansen, manuscript in preparation). The *A. calcoaceticus gdh*A gene is able to restore the GDH activities of all three strains.

Cloning of the *gdh*B gene. For cloning the *gdh*B gene of *A. calcoaceticus* we used the purified GDH-B protein. First the N-terminal aminoacid sequence (23 residues) of GDH-B was determined by Edman degradation (Edman & Begg, 1967). A pool of 20 mer oligonucleotides was synthesized from which the nucleotide sequence corresponds to a part of this aminoacid sequence. This mixture of oligonucleotides was used as a probe to screen the same genomic library of *A. calcoaceticus*. One positive clone was found and DNA sequencing of part of the inserted DNA fragment of this plasmid revealed the presence of a sequence from which an aminoacid sequence could be derived which corresponded to the complete aminoacid sequence as determined by Edman degradation.

In vitro transcription-translation of *gdh*A and *gdh*B. The size of the proteins encoded by the *gdh*A and *gdh*B genes was defined using a commercially available 'Prokaryotic DNA-directed translation kit' (Amersham UK). As this kit contains extracts of *E. coli* the *gdh* genes were cloned under control of the strong *lac* promotor from pUC19. The proteins encoded by these plasmids were visualised on a denaturing polyacrylamide gel which shows that *gdh*A codes for a protein of about Mr 85,000. The product of *gdh*B is estimated on Mr 50,000.

DNA sequences of gdhA and gdhB. The complete nucleotide sequence of the gdhA gene was determined (Cleton-Jansen et al., 1988b) using the dideoxy method (Sanger et al., 1977). It codes for a protein of M_r 86,956. The 140 residues N-terminal part of the GDH-A protein is very hydrophobic. This region contains five hydrophobic domains flanked by polar residues, which are putative membrane spanning segments. It suggests that the N-terminal tail is the domain that anchors the protein in the cytoplasmic membrane.

The gdhB gene was also sequenced (Cleton-Jansen et al., submitted for publication). The M_r of the translated protein is 52,772. The N-terminal aminoacid sequence which was determined by Edman degradation is preceded by a 24 aminoacids leader sequence, which has the characteristics of a signal peptide. Apparently this peptide is cut from the protein after translation between alanaine and aspartate resulting in the mature GDH-B with a M_r of 50,237.

We compared the DNA- and aminoacid sequences of both A. calcoaceticus gdh genes. No obvious homology between gdhA and gdhB could be detected.

Insertion mutants of gdhA and gdhB. Mutants of both gdh genes in A. calcoaceticus were constructed by gene disruption as described in the Material & Methods section. Strain PP2403 carries a kan insertion in the gdhB gene and PP2407 contains a kan insertion in gdhA. PP2410 has a kan insertion in gdhA and a tetracycline encoding fragment in gdhB. On MacConkey agar containing D-glucose it was shown that the gdhB mutant PP2403 still has in vivo GDH activity. PP2407 the gdhA mutant and PP2410, the gdhA gdhB double mutant have no in vivo GDH activity.

Table: In vitro GDH activities
GDH activities are expressed in Units/ml cell free extract.

Strain	glucose	lactose	maltose	2-deoxyglucose
LMD79.41	2.3	0.5	0.9	0.9
PP2403	0.4	0.0	0.0	0.5
PP2407	1.3	0.5	0.6	0.0
PP2410	0.0	0.0	0.0	0.0

The in vitro GDH activity of these mutants was compared with wildtype A. calcoaceticus using several substrates. The results are presented in the table. Both gdhA and gdhB mutants still showed oxidation when D-glucose was used as a substrate, implicating that both GDH enzymes are active in this in vitro system. In PP2410, where both gdh genes are disrupted glucose oxidation could no longer be shown. The disaccharides lactose and maltose are oxidized on wildtype level when gdhA is mutated, but in PP2403 disaccharides are no substrates. This is implicating that GDH-B is responsible for the in vitro oxidation of these sugars. GDH-A is specific for 2-deoxyglucose oxidation because this substrate can only be converted by the gdhB mutant.

Phenotypic in vivo effects of the gdhB mutation could not be detected. Neither the in vivo GDH activity of PP2403, nor the growth rate, or the colony morphology differed from wild type.

DISCUSSION

This report describes the isolation of the genes coding for two different GDH enzymes, designated GDH-A and GDH-B, from *A. calcoaceticus*. The clone which is able to complement *in vivo* GDH⁻mutants of *A. calcoaceticus*, *P. auruginosa* and *E. coli* codes for GDH-A and contains an open reading frame resulting in a protein with M_r 86,956. This was shown by *in vitro* transcription-translation and DNA sequencing.

The second *gdh* gene was isolated by hybridisation with a 20 mer oligonucleotide probe derived from the N-terminal aminoacid sequence of the purified GDH-B protein. The M_r of the protein product of this *gdh*B gene was shown to be 52,772. The mature GDH-B is preceded by a 24 aminoacid leader sequence which is cut from the protein between alanine and aspartate. The resulting protein has a M_r of 50,237. This is in accordance with the M_r of the purified protein (Dokter et al., 1986, Geiger & Görisch, 1986).

The leader sequence of GDH-B has the characteristics of a signal sequence that probably translocates the GDH-B through the plasma membrane into the periplasmic space. GDH-A is also acting in this space but lacks a signal sequence. GDH-A, however, has a 140 residues N-terminus of high hydrophobicity which suggests that the protein is anchored firmly in the cytoplasmic memebrane. We can distinguish five hydrophobic domains which are separated by hydrophilic residues. These are possible membrane spanning segments. GDH-B does not contain a hydrophobic region and is probably not, or only loosely associated with the cytoplasmic membrane.

Duine et al. (1982) reported the presence of a soluble GDH and a membrane asociated GDH in *A. calcoaceticus*. The soluble form is most probably GDH-B, while GDH-A which is strongly anchored in the cytoplasmic membrane with the hydrophobic N-terminal domain is the membrane associated form. The completely different DNA sequences of *gdh*A and *gdh*B show that these two forms of the GDH activity are different proteins. The total lack of homologous regions between the *gdh* genes is striking, considering the fact that both enzymes perform the same reaction and use PQQ as cofactor.

The *in vitro* substratespecificity enables us to discriminate between the activities of GDH-A and GDH-B. The results of insertion mutants in both *gdh* genes show that D-glucose can be used as a substrate by both GDH-A and GDH-B. However, 2-deoxyglucose is a specific GDH-A substrate and disaccharides, like lactose and maltose are only oxidized by GDH-B.

GDH-A is, with our laboratory conditions the only *in vivo* active GDH. GDH-B activity can only be shown *in vitro* when an artificial electron acceptor is present. It has already been postulated (Beardmore-Gray & Anthony, 1986) that the soluble GDH of *A. calcoaceticus* does not play a role in glucose oxidation. It is not very likely that GDH-B acts on another substrate than sugars. Therefore it seems that *A. calcoaceticus* lacks a factor which is necessary in the electrontransport chain between GDH-B and the cytochromes. This factor could be lost during the evolution or be present in the natural surroundings of *A. calcoaceticus*.

ACKNOWLEDGEMENTS
We thank P. Dokter for his gift of purified GDH-B protein and R. Amons for determining the N-terminal aminoacid sequence.
The investigations were supported by the Netherlands Technology Foundation (STW) as part of the joint venture Biotechnology Delft Leiden (BDL).

REFERENCES

Ameyama M Shinagawa E, Matsushita K, Adachi O (1981) D-glucose dehydrogenase of *Gluconobacter suboxydans*: solubilization, purification and characterization. Agric. Biol. Chem. 45: 851-861.

Ameyama M, Nonobe N, Shinagawa E, Matsushita K, Taikimoto K, Adachi O (1986) Purification and characterization of the quinoprotein D-glucose dehydrogenase apoenzyme from *Escherichia coli*. Agric. Biol. Chem. 50: 49-57.

Beardmore-Gray M, Anthony C (1986) The oxidation of glucose by *Acinetobacter calcoaceticus*: interaction of the quinoprotein glucose dehydrogenase with the electron transport chain. J. Gen. Microbiol. 132: 1257-1268

Cleton-Jansen AM, Goosen N, Wenzel TJ, Van de Putte P (1988a) Cloning of the gene encoding quinoprotein glucose dehydrogenase from *Acinetobacter calcoaceticus*: evidence for the presence of a second enzyme. J. Bacteriol. 170:2121-2125

Cleton-Jansen AM, Goosen N, Odle G, Van de Putte P (1988b) Nucleotide sequence of the gene coding for quinoprotein glucose dehydrogenase from *Acinetobacter calcoaceticus*. Nucl. Acid. Res. 16:6228

Deveraux J, Haeberli P, Smithies O (1984) A comprehensive set of sequence analysis programs for the VAX. Nucl. Acids Res. 12:443-446

Dokter P, Frank J, Duine JA (1986) Purification and characterization ofquinoprotein glucose dehydrogenase from *Acinetobacter calcoaceticus* LMD 79.41. Biochem. J. 239:163-169

Duine JA, Frank J, Van Zeeland K (1979) Glucose dehydrogenase from *Acinetobacter calcoaceticus*: a "quinoprotein". FEBS Lett. 108:443-446

Duine JA, Frank J, Van der Meer R (1982) Different forms of quinoprotein aldose (glucose)-dehydrogenase in *Acinetobacter calcoaceticus*. Arch. Microbiol. 131: 27-31

Edman P, Begg G (1967) A protein sequenator. Eur. J. Biochem. 1:80-91

Geiger O, Gorisch H (1986) Crystalline quinoprotein glucose dehydrogenase from *Acinetobacter calcoaceticus*. Biochemistry 25:6043-6048

Goosen N, Vermaas DAM, Van de Putte P (1987) Cloning of the genes involved in the synthesis of coenzyme pyrrolo-quinoline quinone from *Acinetobacter calcoaceticus*. J. Bacteriol. 169:303-307

Hommes RJW, Postma PW, Neijssel OM, Tempest DW, Dokter P, Duine JA (1984) Evidence of a quinoprotein glucose dehydrogenase apoenzyme in several strains of *Escherichia coli*. FEMS Microbiol. Lett. 24: 329-333.

Hauge JG (1966) Glucose dehydrogenase: *Pseudomonas* sp. and Bacterium anitratum. Methods Enzymol. 9:92-98

Maniatis T, Fritsch EF, Sambrook J (1982) Molecular cloning: a laboratory manual. Cold Spring Harbor Laboratory, Cold Spring Harbor, N.Y.

Matsushita K, Ohno Y, Shinagawa E, Adachi O, Ameyama M (1980) Membrane bound D-glucose dehydrogenase from *Pseudomonas* sp.: solubilization, purification and characterization. Agric. Biol. Chem. 44: 1505-1512.

Matsushita K, Shinagawa E, Inoue T, Adachi O, Ameyama M. (1986) Immunological evidence for two types of PQQ dependent D-glucose dehydrogenase in bacterial membranes and the location of the enzyme in *Escherichia coli*. FEMS Microbiol. Lett. 37: 141-144

Messing J, Crea R, Seeberg PH (1981) A system for shotgun DNA sequencing. Nucl. Acids Res. 9:309-321

Midgley M, Dawes EA (1973) The regulation of glucose and methyl-glucoside uptake in *Pseudomonas aeruginosa*. Biochem. J. 132:141-154

Sanger F, Nicklen S, Coulson AR (1977) DNA sequencing with chain-termination inhibitors. Proc. Natl. Acad. Sci. USA 74:5463-5467

Investigations on the Active Site of Glucose Dehydrogenase from Pseudomonas fluorescens

YUJIRO IMANAGA
Department of Chemistry, Nara Women's University, Kitauoyanishi-Machi, Nara, Nara 630, Japan

Abstract. Effects of several inhibitors on the glucose dehydrogenase (EC 1. 1.99.17) from Pseudomonas fluorescens were studied. 2,3-Butanedione (under room light), 8-anilinonaphthalenesulfonate, ethoxyformic anhydride and rosebengal (under irradiation with tungsten lamp) inactivated especially apoenzyme, and Arg and His residue are presumed to participate in the formation of holoenzyme complex. p-chloromercuribenzoate and p-chloromercuribenzenesulfonate also inactivated apoenzyme, but the inactivation was prevented by preincubation with Ca^{2+} alone. N-acetylimidazole showed similar effects on apoenzyme, the activity being restored by hydroxylamine-treatment of the inactivated apoenzyme. These results suggest that both Cys and Tyr residues of apoenzyme bind directly to Ca^{2+} in holoenzyme complex. N-ethylmaleimide and 7-chloro-4-nitrobenzo-2-oxa-1,3-diazole, on the contrary, inactivated especially holoenzyme, and both inactivations were accelerated in the presence of glucose. These results suggest that another type of Cys residue participates directly in the enzyme reaction. Phenylhydrazine inactivated holoenzyme and glucose protected holoenzyme against the inactivation. CN^- inhibited holoenzyme reversibly and competed with substrate glucose in the enzyme reaction. These results suggest the interaction of C(5)-carbonyl group of pyrroloquinoline quinone (PQQ) with glucose. Ca^{2+} promoted the non-enzymic reduction of PQQ by glucose, but was rather inhibitory to the reoxidation process. An active site model and a possible reaction mechanism are proposed.

INTRODUCTION

Membrane-bound glucose dehydrogenase (GDH) from Pseudomonas fluorescens requires PQQ and an essential metal ion such as Ca^{2+} and Mg^{2+}, and is easily converted to the apoform by EDTA-treatmet and following dialysis in contrast with that from Gluconobacter suboxydans where PQQ is tightly attached to the enzyme protein and EDTA has no effect on the reaction(Imanaga et al.1979;Tanaka et al.1981). On the other hand, GDH from Acinetobacter calcoaceticus is converted to the apoform by dialysis against 0.1M acetate buffer (pH 4.5) containing 3M KBr(Duine et al.1979). In P. fluorescens GDH cofactor PQQ seems to bind to the apoprotein in a complex with such metal ion (Imanaga et al.1979;Ameyama et al.1985). But, the amino acid residues of the enzyme protein participating in such complexation are still unknown.

In this paper the effects of several inhibitors such as specific protein modifiers on the enzyme reaction were studied to clarify the binding mode of PQQ-Ca^{2+} and substrate glucose to the protein, and also the reaction mechanism of P. fluorescens GDH, using partially purified enzyme(Imanaga et al.1985;Yokoyama et al.1986;Tsutsumi et al.1987;Imanaga 1987). Non-enzymic reactions of glucose with PQQ in the presence and absence of Ca^{2+} were also

J. A. Jongejan and J. A. Duine (eds.), PQQ and Quinoproteins, 87–95.

studied as a model system.

MATERIALS AND METHODS

Chemicals

PQQ was prepared from the culture medium of P. fluorescens applying Sephadex G-25 gel filtration in stead of PPC and TLC (Imanaga et al. 1979). Commercial preparation (Kanto Chemicals, Tokyo) was also used in non-enzymic experiments. 2,6-Dichlorophenolindophenol (DCIP) was from Wako Pure Chemical Industries, Osaka. Sepharose CL-6B was from Pharmacia Fine Chemicals. Other reagent quality chemicals were from Nakarai Chemicals Kyoto. 2,3-Butanedione was distilled and 1-anilinonaphthalene was recrystallized from ethanol.

Preparation of apoenzyme

All procedures were carried out below 4 °C and 5 mM veronal-HCl buffer (pH 7.0)(buffer I) was used also to make reagent solutions.(Imanaga et al. 1979)

Bacterial membrane fraction was mixed with an equal volume of 2% cholate solution, stirred for 2 h and once frozen. The mixture was centrifuged at 10^5xg for 40 min, the precipitate was suspended in buffer I (1:2 by weight) and treated with an equal volume of 1% deoxycholate solution. The mixture was centrifuged at 10^5xg for 40 min and the supernatant was slowly mixed with a 1/3 volume of 0.1 M $CaCl_2$ solution. The mixture was kept at 0 °C for 1 h and centrifuged. The supernatant was mixed with 0.1 M EDTA solution, keeping the pH in a range of 6-7. After the decrease in pH had stopped, 10 mM excess of EDTA was added. The precipitate obtained by $(NH_4)_2SO_4$ fractionation (15-40%) was dissolved in buffer I (1:5 by weight) and dialyzed against buffer I. The dialyzate [absorbance at 260 nm(A_{260})=a, b ml] was mixed with 1% streptomycin solution (6.5xaxb μl)to remove nucleic acid and the mixture was centrifuged at 10^4xg for 20 min. To the supernatant, glycerol was added to 10% (v/v), and the solution was concentrated by centrifugation in a Centriflo ultrafiltration cone (Amicon). The concentrate was applied to Sepharose CL-6B column (2x125 cm) preequilibrated with buffer I containing 10% glycerol, and gel-filtration chromatography was performed using the same buffer. The main part of the eluate was concentrated and rechromatographed. Specific activity, 2-3; apoenzyme content, >91%.

Enzyme assay

The enzyme activity was assayed by measuring the reduction of A_{600} of DCIP in 5 min at 20°C, using the cuvette with 1 cm light path. The complete mixture in the cuvette contained enzyme solution, 1 mM $CaCl_2$, 40 nM PQQ, 71 μM DCIP, 4.3 mM glucose and 25 mM veronal-HCl buffer(pH 7.0) (buffer II) in a total volume of 3.5 ml. Apoenzyme was incubated in buffer II with Ca^{2+} and PQQ for 25 min to reconstitute holoenzyme. The reaction was started by adding glucose. Reference cuvette contained only DCIP in buffer II. In the case of the inhibition studies, enzyme (holo or apo) was incubated with an inhibitor in buffer II at 20°C (in the case of ethoxyformic anhydride or organic mercurials, 2 mM phosphate buffer(pH 7.0) was used as they react with buffer II) and a sample was taken for enzyme assay at regular time intervals; apoenzyme was instantaneously converted to holoenzyme by adding Ca^{2+} and 10-fold amount of PQQ (0.4 μM).

Non-enzymic reaction of glucose with PQQ and PQQ-Ca^{2+}.

A Thunberg-type cuvette with two side-arms contained 20 μM PQQ and 2 mM CaCl$_2$ in 2.5 ml of 20 mM acetate buffer (pH 6.7) in one arm and 45 mg of glucose in the other. CaCl$_2$ was omitted from control cuvette. PQQ solution was degassed by four freeze-pump-thaw cycles on a vacuum line (below 10^{-4} mmHg, and then sealed into cuvette. The reaction was started by dissolving glucose in PQQ solution and the mixture in the cuvette was kept at 37°C. Spectral change of the solution was followed using Hitachi-200 spectrophotometer. After opening the cuvette to air spectral change was followed again at 23°C.

RESULTS

Inactivation of GDH by 2,3-butanedione (BD)

Incubation of apoenzyme with BD resulted in a progressive loss of enzyme activity and the inactivation followed a pseudo-first order kinetics. When apoenzyme had been preincubated with Ca^{2+} and PQQ (reconstitution of holo-enzyme), the rate of inactivation by BD was greatly reduced (Figure 1), indicating that binding of PQQ-Ca^{2+} to the apoprotein protected the enzyme against BD-attack. A plot of log k_{app} (apparent first order rate constant) versus log BD concentration resulted in a straight line with a slope 1.1 (Figure 1, inset), indicating that 1 mol of BD per mol of enzyme was required to produce inactivation. Arg residue of the enzyme, the possible target of modification by BD, was anticipated to be a

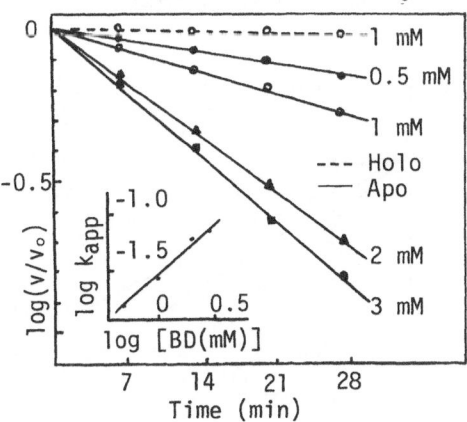

Figure 1. First-order plots of the inactivation of GDH by BD.

linkage site to a carboxyl group of PQQ. But, inactivation of apoenzyme by BD was not observed in the dark. Radical mechanism of the reaction and modification of several amino acid residues including Arg have been reported in such photochemical process (Mäkinen et al. 1982). Though 2-hydroxy-5-nitrobenzyl bromide, Trp-modifying reagent, was without effect, the amino acid residue responsible for BD-inactivation awaits further investigation.

Inhibition by 8-anilino-1-naphthalenesulfonate (ANS)

ANS has an analogous structure to PQQ. ANS (0.8 mM, 15 min) inhibited apoenzyme, even in the presence of Ca^{2+}, more effectively than holoenzyme, but a part of the activity was recovered from ANS-inactivated apoenzyme during assay of enzyme activity (Figure 2). 1-Anilinonaphthalene (AN, 0.4 mM, 15 min) which lacks sulfonic acid group showed no inhibitory effect on holoenzyme and only a weak irreversible inactivation effect (32%) on apoenzyme. These results suggest that ANS binds to a positively-charged site (possibly Arg residue) of apoenzyme which may normally participate in binding to PQQ-COO$^-$ in holoenzyme, and the bound ANS is gradually replaced by PQQ(0.4 μM) during assay of enzyme activity:formation of active holoenzyme.

Inactivation by N-acetylimidazole (NAI)

NAI inactivated apoenzyme following a pseudo-first order kinetics and the protection against inactivation was observed in the case of holoenzyme (Figure 3a). The kinetic order of inactivation was 0.92, indicating that modification of one mol of amino acid residue was responsible for inactivation. Further, the inactivation (15 mM NAI) could be partially prevented by preincubation of apoenzyme with Ca^{2+} (67 mM) alone (Figure 3b). When the inactivated apoenzyme was treated with 0.1 M hydroxylamine about 50% of the original activity was recovered. These results suggest that $Tyr-OH$ rather than $LYS-NH_2$ binds to Ca^{2+} in holoenzyme complex.

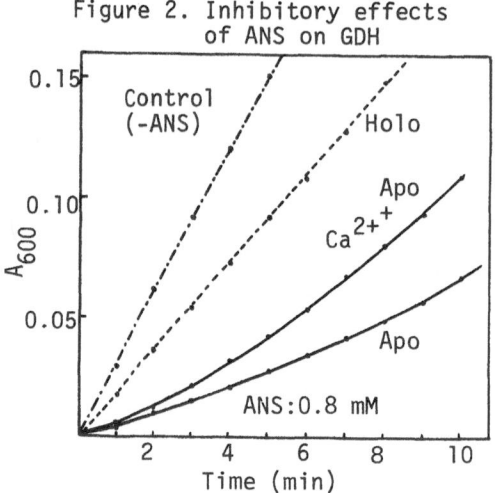

Figure 2. Inhibitory effects of ANS on GDH

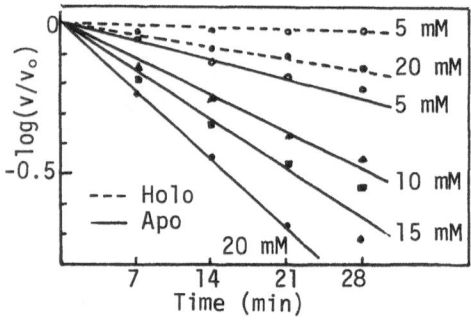

Figure 3a. First-order plots of the inactivation of GDH by NAI.

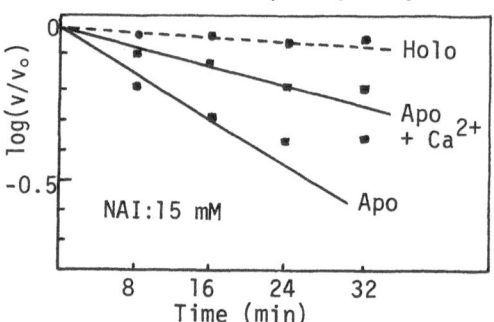

Figure 3b. Effects of Ca^{2+} on the inactivation of apoenzyme by NAI.

Inactivation by p-chloromercuribenzoate (PCMB) and p-chloromercuribenzenesulfonate (PCMBS)

PCMB inactivated apoenzyme more effectively than holoenzyme, but the inactivation (1.6 mM) could be prevented by preincubation of the apoenzyme with Ca^{2+} (1.3 mM) alone (Figure 4a). Similar protective effect of Ca^{2+} (33 mM) was also observed in the case of PCMBS (3 mM)-inactivation (Figure 4b). So, these organic mercurials may react with the essential Cys residue for Ca^{2+}-binding to the enzyme protein.

Inactivation by ethoxyformic anhydride (EFA)

EFA inactivated apoenzyme more effectively than holoenzyme, following a pseudo-first order kinetics. The kinetic order of inactivation was 1.1, indicating that the reaction of single amino acid residue per enzyme leads to inactivation of the enzyme (Figure 5). Inactivated enzyme showed an absorption peak at 240 nm corresponding to the modification of His residue of the protein. But, in this case, Ca^{2+} (1.3 mM) alone could not prevent the inactivation. Photochemical inactivation of the apoenzyme by rosebengal (3.1 nM to 12 nM; pseudo-first order reaction; kinetic order of inactivation, 0.82) was also observed. Though no recovery of the enzyme activity was observed by hydroxylamine-treatment of the EFA-inactivated enzyme, His residue still remains as a possible site in further investigation for direct binding to PQQ to form holoenzyme.

Figure 4. Inactivation of GDH by PCMB and PCMBS. Effects of Ca^{2+}.

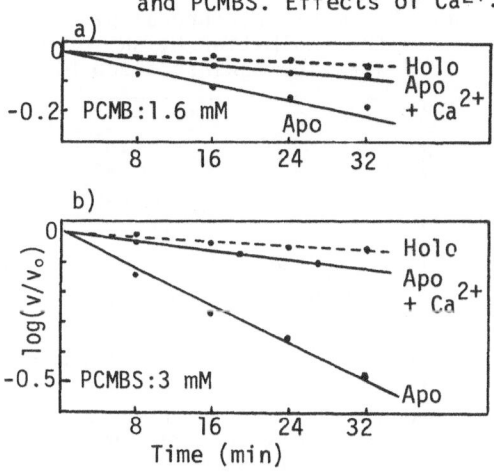

Figure 5. First-order plots of the inactivation of GDH by EFA.

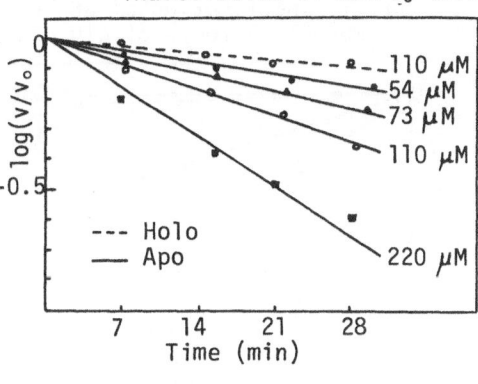

Inactivation by N-ethylmaleimide (NEM) and 7-chloro-4-nitrobenzo-2-oxa-1,3-diazole (NBD-Cl)

Contrary to the inactivation of GDH by the inhibitors mentioned above NEM inactivated holoenzyme more effectively than apoenzyme. The inactivation followed a pseudo-first order kinetics and the kinetic order of inactivation was 0.9 (Figure 6), suggesting the involvement of one Cys residue of holoenzyme in the inactivation process. Further, inactivation of holoenzyme by NEM (1 mM) was accelerated (1.5-fold) in the presence of substrate glucose (4.3 mM) (enzyme assay was started by the addition of DCIP). Similar phenomena were observed using NBD-Cl as an inhibitor. When holoenzyme was treated with 1 μM NBD-Cl in the presence of 23 mM glucose the inactivation was accelerated

Figure 6. First-order plots of the inactivation of GDH by NEM.

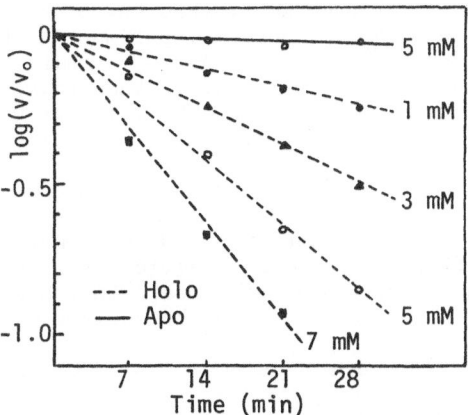

2.2-fold, but glucose showed no effect in the case of apoenzyme (Figure 7) (1.5-fold acceleration in the case of 0.5 μM NBD-Cl and 4.3 mM Glucose). These results suggest that the essential sulfhydryl group which becomes NEM- and NBD-Cl-reactive after reconstitution of the holoenzyme may play a role in catalytic reaction mechanism rather than in substrate binding.

92

Inactivation by phenylhydrazine (PH)

PH (3.5 μM) progressively inactivated holoenzyme, not apoenzyme, and the inactivation rate, contrary to the case of inactivation by NEM or NBD-Cl, was greatly reduced by preincubation (5 min) of holoenzyme with substrate glucose (4.3 mM) (data not shown). These results suggest that PH reacts with C(5)-carbonyl group of PQQ, the reaction being prevented by reduction of PQQ by glucose.

Figure 7. Inactivation of GDH by NBD-Cl. Effects of glucose.

Figure 8. Double-reciprocal plots of the inhibition by CN$^-$ of GDH activity against glucose.

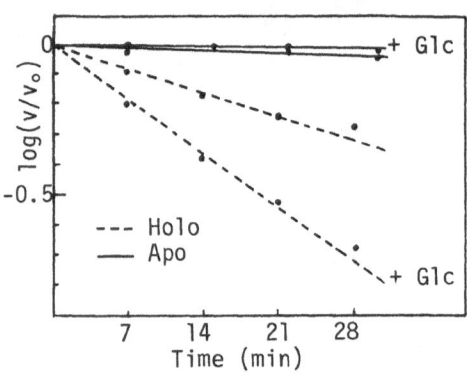

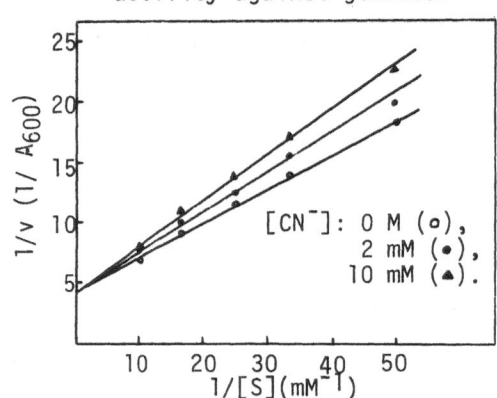

Inhibition by CN$^-$

When holoenzyme was incubated with CN$^-$ (35 mM, 20 min) the activity was reduced to 65% of the original and the activity was restored to 81% by dialysis of the inactivated enzyme against 5 mM veronal buffer containing 1 mM Ca^{2+} (in the control experiment without CN$^-$, 86% of the activity was recovered after dialysis). CN$^-$-inhibition was time-independent. Further, CN$^-$ competitively inhibited the enzyme reaction with substrate glucose (Figure 8), suggesting that CN$^-$ competes with glucose for C(5)-carbonyl group of PQQ in holoenzyme.

Effect of glucopyranose analogs

High concentration of both methyl α- and β-glucoside (57 mM) showed no effect on the GDH reaction even in a lower concentration of substrate glucose (0.14 mM). It appears that hydroxyl group at C(1) of glucose is essential for binding to holoenzyme much more than the pyranose ring moiety.

Non-enzymic reduction of PQQ by glucose

On incubation of a mixture of PQQ(20 μM) and glucose (0.1 M) in 20 mM acetate buffer (pH 6.7) under highly anaerobic condition at 37°C spectral change was observed, owing to the disappearance of PQQ (decrease of A248) and the formation of PQQH$_2$ (increase of A302)(Figure 9a). Addition of Ca^{2+} (2mM) to PQQ solution resulted in a 10 nm red shift of the 330 nm absorption maximum of the PQQ spectrum, suggesting a complexation between them. Reduction of PQQ was promoted in the presence of Ca^{2+} especially with regards to the decrease of A_{248} (Figure 9a). On exposure to air spectra of both solutions returned to that of PQQ (Figure 9b): Ca^{2+} was stimulatory to the decrease of A_{302}, but finally was inhibitory to the increase of A_{248} to recover a complete PQQ spectrum (cf. spectra at 40 min, curve 3). At pH 6.0 both the reduction of PQQ and reoxidation of $PQQH_2$ were much slower, but the effect of Ca^{2+} was more eminent in both processes: stimulatory to reduction and inhibitory to reoxidation. It appears that complexation between Ca^{2+} and PQQ plays a role in the reaction of PQQ with glucose.

DISCUSSION

C(9)-carboxyl group of PQQ has been reported to be essential in the formation of active GDH of Escherichia coli (Shinagawa et al. 1986). Although the real target of BD-modification is still obscure, a positively-charged site of apoprotein (possibly Arg residue) which binds reversibly to ANS is supposed to contribute to the binding to such PQQ-carboxyl in holoenzyme.

9-decarboxy PQQ, like α-picolinate forms 1:1 complex with Cd^{2+}, while 7,9-didecarboxy PQQ does not (Noar et al. 1985). Further, as a model for the active site of copper-containing amine oxidases, the formation of Cu^{2+}-PQQ complex (1:1) (Jangejan et al.1987) and the preparation of ternary Cu^{2+} complex containing PQQ and 2,2':6',2''-terpyridine (Suzuki et al.1988) have been reported. Addition of Ca^{2+} to PQQ solution resulted in a 10 nm red shift of the 330 nm absorption maximum of the PQQ spectrum, suggesting that Ca^{2+} also binds to α-picolinate part of PQQ [N(6) and C(7)-carboxyl group]. Apo-GDH of P. fluorescens requires Ca^{2+} for the formation of active holoenzyme (Imanaga et al. 1979) Both Tyr and Cys residues of apoprotein are presumed to bind directly to Ca^{2+} in the holoenzyme complex.

Figure 9. Non-enzymic reduction of PQQ by glucose
a) Reduction of PQQ   b) Reoxidation of $PQQH_2$

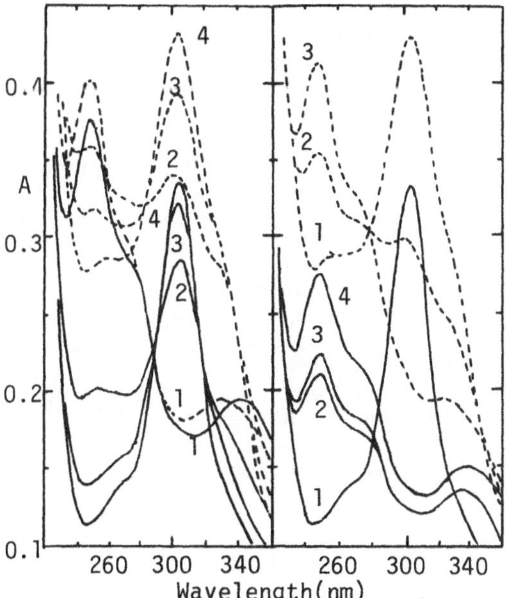

$\underline{\hspace{1cm}}$:$+Ca^{2+}$, $----$:$-Ca^{2+}$
1, 0; 2, 9.5; 3, 15; 4, 20h.   1, 0; 2, 15; 3, 40; 4, 150 min.

CN$^-$ reversibly inhibited holoenzyme and the inhibition was competitive with substrate glucose. Since it is well known that C(5)-carbonyl group of PQQ is very reactive towards nucleophiles and reversible adduct formation takes place with H_2O, NH_3, HCN and alcohol (Dekker et al.1982), the inhibition by CN$^-$ of holoenzyme appears to depend on the above reaction of CN$^-$ with the cofactor. Thus, it is supposed that substrate glucose reacts at first with C(5)-carbonyl group of cofactor PQQ to form an intermediate adduct at the active site of GDH, CN$^-$ inhibiting this process by competing with glucose, and then intramolecular electron-transfer in the intermediate results in the formation of products, δ-gluconolactone and $PQQH_2$.

Cys residue which becomes reactive with NEM and NBD–Cl after reconstitution of holoenzyme may play a role in catalytic reaction rather than in substrate binding, probably withdrawing proton from anomeric OH to promote the addition of glucose to PQQ. Participation of glucose anion in oxidation of glucose with bromine has been discussed (Capon 1969).

Glucose reduced PQQ, though the rate was low, in mild, bioligical conditions, and Ca^{2+} promoted the reaction. A catalytic role of metal ion in the reduction of PQQ has been suggested, since metal ion binding to $PQQH_2$ [tridentate, (α-picolinate + 8-hydroxyquinoline) type] should be much greater than to PQQ (bidentate, α-picolinate type) (Noar et al. 1985). It has been reported that ternary Cu^{2+} complex containing PQQ and 2,2':6',2"-terpyridine, whose o-quinone moiety of PQQ being activated by the electron-withdrawing effect of copper, has a much higher catalytic activity in oxidation of benzylamine than PQQ (Suzuki et al. 1988). Ca^{2+} may promote the reaction between glucose and PQQ at the active site of GDH: 1) promotion of the intermediate formation by electron-withdrawing from C(5)-carbonyl oxygen---activation of C(5)-carbonyl group towards nucleophiles; 2) promotion of the intramolecular electron-transfer in the intermediate adduct (six-membered ring structure has been suggested: Kieboom, Duine 1988) to form oxidation-reduction products. Summarizing above considerations, Figure 10 presents an active site model and a possible reaction mechanism as a working hypothesis. Identification of the participating amino acid residues and the elucidation of the reaction mechanism await further investigation.

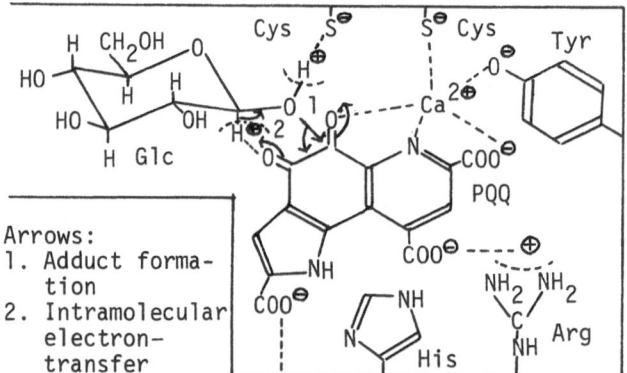

Figure 10. Active site model and possible reaction mechanism of GDH

Arrows:
1. Adduct formation
2. Intramolecular electron-transfer

ACKNOWLEDGEMENT

The author wish to thank to Misses Asako Itaya, Sonoko Suzuki, Akino Yokoyama, Ritsuko Tsutsumi, Hiromi Nishikawa, Naoko Shiozaki and other students for their assistance. The author is also indebted to Drs. Fukuo Takemura and Kaoru Iwai (Laboratory of Polymer Chemistry, Department of Chemistry) for their technical assistance in high vacuum experiments.

REFERENCES

Ameyama M.,Nonobe M.,Hayashi M.,Shinagawa E.,Matsushita K. and Adachi O. (1985) Mode of binding of pyrroloquinoline quinone to apo-glucose dehydrogenase. Agr.Biol.Chem.49:1227-1231.
Capon B.(1969) Mechanism in carbohydrate chemistry. XV. Oxidation of aldoses with bromine. Chem. Rev.69:493-496
Dekker R.H.,Duine J.A.,Frank J.K.N. J.,Verwiel D.E.J. and Westerling J. (1982) Covalent addition of H_2O, enzyme substrates and activators to

pyrroloquinoline quinone, the coenzyme of quinoproteins. Eur.J.Biochem. 125:69-73

Duine J.A.,Frank J.Z.N. J. and Van Zeeland J.K.(1979) Glucose dehydrogenase from Acinetobacter calcoaceticus. A 'quinoprotein'. FEBS Letters 108:443-446

Imanaga Y.,Hirano-Sawatake Y., Arita-Hashimoto Y., Itou-Shibouta Y. and Katou-Semba R.(1979) On the cofactor of glucose dehydrogenase of Pseudomonas fluorescens. Proc. Japan Acad.55(B):264-269

Imanaga Y.,Suzuki S.,Yoshitomi J.,Murase H. and Minami Y.(1985) On the active site of membrane-bound glucose dehydrogenase from Pseudomonas fluorescens. Effects of inhibitors. J.Biochem.Soc.Japan(Seikagaku)57:1065 (Abstract in Japanese)

Imanaga Y.(1987) On the active site of membrane-bound glucose dehydrogenase from Pseudomonas fluorescens. Effects of inhibitors.Biochemistry of PQQ and Quinoproteins (Yabuta Seminar:Agr.Chem.Soc.Japan, Oct.22, Kyoto)37-40

Jongejan J.A.,Van der Meer R.A.,van Zylen and Duine J.A.(1987) Spectrophotometric studies on pyrroloquinoline quinone-copper(II) complexes as possible models for copper-quinoprotein amine oxidases.Rec.Trav.Chim. Pays-Bas106:365

Prof. Dr. Kieboom and Prof. Dr.Duine J.A.(1987) Personal communication.

Mäkinen K.K., Mäkinen P.-L.,Wilkes S.H.,Bayliss M.E. and Brescott J.M.(1982) Photochemical inactivation of Aeromonas aminopeptidase by 2,3-butanedione. J. Biol.Chem.257:1765-1772

Noar J.B.,Rodriquez E.J. and Bruice J.C.(1985) Synthesis of 9-decarboxymethoxatin. Metal complexation of methoxatin as a possible requirement for its biological activity. J. Am. Chem. Soc. 107: 7198-7199

Shinagawa E.,Matsushita k.,Nonobe M.,Adachi O.,Ameyama M.,Ohshiro Y.,Itoh S. and Kitamura Y.(1986) The 9-carboxyl group of pyrroloquinoline quinone, a novel prosthetic group, is essential in the formation of holoenzyme of D-glucose dehydrogenase. Biochem.Biophys.Res.Commun.139:1279-1284

Suzuki S.,Sakurai T.,Itoh S. and Ohshiro Y.(1988) Preparation and Characterization of ternary copper(II) complexes containing coenzyme PQQ and bipyridine or terpyridine. Inorg. Chem.27:591-592

Tanaka N. and Imanaga Y.(1981) Purification and properties of glucose dehydrogenase from Gluconobacter suboxydans. J.Biochem.Soc.Japan(Seikagaku) 53:972(Abstract in Japanese)

Tsutsumi R.,Ikawa K. and Imanaga Y.(1987) On the active site of membrane-bound glucose dehydrogenase from Pseudomonas fluorescens. Effects of inhibitors.J.Biochem.Soc.Japan(Seikagaku)59:503(Abstract in Japanese)

Yokoyama A.,Nakata Y. and Imanaga Y.(1986)On the active site of membrane-bound glucose dehydrogenase from Pseudomonas fluorescens. Effects of inhibitors. J.Biochem.Soc.Japan(Seikagaku)58:1338(Abstract in Japanese)

PRODUCTION OF GLUCONIC AND GALACTONIC ACIDS FROM WHEY.

N. Van Huynh, M. Decleire, J.C. Motte and X. Monseur.
Institute for Chemical Research, Museumlaan 5, 1980 Tervuren - Belgium.

Key words : whey ; gluconic acid ; galactonic acid.

ABSTRACT

Whey was hydrolysed by *Kluyveromyces bulgaricus* cells immobilized in either alginate or chitosan beads. The hydrolysate was used to grow *Gluconobacter oxydans*. The identification of the acids produced was carried out by Gas Capillary Chromatography. Glucose was oxidised to gluconic acid in less than 24 hours. Galactose was oxidised to galactonic acid at a lower rate.

INTRODUCTION

Whey is a relatively rich by-product of the milk industry : it contains lactose, proteins and minerals. It is used on a small scale for the production of single cells and lactic acid. Its wide use is hampered by the inability of most microorganisms to grow on lactose. This difficulty can be alleviated either by a preliminary hydrolysis of lactose into glucose and galactose or by cloning the lactose transport and hydrolysis systems and if required, the galactose epimerization pathway. We are being involved in exploiting the first alternative. The hydrolysis of whey was carried out by immobilized *Kluyveromyces bulgaricus* whole cells (Decleire et al., 1985). The hydrolysate was used for the production of acids by *Gluconobacter oxydans* (Van Huynh et al., 1986). The determination of the acids produced by fermentation is usually carried out by HPLC (Hommes et al., 1986; Van Huynh et al., 1986). Although this method is extremely convenient to use routinely, special techniques are required for the monitoring of substances that have very similar structures, such as isomers. This paper deals with the production of acids from whey hydrolysate and the identification of the acids by Gas Capillary Chromatography.

MATERIALS AND METHODS

Microbial strains.

The *Kluyveromyces bulgaricus* IRC 101 used for whey hydrolysis was already described (Van Huynh and Decleire, 1982). *Gluconobacter oxydans* ATCC 621 H was grown at 30°C in a 1-L fermenter.

Determination of sugars and acids.

This was carried out by High Performance Liquid Chromatography (HPLC) and Gaz Capillary Chromatography (GCC). The HPLC system comprised a 6000 A pump, a "Wisp" intelligent automated injector, a R-401 refractometer, a UV 411 detector (set at 214 nm), all from Waters. The results were recorded with a Trilab 2500 from Trivector

97

J. A. Jongejan and J. A. Duine (eds.), PQQ and Quinoproteins, 97–99.

(Sandy, Bedfordshire, UK). Two columns were used: Bio Rad Aminex HPX 87 H (elution with 0.01 N H_2SO_4 at 0.7 ml/mn and 70°C) and Ionex HPIC -AS5 fitted with a HPIC -AG5 precolumn (elution with 0.01 N NaOH at 0.7 ml/mn and 25°C). GCC was performed with a Carlo Erba Vega Model equipped with a OV 17 column (length: 20 m, internal diameter: 320 microns). The elution conditions were: 10 min from 60°C to 150°C and 10 min from 150°C to 300°C.

RESULTS AND DISCUSSION

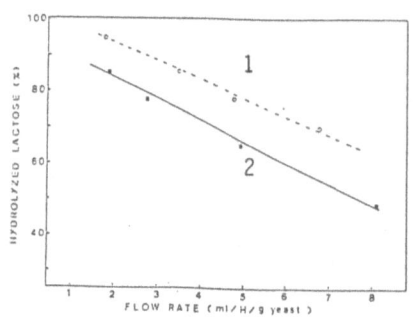

Fig. 1 : Lactose hydrolysis by gel-
entrapped *K. bulgaricus*
cells (1.3 U β-galactosi-
dase/mg).
1. 1.5% alginate gel;
2. 3% chitosan gel.

Preliminary assays were carried out with cells immobilized in either alginate or chitosan beads. The hydrolysis rate of whey by *K. bulgaricus*
immobilized in Ca alginate beads was higher than that carried out in the
chitosan column (Fig. 1). Whey was continuously hydrolysed for over 3
months in a alginate column.

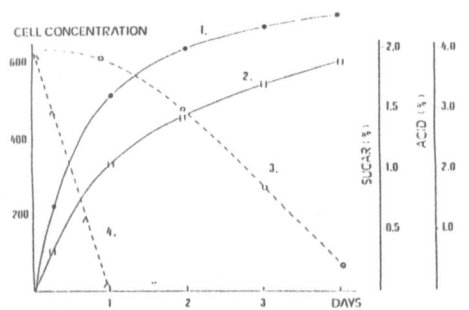

Fig. 2 : Production of acids from
whey hydrolysate.
1. *G. oxydans* cell concentration
(million/ml);
2. Acid production (%);
3. Galactose (%);
4. Glucose (%).

Gluconobacter oxydans was grown on the whey hydrolysate. Glucose was consumed in 18 hours and galactose in 3 days with the production of acids
(Fig. 2).

The identification of the acids produced was carried out by HPLC and GCC.
With the first technique it was not possible to discriminate the acids
while GC allowed the separation of gluconic and galactonic acids. So *G.
oxydans* oxidised glucose to gluconic acid and galactose to galactonic
acid but at a much lower rate.

Although, glucose dehydrogenase purified from *G. oxydans* is devoid of any
activity towards galactose (Ameyama et al., 1981), the strain 621 H can
grow (very slowly) in galactose-based medium with the production of galacto-

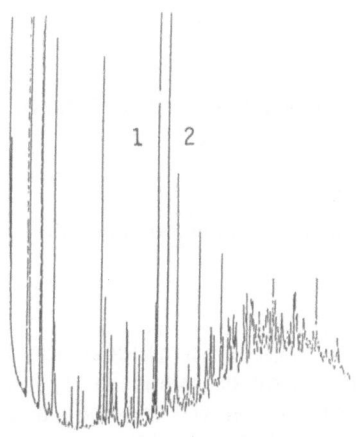

Fig. 3 : Identification of the acids by GCC.
1. Gluconic acid;
2. Galactonic acid.

nic acid. This in vivo oxidation of galactose could be attributed to the activity of glucose dehydrogenase since the same authors reported a faint activity of the membrane fraction. In addition to the discrepancy of activity due to localization of the enzyme, activity is also strain-dependent. Glucose dehydrogenase purified from *Escherichia coli* catalyses the oxidation of galactose and other sugars (Ameyama et al., 1986) and it seems that the oxidation of galactose by *Salmonella typhimurium* could be attributed to the same enzyme (Hommes et al., 1986). Galactose is also oxidised by *Acitenobacter calcoacticus* cells and by glucose dehydrogenase purified from the same strain, but at different rates and affinities (Dokter et al., 1987).

REFERENCES

Ameyama M, Shinagawa E, Matsushita K and Adachi O, 1981. D-glucose dehydrogenase of *Gluconobacter oxydans*: solubilization, purification and characterization. Agric. Biol. Chem. 45: 851-861.

Ameyama M, Nonobe M, Shinagawa E, Matsushita K, Takimoto K and Adachi O, 1986. Purification and characterization of the quinoprotein D-glucose dehydrogenase apoenzyme from *Escherichia coli*. Agric. Biol. Chem. 50: 49-57.

Decleire M, Van Huynh N, Motte JC and De Cat W, 1985. Hydrolysis of whey by whole cells of *Kluyveromyces bulgaricus* immobilized in calcium alginate gels and in hen egg white. Appl. Microbiol. Biotechnol. 22: 438-441.

Dokter P, Pronk JT, Van Schie BJ, Van Dijken JP and Duine JA, 1987. The in vivo and in vitro substrate specificity of quinoprotein glucose dehydrogenase of *Acinetobacter calcoaceticus*. FEMS Microbiology Letters 43: 195-200.

Hommes RWJ, Loenen WAM, Neijssel OM and Postma PW, 1986. Galactose metabolism in gal mutants of *Salmonella typhimurium* and *Escherichia coli*. FEMS Microbiology Letters 36: 187-190.

Van Huynh N and Decleire M, 1982. Les cellules entières de *Kluyveromyces bulgaricus* comme source de β-galactosidase. Revue des Fermentations et des Industries Alimentaires. 37: 153-157.

Van Huynh N, Decleire M, Voets M, Motte JC and Monseur X, 1986. Production of gluconic acid from whey hydrolysate. Process Biochemistry 21: 31-32.

MODE OF BINDING OF PYRROLOQUINOLINE QUINONE TO GLUCOSE DEHYDROGENASE FROM *ACINETOBACTER CALCOACETICUS*

Helmut GÖRISCH , Otto GEIGER and Martin ADLER
Institut für Mikrobiologie, Universität Hohenheim, Garbenstraße 30,
D-7000 Stuttgart 70, Federal Republic of Germany

In a number of Gram-negative bacteria a quinoprotein glucose deydro-
genase is found, which will loose its prosthetic group, when dialysed
against buffer containing EDTA. The apoform of the enzyme can be converted
to active holoenzyme by pyrroloquinoline quinone (PQQ) in the presence of
Mg^{2+} or Ca^{2+} ions. In contrast the quinoprotein glucose dehydrogenase from
Gluconobacter suboxidans (Ameyama et al., 1981) and from *Acinetobacter
calcoaceticus* (Dokter et al., 1986; Geiger and Görisch, 1986) are not
inactivated by dialysis against EDTA-containing buffers. However glucose
dehydrogenase from *A. calcoaceticus* is reversibly inactivated by heat
treatment, and restoration of active enzyme depends on the presence of PQQ
and Ca^{2+} ions.

Glucose dehydrogenase from A. calcoaceticus is inactivated
at temperatures above 35°C. At a given temperature the enzyme is not inactivated completely. Between 35 and 48°C an equilibrium is established
between active and inactive enzyme, which depends on the inactivation
temperature and the concentration of enzyme used. At 50°C the enzyme is
inactivated completely. When afterwards the temperature is shifted to 40°C
the activity of the sample increases again. The final residual activity
reached is the same as that of an identical sample, which was simply inactivated at 40°C, Fig 1. Glucose dehydrogenase is stabilized against
thermal inactivation by Ca^{2+} ions; even though at room temperature the
enzyme is not inactivated in the presence of 10 mM EDTA, nor by prolonged

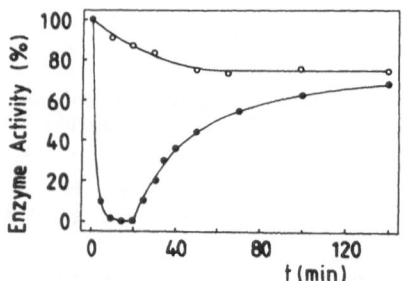

Fig. 1. Eqilibrium between native
and heat-inactivated glucose dehydrogenase.
Samples of glucose dehydrogenase
with 20 units/ml were incubated at
either 40°C (o-o-o) or at 50°C for
20 min and afterwards at 40°C
(●-●-●). Residual activities were
determined at the times indicated.

J. A. Jongejan and J. A. Duine (eds.), PQQ and Quinoproteins, 100–102.
© 1989 by Kluwer Academic Publishers.

dialysis against buffer containing EDTA. X-ray fluorescence was used to determine the metal ion content. Based on a subunit molecular weight of 54 000, two Ca^{2+} ions were found per subunit of the enzyme.

Upon heat inactivation glucose dehydrogenase of *A. calcoaceticus* dissociates into apoenzyme and the prosthetic group PQQ. When PQQ is removed from the reaction mixture by binding to an anion exchange resin no activity reappears. When Ca^{2+} is removed from the reaction mixture by EDTA also no reactivation occurs. The velocity of reactivation depends on the Ca^{2+} ion concentration, the concentration of PQQ and the squared protein concentration. The apoform of glucose dehydrogenase can be isolated, and in the presence of PQQ and Ca^{2+} ions active holoenzyme is formed. Mn^{2+}, Cd^{2+} and Sr^{2+} can replace Ca^{2+} and allow the formation of active glucose dehydrogenase after heat inactivation. Only divalent metal ions with an ionic radius between 0.82 and 1.16 Å will support the formation of enzymically active glucose dehydrogenase.

When Ca^{2+} is added to a solution of PQQ a spectral shift of the absorption maxima occurs. A similar shift is observed with Mn^{2+}, Cd^{2+}, Sr^{2+} and with $HgCl_2$, indicating complex formation between PQQ and the divalent metal ions. The association constant for Ca^{2+} has been determined, $K_{ass} = 103\ M^{-1}$. The titration data suggest a complex with a stoichiometry of 1:1.

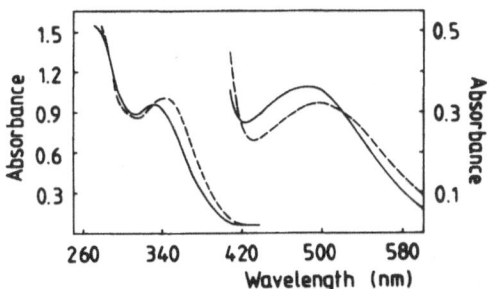

Fig. 2. Absorbtion spectrum of PQQ and the Ca^{2+}/PQQ-complex.
Spectrum of PQQ in 100 mM sodium acetate buffer, pH 6, (———).
Spectrum of PQQ in 100 mM sodium acetate buffer, pH 6, in the presence of 200 mM $CaCl_2$ (- - -).
The ionic strength of the samples was adjusted to $I = 1$ by the addition of $NaClO_4$. Left: 100 µM solution; right: 500 µM solution of PQQ.

The absorption maxima of the complex between Ca^{2+} and PQQ are shifted to longer wavelengths with respect to the absorption spectrum of PQQ. The maximum at 333nm is shifted to 344 nm, and the maximum with a low extinction coefficient at 478 nm is shifted to 499 nm, Fig 2. Based on the finding, that the apoenzymes of quinoprotein glucose dehydrogenase which occur in a variety of Gram-negative bacteria, can be reconstituted to active enzyme with PQQ in the presence of Mg^{2+} or Ca^{2+} ions, it has been suggested, that the prosthetic group is bound at the active site via a Ca^{2+} or Mg^{2+} ion bridge (Ameyama et al., 1985).

Since quinoprotein glucose dehydrogenase from *A. calcoaceticus* shows an absorption spectrum similar to that of the Ca^{2+}/PQQ-complex it is proposed, that the prosthetic group is bound as a complex with Ca^{2+} at the active site of the enzyme.

Ameyama M, Shinagawa E, Matsushita K and Adachi O, 1981. D-Glucose Dehydrogenase of *Gluconobacter suboxidans*: Solubilization, Purification and
Characterization. Agricultural and Biological Chemistry 45: 851-861

Ameyama M, Nonobe M, Hayashi M, Shinagawa E, Matsushita K and Adachi O,
1985. Mode of Binding of Pyrroloquinoline Quinone to Apo-glucose Dehydrogenase. Agricultural and Biological Chemistry 49: 1227-1231

Dokter P, Frank Jzn J and Duine JA, 1986. Purification and characterization
of quinoprotein glucose dehydrogenase from *Acinetobacter calcoaceticus*
L.M.D. 79.41. Biochemical Journal 239: 163-167

Geiger O and Görisch H, 1986. Crystalline Quinoprotein Glucose Dehydrogenase from *Acinetobacter calcoaceticus*. Biochemistry 25: 6043-6048

GLUCOSE DEHYDROGENASE FROM *ZYMOMONAS MOBILIS*:
EVIDENCE FOR A QUINOPROTEIN

M. Strohdeicher[1], S. Bringer-Meyer[1], B. Neuß[1], R. van der Meer[2], J.A. Duine[2], and H. Sahm[1]

[1]Institut für Biotechnologie, Kernforschungsanlage Jülich GmbH, Postfach 1913, D-5170 Jülich, Federal Republic of Germany
[2]Department of Microbiology & Enzymology, Delft University of Technology, Julianalaan 67, 2628 BC Delft, The Netherlands

KEY WORDS: *Zymomonas mobilis*, glucose dehydrogenase, quinoprotein, phenylhydrazine

ABSTRACT

A membrane bound glucose dehydrogenase was detected in the gram-negative anaerobic bacterium *Zymomonas mobilis*. The enzyme-activity was measured with artificial electron acceptors and the enzyme was partially purified. Furthermore, by the phenylhydrazine method pyrroloquinoline quinone was shown to be the natural cofactor of this glucose dehydrogenase.

INTRODUCTION

The anaerobic gram-negative bacterium *Zymomonas mobilis* is well known as an efficient ethanol producer on glucose based media (Rogers et al. 1982). However, besides the main catabolic enzymes involved in ethanol-generation only little information is available about other enzymes in *Z. mobilis*. Exceptions are the activities which are involved in the sorbitol- and gluconate- formation. Recently it was shown that both reactions are simultaneously performed by a cytosolic glucose-fructose oxidoreductase with tightly bound NADP as cofactor (Zachariou and Scopes 1986).
Furthermore, a membrane-bound glucose dehydrogenase was detected with artificial electron acceptors (Leigh et al. 1984, Strohdeicher et al. 1988). The role of this enzyme is still unclear. To get insight into its function in *Z. mobilis* we tried to purify and characterize it. Since the glucose dehydrogenase seemed to be similar to recently described quinoproteins (Duine et al. 1986), we tried to elucidate whether PQQ is the natural cofactor also for this dehydrogenase.

MATERIALS AND METHODS

Microorganism and culture conditions. Z. mobilis ATCC 29191 was anaerobically cultivated as described previously (Bringer et al. 1985).
Preparation of cell free extracts and membrane fractions. Cell-free extracts were prepared according to Strohdeicher et al. (1988). Ultracentrifugation I (140,000 x g, 60 min.) of the extract resulted in a precipitate, referred to as "membrane fraction".

J. A. Jongejan and J. A. Duine (eds.), PQQ and Quinoproteins, 103–105.

Enzyme-purification steps. The enzyme could be detached and separated from the membranes by treatment with 0.7 M KCl + 1% TRITON X-100 and subsequent ultracentrifugation II (140,000 x g, 60 min.). The buffer was changed to 150 mM potassium phosphate, pH 6.5 + 0.3 % TRITON X-100 (KPB) on a Sephadex G25- column. Afterwards the sample was applied to a hydroxylapatite column which was equilibrated with the same buffer. After washing the column with starting buffer the GDH was eluted with 350 mM KPB. Active fractions were pooled and separated by gel filtration on Sephacryl S-200.
Identification of PQQ. For PQQ-detection the recently described phenylhydrazine method (Meer et al. 1987) was used.
Enzyme assays. Glucose dehydrogenase was assayed with Wurster's Blue as electron acceptor as described by Strohdeicher et al. (1988).

RESULTS

The purification steps led to a 40fold overall enrichment of the glucose dehydrogenase, but the enzyme was not pure as could be seen by SDS-polyacrylamide gelelectrophoresis (data not shown).
As recently demonstrated by Meer et al. (1987), PQQ can be detected by phenylhydrazine-treatment of the samples. PQQ and phenylhydrazine form the $C(5)$-hydrazone which can be identified by HPLC on C_{18}-reversed phase columns and by the characteristic absorption spectrum, which is obtained by photodiode-array detection.
Since our enzyme-preparation was not completely purified, aliquots of the pooled GDH-active and -inactive fractions from hydroxylapatite- and gelfiltration chromatographies were analysed. HPLC analysis of the phenylhydrazine- and pronase-treated active fraction from the hydroxylapatite chromatography showed a peak at the same retention time as the authentical $C(5)$-hydrazone. The spectrum of this peak (**Fig. 1a**, arrow) was compared with the spectrum of the $C(5)$-hydrazone (**Fig. 1b**) and judged as identical. Analysis of the active gel filtration sample gave the same results, while inactive fractions were negative in the test.
These results indicate that the membrane-bound glucose dehydrogenase from *Z. mobilis* is a quinoprotein.

Figure 1:

a) HPLC-Chromatogramm of a phenylhydrazine- and pronase-treated sample of the *Z. mobilis* glucose dehydrogenase on a C_{18}- column. The arrow indicates the analysed peak.

b) Absorption spectra of authentical hydrazone **(r)** and of the sample **(s)**.

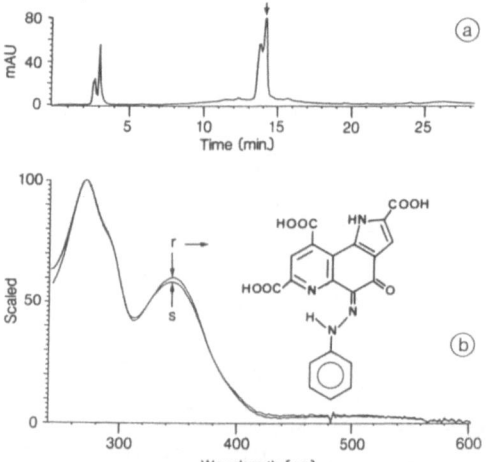

CONCLUSIONS

The results have shown that *Zymomonas mobilis* possesses a membrane bound glucose dehydrogenase which is comparable to a wide variety of glucose dehydrogenases from aerobic bacteria (Duine et al. 1986). In these bacteria the quinoenzymes are coupled with the electron transport chain and oxygen is the terminal electron acceptor. Furthermore, the oxidation of glucose seems to be an energy-yielding process (Schie et al. 1985).
Since *Z. mobilis* is an anaerobic bacterium the nature of the terminal electron acceptor is still unclear. To get insight into the physiological role of the glucose dehydrogenase in *Z. mobilis*, the analysis of possible interactions between the enzyme and other membrane components, i.e. quinones and cytochromes, will be subject of our further investigations.

REFERENCES

Bringer S, Härtner T, Poralla K, and Sahm H (1985). Influence of ethanol on the hopanoid content and the fatty acid pattern in batch and continuous cultures of *Zymomonas mobilis*. Arch Microbiol 140: 312-316

Duine JA, Frank Jzn J, and Jongejan JA (1986). PQQ and quinoprotein enzymes in microbial oxidations. FEMS Microbiol Rev 32: 165- 178

Leigh D, Scopes RK, and Rogers PL (1984). A proposed pathway for sorbitol production by *Zymomonas mobilis*. Appl Microbiol Biotechnol 20: 413-415

Meer RA van der, Jongejan JA, and Duine JA (1987). Phenylhydrazine as probe for cofactor identification in amine oxidoreductases. FEBS Lett 221: 299-304

Rogers PL, Lee KJ, Skotnicki ML, and Tribe DE (1982). Ethanol production by *Zymomonas mobilis*. Adv Biochem Eng 23: 37-84

Schie BJ van, Hellingwerf KJ, van Dijken PJ, Elferink MGL, van Dijl JM, Kuenen JG, and Konings W.N. (1985). Energy transduction by electron transfer via pyrroloquinoline quinone-dependent glucose dehydrogenase in *Escherichia coli, Pseudomonas aeruginosa,* and *Acinetobacter calcoaceticus* (var. *lwoffi*). J Bact 163: 493-499

Strohdeicher M, Schmitz B, Bringer-Meyer S, and Sahm H (1988). Formation and degradation of gluconate by *Zymomonas mobilis*. Appl Microbiol Biotechnol 27: 378-382

Zachariou M and Scopes RK (1986). Glucose-fructose oxidoreductase, a new enzyme isolated from Zymomonas mobilis that is responsible for sorbitol production. J Bacteriol 167: 863-869

HAEM-CONTAINING PROTEIN COMPLEXES OF *ACINETOBACTER CALCOACETICUS* AS SECONDARY ELECTRON ACCEPTORS FOR QUINOPROTEIN GLUCOSE DEHYDROGENASE

A. Geerlof, P. Dokter, J.E. van Wielink and J.A. Duine

Department of Microbiology & Enzymology, Delft University of Technology
Julianalaan 67, 2628 BC Delft, The Netherlands

INTRODUCTION

Some strains of *Acinetobacter calcoaceticus* convert glucose with oxygen into gluconolactone and further to gluconic acid as end product. This incomplete oxidation can provide the cell with useful energy (De Bont *et al.*, 1984; Van Schie *et al.*, 1985, 1987; Kitagawa *et al.*, 1986; Müller and Babel, 1986).

Two complete different quinoprotein glucose dehydrogenases (GDH) have been isolated from *A. calcoaceticus* LMD 79.41: (1) a membrane-bound GDH (m-GDH) (Matsushita *et al.*, this Proceedings), related to the GDH's in *Escherichia coli* and *Pseudomonas* species (Ameyama *et al.*, 1986). M-GDH, localized at the periplasmic side of the cytoplasma membrane, donates its reducing equivalents directly to ubiquinone-9 (Ameyama *et al.*, 1986; Beardmore-Gray and Anthony, 1986; Matsushita *et al.*, 1987); (2) a soluble GDH (s-GDH) (Dokter *et al.*, 1986; Geiger and Görisch, 1986) located in the periplasm and closely associated with the soluble cytochrome *b*-562, which is able to accept electrons from s-GDH (Dokter *et al.*, 1988). It should be noted, however, that the physiological role of s-GDH is still unknown as it is unable to oxidize glucose *in vivo*.

To elucidate the significance of both enzymes for this bacterium, more knowledge on the components of the electron transport chain is indispensible. Here we report the (partial) purification and characterization of the membrane-bound *b*-type cytochrome complexes of *A. calcoaceticus* LMD 79.41.

MATERIALS AND METHODS

A. calcoaceticus LMD 79.41 was grown in batch cultures or in oxygen-limited fed-batch cultures on a mineral medium (Van Schie *et al.*, 1984) with ethanol or malate as a carbon and energy source.

Cells were disrupted in a French pressure cell (4000 psi). Cytoplasmic membranes were isolated by ultracentrifugation (2 hours at 100,000 x g). The cytochrome-containing complexes were detached from the cytoplasmic membrane by incubation with detergent (Triton X-100 or Sulfobetain (SB_{12})). Further purification of these extracts was accomplished by means of ion-exchange chromatography (DEAE-Sepharose) and hydroxylapatite chromatography (Bio-Gel HT).

M_r values of the proteins were determined by gelfiltration chromatography (FPLC Superose™12) or native- and SDS-PAGE on gradient gels (Pharmacia 4-30%). Potentiometric titrations were performed at pH 7.0 according to Van Wielink *et al.* (1982).

RESULTS AND DISCUSSION

The cytoplasmic membranes of *A. calcoaceticus* appeared to contain three cytochrome *b*-containing complexes. Under conditions of high aeration a cytochrome *o*-containing oxidase and a cytochrome b_{554} were the predominant species. These were separated and partly purified (Table 1).

106

J. A. Jongejan and J. A. Duine (eds.), PQQ and Quinoproteins, 106–109.
© 1989 by Kluwer Academic Publishers.

Under oxygen-limited growth conditions only the cytochrome d-containing oxidase was present. This complex was purified to almost homogeneity (Table 2).

Comparison of the cytochrome-containing complexes with those of *E. coli* (Table 3) revealed similarities and dissimilarities. The cytochrome o-containing oxidase is rather similar to the corresponding oxidase of *E. coli* (Kita *et al.*, 1984a; Matsushita *et al.*, 1986). The cytochrome d-containing oxidase shows some different characteristics in comparison with the cytochrome d-containing oxidase from *E. coli* (Kita *et al.*, 1984b; Miller and Gennis, 1983): (1) It consists of a greater number of subunits (5 vs. 2). (2) Cytochrome b_{595} (formerly cytochrome a$_1$) was not found. Both cytochrome complexes were able to oxidize ubiquinol-0 and duroquinol and function as terminal oxidases in the electron transport chain. Cytochrome b_{554} is quite different from cytochrome b_{556} found in *E. coli* (Kita *et al.*, 1978; Murakami *et al.*, 1985). It's physiological role is unclear. Addition of cytochrome b_{554} to the cytochrome o-containing oxidase preparation

TABLE 1. Purification of the cytochrome o-containing oxidase and cytochrome b_{554}*.

Step	Protein (mg)	Cytochrome b (nmol)	Cyt. b /protein (nmol/mg)	Yield (%)
Cell-free extract	4292	1299	0.3	100
Cytoplasmic membranes	1229	947	0.8	73
Triton X-100 extract	690	573	0.8	44
Cytochrome o-containing oxidase				
1st DEAE-Sepharose	73	256	3.5	20
2nd DEAE-Sepharose	45	223	5.0	17
Cytochrome b_{554}				
1st DEAE-Sepharose	134	174	1.3	13

*) the name refers to the α-band in the spectrum of the reduced protein at 77K.

TABLE 2. Purification of the cytochrome d-containing oxidase.

Step	Protein (mg)	Cytochrome d (nmol)	Cyt. d/protein (nmol/mg)	Yield (%)
Cell-free extract	1710	275	0.16	100
Cytoplasmic membranes	288	117	0.41	43
Sulfobetain extract	207	122	0.59	44
DEAE-Sepharose	9.8	34.5	3.50	13
Hydroxylapatite	2.2	10.8	4.90	4

didn't stimulate the quinol oxidation rate.

Based on former (Dokter *et al.*, 1988) and present data, the linkage of components in the respiratory chain of *A. calcoaceticus* LMD 79.41 can be indicated as depicted in Scheme 1.

TABLE 3. Properties of the cytochrome-containing complexes from *A. calcoaceticus* and *E. coli*.

	Cytochrome *o* oxidase		Cytochrome *d* oxidase		Cytochrome b_{554} b_{556}	
	A. calc.	*E. coli*	*A. calc.*	*E. coli*	*A. calc.*	*E. coli*
M_r (native) (kD)	150	140	110	77-100	70	oligomer
M_r (denatured)	55 (h)	51-66	55 (h)	51- 57	70 (h)	14-17.5
(kD)	41	28-36	39	26- 43		
	33	18-22	34			
	17	13-17	22 (h)			
			15 (h)			
λ_{max} (reduced)	426	427	426	429-30	422	425
at 25 °C (nm)	530	530	530	532	530	529
	564	560	560 (*b*)	560-62	556	558
	558 (s)			594-95		
			626 (*d*)	628-29		
E_o' (mV)	*b*: 160*	150*	*b*: 170**	10*	*b*: 100*	-45/+35*
	o: ?	150*	*d*: ?	240*	50**	

*, measured after purification; **, measured in membranes; *b*, cytochrome *b*; *d*, cytochrome *d*; h, haem group; s, shoulder.

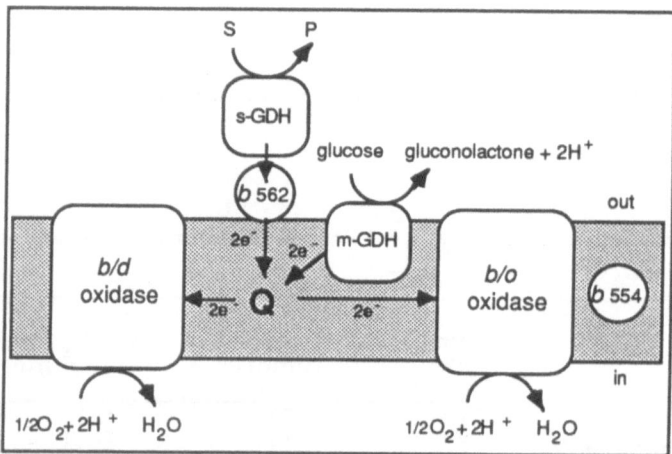

SCHEME 1. The linkage of GDH to the respiratory chain of *A. calcoaceticus*.

REFERENCES

Ameyama M, Nonobe M, Shinagawa E, Matsushita K, Takimoto K and Adachi O, 1986. Agric Biol Chem 50: 49-57.
Beardmore-Gray M and Anthony C, 1986. J Gen Microbiol 132: 1257-1268.
De Bont JAM, Dokter P, Van Schie BJ, Van Dijken JP, Frank JJzn and Kuenen JG, 1984. Antonie van Leeuwenhoek 50: 76-77.
Dokter P, Frank JJzn and Duine JA, 1986. Biochem J 239: 163-167.
Dokter P, Van Wielink JE, Van Kleef MAG and Duine JA, 1988. Biochem J 254: 131-138.
Geiger O and Görisch H, 1986. Biochemistry 25: 6043-6048.
Kita K, Yamato I and Anraku Y, 1978. J Biol Chem 253: 8910-8915.
Kita K, Konishi K and Anraku Y, 1984a. J Biol Chem 259: 3368-3374.
Kita K, Konishi K and Anraku Y, 1984b. J Biol Chem 259: 3375-3381.
Kitagawa K, Tateishi A, Nakano F, Natumoto T, Morohoshi T, Tanino T and Usui T, 1986. Agric Biol Chem 50: 1453-1457.
Matsushita K, Patel L and Kaback HR, 1986. Methods in Enzymology 126: 113-122.
Matsushita K, Nonobe M, Shinagawa E, Adachi O and Ameyama M, 1987. J Bacteriol 169: 205-209.
Miller MJ and Gennis RB, 1983. J Biol Chem 258: 9159-9165.
Müller RH and Babel W, 1986. Arch Microbiol 144: 62-66.
Murakami H, Kita K, Oya H and Anraku Y, 1985. FEMS Microbiol Lett 30: 307-311.
Van Schie BJ, Hellingwerf KJ, Van Dijken JP, Elferink MGL, Van Dijl JM, Kuenen JG and Konings WN, 1985. J Bacteriol 163: 493-499.
Van Schie BJ, Pronk JT, Hellingwerf KJ, Van Dijken JP and Kuenen JG, 1987. J Gen Microbiol 133: 3427-3435.
Van Schie BJ, Van Dijken JP and Kuenen JG, 1984. FEMS Microbiol Lett 24: 133-138.
Van Wielink JE, Oltmann LF, Leeuwerik FJ, De Hollander JA and Stouthamer AH, 1982. Biochim Biophys Acta 681: 177-190.

IDENTIFICATION AND QUANTIFICATION OF PQQ

Robert A. van der Meer, Jacob A. Jongejan and Johannis A. Duine

Department of Microbiology and Enzymology, Delft University of Technology, Julianalaan 67, 2628 BC DELFT, The Netherlands.

1. SUMMARY

PQQ can occur in free form, as a C_5-addition compound, as a condensation product with amino acids (e.g. an oxazole) and in an extractable or covalently bound form. Free, underivatized PQQ can be quantified by several chromatographic methods and biological assays. However, alternative analytical procedures had to be developed for the other cases. The so-called hydrazine method has proven to be a powerful tool in the detection and determination of PQQ in enzymes where it is covalently bound. The conditions established also give some insight into the inhibition mechanism of quinoproteins with hydrazines and allow isolation of peptides to which PQQ is still attached in a non-reactive form. On the other hand, the so-called hexanol extraction procedure is suited to detect PQQ in enzymes for which the hydrazine method fails, and in its condensation products. Since PQQ is very reactive with nucleophilic compounds, the occurrence of free PQQ in complex biological systems is very unlikely so that the latter method could be very useful in studies on the biosynthesis and the possible vitamin role of PQQ in mammals.

2. INTRODUCTION

A steadily growing number of enzymes (quinoproteins), from prokaryotic [Duine et al., 1986] as well as eukaryotic [Duine et al., 1987] origin, appear to contain PQQ. Since some of these enzymes are vital for cellular reproduction or regulation of physiological processes in mammals [Duine, this Proceedings], it is obvious that the enzymology of quinoproteins and the biosynthesis and distribution of PQQ will become important topics of research in the near future. To study these topics, the availability of reliable and sensitive methods of PQQ analysis will be crucial.

Originally, PQQ was discovered in several bacterial dehydrogenases from which it could be removed by denaturing the protein. In analogy with other cofactors, methods were developed to quantify PQQ with biological and chromatographic (h.p.l.c.) assays. Soon thereafter, several mammalian enzymes were found to

111

J. A. Jongejan and J. A. Duine (eds.), PQQ and Quinoproteins, 111–122.
© *1989 by Kluwer Academic Publishers.*

contain covalently bound PQQ and nowadays the number of enzymes of the latter group has surpassed that of the former.

A characteristic feature of the cofactor is its reactivity with nucleophiles, leading to biological inactive products [van Kleef et al., these Proceedings]. In view of this property it cannot be expected that complex biological samples containing nucleophiles (e.g. amino acids, SH-compounds, etc.), will show the presence of free PQQ. Therefore, development of methods suited for the analysis of the cofactor in all its forms should be directed to prevent uncontrolled nucleophilic attack (e.g. during detachment of PQQ from proteins where it is covalently bound) and to convert existing products (e.g. oxazoles) into a single, analyzable product. For these reasons, emphasis will be laid in this paper on those methods having the required capabilities.

3. FREE, NON-COVALENTLY BOUND, AND PRODUCTS OF PQQ

Free PQQ has sofar only been detected in culture media of certain bacteria (methylotrophs, ethanol grown *Pseudomonas* species), cultured on an inorganic medium. Two types of analysis procedure have been developed: biological assays based on reconstitution of a quinoprotein apo-enzyme (see Section 3.1) and, if PQQ concentrations are high and contamination is low, h.p.l.c. methods (see Section 3.2), using a reversed phase column. In complex samples, product formation has already occurred in an uncontrolled way. Therefore, although the presence of PQQ has been claimed in many biological materials [Ameyama et al., 1984], the detection methods used (a biological assay, fluorescence spectroscopy) seem unsuitable for that purpose since the products formed with amino acids are inactive in a biological assay [van Kleef et al., this proceedings] and it is difficult to discriminate the fluorescence of PQQ from that of other compounds [van Kleef et al., 1987].

Non-covalently bound PQQ has only been detected in certain bacterial dehydrogenases (e.g. methanol dehydrogenase (MDH), ethanol dehydrogenase (EDH) and glucose dehydrogenase (GDH)). Both methods indicated for the determination of free PQQ can be applied provided that the extraction method is able to detach PQQ quantitatively from the enzymes and no reaction occurs with nucleophilic residues of the protein. No systematic studies have been performed on these aspects, but in our laboratory, extraction by adding methanol (90 %) and acidification to pH 2, seems to work satisfactory (it should be noted, however, that $PQQH_2$ is rather stable under these conditions so that if the cofactor occurs in a reduced form in the enzyme, the total amount of cofactor is not represented by the PQQ-peak in the chromatogram). For unknown reasons, removal of PQQ from MDH was sometimes variable but quantitative results were obtained by extraction with SDS (sodium dodecyl sulphate) at pH 7 as checked with h.p.l.c. gel filtration and photodiode array detection [Dijkstra et al., 1984].

If there is reason to believe that PQQ is present in derivatized form, a degradative conversion procedure via $PQQH_4$, leading to a highly fluorescing compound after oxidation with sodium periodate (Fig. 1A), could be tried [Duine et al., 1981]. A second possibility in such a case is an alkaline treatment (50 % KOH) of the sample at 100 °C, leading to another highly fluorescing compound (Fig. 1B) [Duine et al., 1987]. Both degradative procedures are able to detect PQQ only in a qualitative way. In this respect, the hexanol extraction procedure (see Section 4.2) looks more promising.

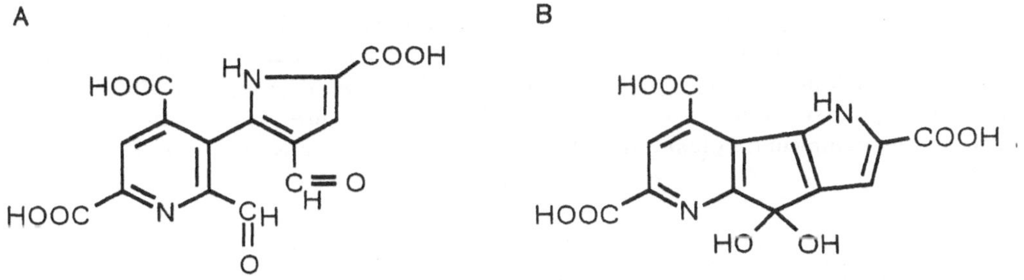

A   B

Figure 1: Proposed structures of the fluorescing products obtained by sodium periodate oxidation (A), and alkaline treatment (B).

3.1. High performance Liquid Chromatography Methods.

PQQ, its reduced forms, and several of its derivatives have been analyzed on a C_{18} reversed phase column by using absorption or fluorescence detection. Ion-suppression chromatography with a CH_3OH gradient in H_3PO_4 (0.4 %) gave satisfying results for PQQ, $PQQH_2$, and several derivatives [Duine et al., 1983]. Ion-pairing chromatography works even better since the peak of PQQ in the chromatogram is much sharper (perhaps because addition of H_2O or CH_3OH to the C(5) group is prevented under this condition). Fluorescence detection is preferred over absorption detection in view of its higher sensitivity and the fact that most of the contaminants do not fluoresce (the latter property makes that the requirements for purity of the samples are less stringent). If samples are heavily contaminated (culture fluids), it is advisable to perform a clean-up with an Amberlyst A21 anion exchanger [van Kleef et al., 1987]. Controlled adduct formation with an aldehyde or ketone can be used to check the identity of a presumed PQQ-peak in the chromatogram [Duine et al., 1983].

3.2. Biological assays.

This method was first performed with GDH apoenzyme from *Acinetobacter cal-coaceticus* [Duine *et al.*, 1979]. Meanwhile, several naturally occurring and self-prepared apo-quinoproteins have been described and these have been used, either in purified form, or present in membrane particles or whole cells (Table 1). Factors that are important to make a choice are: the availability, the ease of preparation, and the stability of the apo-enzyme; absence of background activity; a high turnover number; rapid reconstitution; no interference of other enzymes (especially important if unpurified apoenzymes or whole cells are used).

As is inherent with many biological assays, the high sensitivity requires special care. Thus, we found that chromatographic column materials and laboratory glassware becomes easily contaminated with PQQ, giving high backgrounds so that false positive results are obtained if no checks are made. Since PQQ is a very stable compound, glassware should be thoroughly cleaned [van Kleef *et al.*. 1987].

Table 1: Quinoprotein apo-enzymes in bio-assays for PQQ. The apoenzymes have been used in purified form (enzyme) or in unpurified form occurring in cell-free extract (extract). in cytoplasmic membranes (membranes) or in whole cells (cells). Abbreviations used: GDH, glucose dehydrogenase; QHEDH, alcohol dehydrogenase (quinohaemoprotein).

Quinoprotein	Bacterium	State of the apo-enzyme	Method of activity detection	Ref.[*]
GDH	*A. calcoaceticus*	enzyme	dye reduction	1, 2, 3
GDH	*A. calc.* PQQ⁻ mutant	cells	dye reduction	4
GDH	*Ps. aeruginosa*	enzyme	dye reduction	5
GDH	*E. coli*	cells	gluconate production	6, 7, 8
GDH	*E. coli*	membranes	gluconate production	8
GDH	*Ps. fluorescens*	enzyme	dye reduction	9
QHEDH	*Ps. testosteroni*	enzyme	dye reduction	10

References: 1. Hauge, 1964; 2. Duine *et al.*, 1979; 3. Kilty *et al.*, 1982; 4. Ameyama *et al.*, 1980; 5. Duine *et al.*, 1983; 6. van Kleef *et al.*, 1987; 7. Ameyama *et al.*, 1985; 8. Geiger and Goerisch, 1987; 9. Ameyama *et al.*, 1981; 10. Groen *et al.*, 1986.

4. COVALENTLY BOUND PQQ

"Covalently bound PQQ" is used here in an operational sense, that is the co-factor cannot be detached from the protein by simple dialysis, denaturation or extraction with SDS. Evidence for covalent bonding has been obtained in the case of porcine kidney diamine oxidase [van der Meer *et al.*, this Proceedings] and methylamine dehydrogenase from *Pseudomonas* AM1 [Ishii *et al.*, 1983]. Although the presence of PQQ was observed (in enzymes where it is covalently bound) by protein hydrolysis and fluorescence spectroscopy [Ameyama *et al.*, 1984], as discussed already to our view this approach is unsuited, reason why the hydrazine method and the hexanol extraction procedure were developed.

4.1. The hydrazine method

Procedure

The idea behind this method was to derivatize PQQ in the enzymes with a reagent giving a product which should survive common hydrolysis procedures for proteins so that subsequently the isolated product could be identified and quantified by comparison with a model compound.

In a screening of several carbonyl group reagents, it appeared that 2,4-dinitrophenylhydrazine (DNPH) reacted with free PQQ to a product which behaved satisfactory. The structure of the product, PQQ-DNPH hydrazone at the C(5) position, was solved by X-ray analysis of its ester [van Koningsveld *et al.*, 1986]. The first example on which the method was applied, bovine serum amine oxidase, appeared to be a good choice: DNPH treatment led to a product which was identical (chromatographically and spectrally) to the model compound [Lobenstein-Verbeek *et al.*, 1985]. However, the aim specified, that is to develop a quantitative method, was not met in this case. Based on the assumption of 1 PQQ per enzyme molecule, the yield of hydrazone (6 %) was far too low. The problem was solved, however, as it was found that depending on the applied conditions, PQQ can be quantitatively transformed into either the PQQ-DNPH hydrazone or into a product with quite different properties, here indicated as the PQQ-DNPH azo-compound [van der Meer *et al.*, 1986]. Nowadays, it is clear that also the nature of the enzyme is important whether a product is formed. For the copper-containing amine oxidases, it appears that the azo-compound is immediately formed on addition of the hydrazine whereas hydrazone formation requires long incubation times and high O_2 tensions, as exemplified by diamine oxidase [van der Meer *et al.*, 1986].

The application of a hydrazine is not limited to DNPH. For instance, in the case of methylamine dehydrogenase (MADH) from *Thiobacillus versutus*, phenylhydrazine (PH) was used with success, also yielding two products depending on the conditions applied. The model compounds are more difficult to obtain for this hydrazine since PH (like other hydrazines, exept DNPH) tends to reduce the PQQ. However, using a standard method for reaction of hydrazines with ketones [Stevens and Higginbotham, 1953], and not more than 10 mg of PQQ with a slight molar excess of PH, both products could be obtained by purifying the reaction mixture with reversed phase chromatography on a C_{18} column [van der Meer et al., 1987].

Table 2: Absorption maxima of enzymes derivatized with hydrazines.
Methylamine oxidase (MeAO) from *Arthrobacter* P1, diamine oxidase (PKDAO) from porcine kidney, bovine serum amine oxidase (BSAO), lysyl oxidase (HPLYO) from human placenta, dopamine β-hydroxylase (DBH) from bovine adrenal medulla, methylamine dehydrogenase (MADH) from *Thiobacillus versutus*, and soybean lipoxygenase-1 (SLO) were incubated with DNPH, MH (methylhydrazine) or PH under the following conditions: I, incubation at 40 °C for 16 h at pH 7; II, incubation for 5 min at roomtemperature at pH 7; a, incubation while blowing a stream of nitrogen over the surface of the solution; b, incubation while blowing a stream of oxygen over the surface; c, incubation under air in 0.1 M borate buffer, pH 10; *enzyme treated with diethyldithiocarbamate for copper removal. Spectra were measured after excess reagent was removed by gel filtration.

Enzyme	Reaction condition	Hydrazine	Absorption maxima derivatized enzyme	Reference[**]
			nm	
MeAO	I,a/b	DNPH	410	1
	I,a/b	DNPH	no reaction	1
		MH	383	1
	I,c	PH	365,440	1
PKDAO	I,a	DNPH	454	2
	I,b	DNPH	445	2
	I,a/b*	DNPH	410	2
	I,b	PH	335	2
	II,a	PH	365,440	2
BSAO	I,a	DNPH	454	3
	I,b	DNPH	355,445	3
	I,a/b*	DNPH	410	3
HPLYO	I,b	DNPH	355,445	4
MADH	I,b	PH	340	5
	II,a	PH	365,440	5
DBH	I,a/b	PH	335	6
	I,a*	PH	335	6
	II,a/b	PH	no reaction	–
SLO	I,c	PH	335	7
	II,a/b	PH	no reaction	–

[**]**References:** 1. Van Iersel et al., 1986; 2. van der Meer et al., 1986; 3. Lobenstein-Verbeek et al., 1984; 4. van der Meer and Duine, 1986; 5. van der Meer et al., 1987; 6. van der Meer et al., 1988; 7. van der Meer and Duine, 1988.

The hydrazine method has now been tested with several hydrazines on a variety of enzymes (Table 2). As is apparent from the absorption maxima, the values correspond with the condition applied and the type of product expected. However, although the maxima in the derivatized enzymes are indicative for the presence of PQQ, identification and quantification is only possible with the isolated products. This requires a delicate step, namely the detachment from the protein. On testing several proteases, it appeared that sometimes severe losses of product occurred for reasons unknown sofar. However, application of pronase E (Boehringer, Mannheim) has been proven to be reliable so that this enzyme preparation is routinely used in our laboratory for that purpose. After hydrolysis, the mixture is acidified and applied to a Seppak C_{18} cartridge to remove the amino acids and concentrate the product. Finally, h.p.l.c. on a reversed phase column with a methanol gradient gives the homogeneous product, as judged by monitoring the peaks with a photodiode array detector, used for checking peak homogeneity and comparison with the spectrum of the model compound (see for instance van der Meer *et al.*, 1987).

Products

The assignment of structures (Fig.2) for the products obtained with DNPH having the spectroscopic properties and retention times given in Table 3, is based on the following reasoning. The structure of PQQ-DNPH hydrazone was elucidated by X-ray analysis of its ester. The PQQ-DNPH azo-compound has a some-

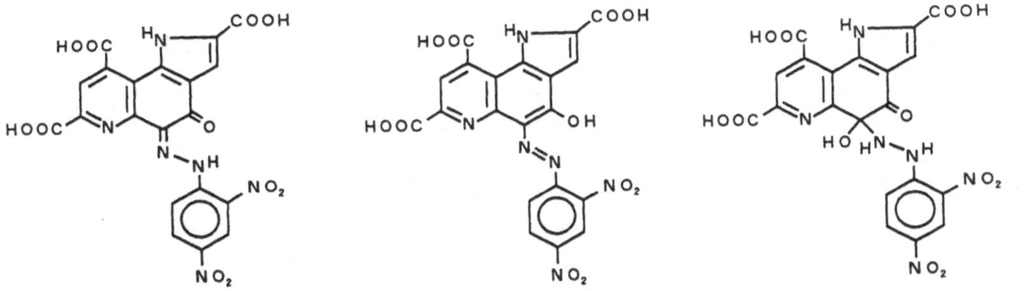

hydrazone
(λ_{max} = 445 nm)

azo-compound
(λ_{max} = 454 nm)

hydrazine adduct
(λ_{max} = 410 nm)

Figure 2: Postulated structures of PQQ-DNPH products.

what higher absorption maximum, suggesting more resonance in the structure (the differences for the products formed with PH under analogues conditions are even higher). Therefore, the structure depicted for the PQQ-DNPH azo-compound is tentatively given to it. The structure is in accordance with the ^{1}H-NMR data while mechanistic arguments have been given elsewhere [van der Meer *et al.*, 1987]. Although both structures are in fact tautomers, they do not easily interconvert into each other. Another product, detected in copper-depleted amine oxidases derivatized with DNPH (λ_{max} = 410 nm, Table 2), is probably the hydrazine adduct. Addition of Cu^{2+} transforms this product immediately into the azo-compound (similarly, the copper depleted amine oxidases are inactive, activity is restored on Cu^{2+} addition). In analogy, products formed with PH or MH under similar conditions as with DNPH have been assigned the structures as those formed with DNPH.

Table 3: Properties of the hydrazine products.

Compound	Type	λ_{max}	ε	Retention time*	¹H-NMR signals
		nm	$M^{-1}.cm^{-1}$	min	ppm
PQQ-PH	hydrazone	335		12.0	7.26, 8.83, 13.90, 16.99, 8.02
PQQ-PH	azo	445		13.0	7.27, 8.83, 16.97, 8.02
PQQ-DNPH	hydrazone	445	31 400	14.6	7.38, 8.72, 13.67, 16.51, 8.97, 8.57, 8.32
PQQ-DNPH	azo	454		18.7	
PQQ-DNPH	hydrazine adduct	410			
PQQ		249		1.9	7.38, 8.54, 12.45
PQQ-MH	azo	383			

*as determined with reversed phase h.p.l.c. described in van der Meer et al., 1987.

Implications

The experience obtained sofar with the hydrazine method, indicates that the conditions during derivatization determine the course of the reaction. This implicates for instance that under the common conditions used for inhibition of amine oxidases, hydrazone formation does scarcely take place (contrary to what is generally believed in older but also in recent literature) but instead of this, the so-called azo-compound is rapidly formed.

Knowledge on the conditions is also relevant in case it is attempted to isolate a peptide to which PQQ is still attached. For obvious reasons, treatment of underivatized enzymes will lead to uncontrolled derivatization of PQQ in the produced peptide. Derivatization with DNPH to the hydrazone (the most stable adduct in our hands) allowed the isolation of a PQQ-containing peptide from porcine kidney diamine oxidase [van der Meer et al., this proceedings].

Evaluation

Derivatization of an enzyme with a suitable hydrazine can already give an indication for the presence of PQQ from the absorption maximum. This indication is only reliable, however, after removal of excess hydrazine has occurred. This was apparently forgotten by Tur and Lerch after derivatizing benzylamine oxidase from *Pichia pastoris* [Tur and Lerch, 1988], where the absorption spectrum given for DNPH-treated enzyme only showes the absorption band of DNPH itself (λ_{max} = 365 nm), since the hydrazone should have an absorption maximum at 445 nm ($\varepsilon_{445\,nm}$ = 31 400 $M^{-1}.cm^{-1}$).

One also has to be careful in which conditions are applied during derivatization, because these conditions define which product (hydrazone or azo-compound) is formed. If these facts are not taken into account, errors in interpretation are inevitable (see for example Williamson *et al.*, 1986 where Raman spectroscopy was used to compare the model compound with the (wrong) product in the enzyme).

The hydrazine method is based on comparison of enzyme-isolated adduct with the corresponding model compound using spectroscopy (UV-VIS and ^{1}H-NMR) and chromatography, methods well accepted for identification of unknown compounds, especially when used in combination. Nevertheless, doubt has been expressed on the validity of this approach for identification of the cofactor [see e.g. Hartman and Klinman, 1988]. It might therefore be reassuring that in all cases we were able to transform adducts isolated from enzymes (as well as model compounds) into PQQ itself by addition of dimethylsulphoxide [van der Meer *et al.*, 1987]. After this transformation, quantification of PQQ is possible with a biological assay [Groen *et al.*, 1987].

In several recent papers [Williamson *et al.*, 1986; Moog *et al.*, 1986 and Knowles *et al.*, 1986], Raman spectroscopy was advocated as the tool to provide evidence for the existence of PQQ in amine oxidases. However, although Raman spectra provide a fingerprint of an unknown compound, identification was based on only this criterium. Subtle differences between the spectra can be ascribed to either interaction with local protein structural elements or to a diverging PQQ structure. Therefore, isolation of the adduct and comparison to a model compound should be performed in order to discriminate between these possibilities.

It may be clear from the foregoing, that the hydrazine method has been a powerful tool in establishing the quinoprotein nature of several enzymes. However, it can be imagined that it will fail in certain cases, for instance if the enzyme-bound cofactor is not in the right redox state, or the active site is unsuited to allow reaction of PQQ with the hydrazine. In this respect, a precedent already exists, namely 3,4-dihydroxyphenylalanine (dopa) decarboxylase of pig kidney, where hydrazone formation is limited to only a few percent [Groen *et al.*, 1988]. The fact that hydrazone formation with pyridoxal phosphate (PLP) under the conditions applied, is also limited, while PLP-PH hydrazone formation under denaturing conditions [Wada and Snell, 1961] is easily achieved, suggests that the active site does not allow the entry of the hydrazine. Furthermore, it is questionable whether the hydrazine method will be able to determine the total number of PQQ molecules in an enzyme in all cases. Enzymes may show half-of-the-site reactivity, that is that after one PQQ has reacted with the hydrazine, derivatization of the second one may become blocked. This may explain the assymmetry of amine oxidases, that is the presence of 2 Cu^{2+} ions and only 1 PQQ detected by the hydrazine method.

4.2. The hexanol extraction procedure

In view of the indicated possible drawbacks of the hydrazine method, the attractiveness of an independent second method for the determination and quantification of covalently bound PQQ, and the need for a procedure able to detect PQQ in its products, we developed the so-called hexanol extraction procedure. The rationale behind this procedure is the following: the higher aliphatic alcohol (hexanol) has a high boiling point so that refluxing creates the high temperatures required for detachment of PQQ (with HCl) and forms an adduct with PQQ; the adduct becomes extracted by the organic solvent so that it escapes undesired attack by other nucleophiles; the strong acid used (6 N HCl) not only detaches PQQ (or decomposes its products) from the protein but also protonates its carboxylic acid groups so that extraction by the alcohol layer becomes possible. In a typical experiment, the following quantities and conditions were used: to a solution of 10 ml enzyme (0.55 mg/ml in 10 mM sodium phosphate buffer, pH 6.5), 10 ml concentrated HCl and 4 ml n-hexanol were added; the mixture was refluxed for 4 h. The product formed has already been characterized and appears to be the PQQ-5,5-dihexyl ketal (see figure 3). Identification and quantification of the isolated product is possible by comparison with the model compound in respect of its characteristics on a reversed phase h.p.l.c. column, absorption spectrum, and ^{1}H-NMR parameters. On the other hand, the product can also easily be hydrolyzed to PQQ (hydrolysis in aqueous sodium carbonate (2 M) at 90 °C for 2 h) so that identification and quantification is also possible with a biological assay. It should be admitted that under certain, as yet unexplained conditions, a slightly different product (slightly different absorption spectrum, insoluble in water and soluble in methanol) is formed. Provisional characterization of this product indicates that it is the corresponding dihexyl ester of the PQQ-5,5-dihexyl ketal.

The method has already been successfully in the case of galactose oxidase (unpublished), dopa decarboxylase [Groen et al., 1988] and glutamate decarboxylase from E. coli (unpublished).

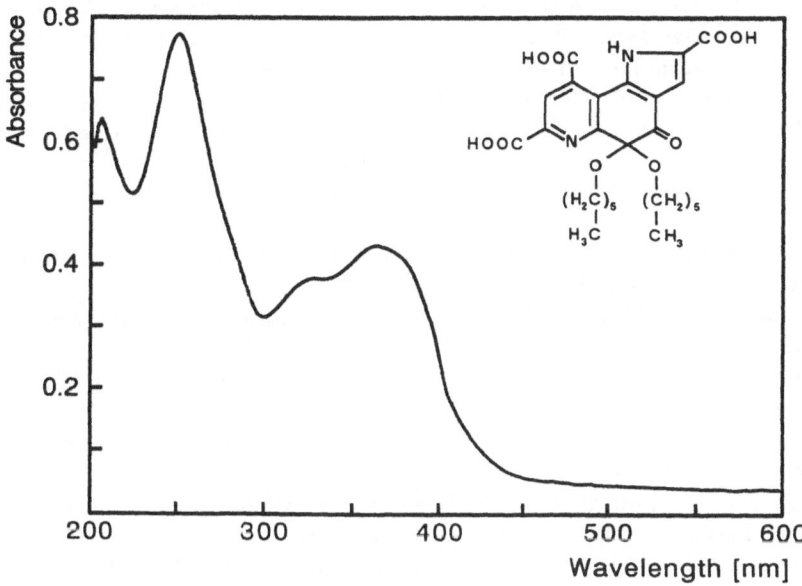

Figure 3: Structure and absorption spectrum of PQQ-5,5-dihexyl ketal.

The hexanol extraction procedure seems also a very valuable tool for the analysis of PQQ in all kinds of biological materials where it occurs in derivatized form. Thus it was found that PQQ-oxazoles, hydrazones/azo-compounds can be transformed into PQQ-5,5-dihexyl ketal with this procedure (unpublished results).

REFERENCES

Ameyama, M., Hayashi, M., Matsushita, K., Shinagawa, E. and Adachi, O. (1984) Agric. Biol. Chem. *48*, 561–565.

Ameyama, M. Matsushita, K., Ohno, Y., Shinagawa, E., and Adachi, O. (1981) FEBS Lett. *130*, 179–183.

Dijkstra, M., Frank, Jzn., J., Jongejan, J.A., and Duine, J.A. (1984) Eur. J. Biochem. *140*, 369–373.

Duine, J.A., Frank Jzn., J. and van Zeeland, J.K. (1979) FEBS Lett. *108*, 443–446.

Duine, J.A., Frank Jzn., J. and Verwiel, P.E.J. (1981) Eur. J. Biochem. *118*, 395–399.

Duine, J.A., Frank Jzn., J. and Jongejan, J.A. (1983) Anal. Biochem. *133*, 239–243.

Duine, J.A., Frank jzn., J. and Jongejan, J.A. (1986) FEMS Microbiol. Rev. *32*, 165–178.

Duine, J.A., Frank jzn., J. and Jongejan, J.A. (1987) Adv. Enzymol. Relat. Areas Mol. Biol. *59*, 169–212.

Groen, B.W., van Kleef, M.A.G. and Duine, J.A. (1986) Biochem. J. *234*, 611-615.

Groen, B.W., van der Meer, R.A. and Duine J.A. (1988) FEBS Lett. *237*, 98-102.

Hartman, C. and Klinman, J.P. (1988) Biofactors *1*, 41-49.

Hauge, J.G. (1964) J. Biol. Chem. *239*, 3630-3639.

Ishii, Y., Hase, T., Fukumori, Y., Matsubara, H., and Tobari, J. (1983) J. Biochem. *93*, 107-119.

Kilty, C.J., Maruyama, K., and Forrest, H.S. (1982) Arch. Biochem. Biophys. *218*, 623-625.

Knowles, P.F., Pandeya, K.B., Rius, F.X., Spencer, C.M., Moog, R.S., McGuirl, M.A. and Dooley, D.M. (1986) Biochem. J. *241*, 603-608.

Van Kleef, M.A.G., Dokter, P., Mulder, A.C. and Duine, J.A. (1987) Anal. Biochem. *162*, 143-149.

Lobenstein-Verbeek, C.L., Jongejan, J.A., Frank jzn., J. and Duine, J.A. (1984) FEBS Lett. *170*, 305-309.

Moog, R.S., McGuirl, M.A., Cote, C.E. and Dooley, D.M. (1986) Proc. Natl. Acad. Sci USA *83*, 8435-8439.

Stevens, F.J. and Higginbotham, D.H. (1953) J. Am. Chem. Soc. *76*, 2206-2207.

Tur, S.S. and Lerch, K. (1988) FEBS Lett. *238*, 74-76.

Van der Meer, R.A. and Duine, J.A. (1986) Biochem. J. *239*, 789-791.

Van der Meer, R.A. and Duine, J.A. (1988) FEBS Lett. *235*, 194-200.

Van der Meer, R.A., Jongejan, J.A., Frank jzn., J. and Duine, J.A. (1986) FEBS Lett. *206*, 111-114.

Van der Meer, R.A., Jongejan, J.A., and Duine, J.A. (1987) FEBS Lett. 221, 299-304.

Van der Meer, R.A., Jongejan, J.A., and Duine, J.A. (1988) FEBS Lett. 231, 303-307.

Van Iersel, J., Frank jzn., J. and Duine, J.A. (1985) Anal. Biochem. *151*, 196-204.

Van Iersel, J., van der Meer, R.A. and Duine, J.A. (1986) Eur. J. Biochem. *161*, 415-419.

Van Koningsveld, H., Jansen, J.C., Jongejan, J.A., Frank Jzn, J. and Duine, J.A. (1985) Acta Cryst. C41, 89-92.

Wada, H. and Snell, E.E. (1961) J. Biol. Chem. *236*, 2089-2095.

Williamson, P.R., Moog, R.S., Dooley, D.M. and Kagan, H.M. (1986) J. Biol. Chem. *261*, 16302-16305.

GAS CHROMATOGRAPHY OF PQQ

O. Suzuki, T. Kumazawa, H. Seno, T. Matsumoto* and T. Urakami**

Department of Legal Medicine, Hamamatsu University School of Medicine, 3600 Handa-cho, 431-31; *First Department of Surgery, Nagoya University School of Medicine, 65 Tsuruma-cho, Nagoya 466; and **Niigata Research Laboratory, Mitsubishi Gas Chemical Company, Inc., 182 Shinwari, Tayuhama, Niigata 950-31, Japan

Abstract. A new method for gas chromatographic analysis of PQQ is presented. A sample was made weakly-alkaline and passed through a Sep-Pak C_{18} cartridge. The eluate was then made acid with HCl and passed through the second Sep-Pak C_{18} cartridge. Finally, PQQ was eluted with pyridine solution; it was evaporated to dryness. The residue was derivatized with a phenyltrimethylammonium reagent and subjected to gas chromatography (GC). The chemical structure of the reaction product was estimated to be 3-[1-methyl-3,5-di(methoxy-carbonyl)-pyrrole-2-yl]-2,4,6-tri(methoxycarbonyl)pyridine by mass spectral analysis. GC was made with SPB-1 fused silica wide-bore and narrow-bore capillary columns, and with flame ionization detection. The detection limit of PQQ was 5-10 ng in an injected volume with the narrow-bore capillary column in the splitless mode. The endogenous PQQ in human urine, plasma, brain and liver, and rat brain and liver, was below the detection limit (less than 100 ng/ml or g).

INTRODUCTION

For analyses of PQQ, enzymatic and high-performance liquid chromato-graphic methods have been reported (Duine et al, 1983; Ameyama et al, 1985; van Kleef et al, 1987; Geiger and Görisch, 1987). However, these studies dealt only with clean aqueous solutions of PQQ and are not applicable to crude biological samples with high protein contents. In this study, we present a new method for gas chromatographic detection of PQQ and try to measure PQQ in crude biological samples.

MATERIALS AND METHODS

Materials. PQQ was obtained from Mitsubishi Gas Chemical Company, Inc. (Niigata); phenyltrimethylammonium (PTMA) hydroxide (20-25% in methanol) from Tokyo Kasei Kogyo Co., Ltd. (Tokyo); and fused silica capillary columns from Supelco, Inc. (Bellefonte, PA).

GC conditions. GC was carried out on a Shimadzu GC-4CM instrument

J. A. Jongejan and J. A. Duine (eds.), PQQ and Quinoproteins, 123–129.
© 1989 by Kluwer Academic Publishers.

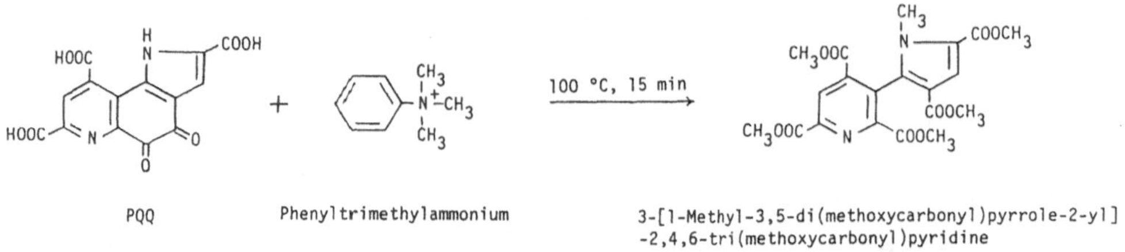

124

FIGURE 1. Reaction of PQQ with phenyltrimethylammonium (PTMA).

with fused silica wide-bore (SPB-1, 15 m x 0.53 mm i.d., film thickness 1.5 μm) and narrow-bore (SPB-1, 15 m x 0.32 mm i.d., film thickness 0.25 μm) capillary columns, with flame ionization detection. The GC conditions were: injection temperature 300°C and column temperature 150-300°C (10°C/min) for both columns. GC was in the splitless mode with the wide-bore capillary column in the whole procedure with nitrogen flow rate of 20 ml/min. In narrow-bore capillary GC, the sample solution was injected into the GC port in the splitless mode at 150°C; it was left for 5 min at this temperature and switched to the split mode with nitrogen flow rate of 3 ml/min during elevation of column temperature.

Mass spectrometry (MS). Mass spectra of the reaction product were recorded on a JMS-D300 GC/MS instrument with a JMA-2000E computer-controlled data analysis system in the positive ion electron impact mode.

Clean-up and derivatization. One-milliliter of human urine or plasma was mixed with 9 ml 1% $NaHCO_3$ solution and passed through a Sep-Pak C_{18} cartridge to remove impurities. The eluate was mixed with 0.15 ml conc. HCl, applied to the second Sep-Pak C_{18} cartridge and washed with 20 ml 0.001 N HCl. PQQ was eluted with 1.5 ml 50% pyridine in water; it was evaporated to dryness in vacuo. A 0.1-ml aliquot of PTMA hydroxide (20-25% in methanol) was added to the residue and heated at 100°C for 15 min; 1-3 μl of it was subjected to GC analysis.

For clean-up of PQQ in brain and liver tissues, deproteinization with acid was added to the above procedure (Suzuki et al, in preparation).

RESULTS

Reaction product. PQQ is a very polar compound and it was impossible to detect PQQ by GC in the underivatized form. Derivatization of PQQ with various reagents, such as bis(trimethylsilyl)trifluoroacetamide, diazomethane and trimethylsilyldiazomethane, were tested, but none of them gave any peak on gas chromatograms. Thus, we tried to derivatize the ortho-quinone group of PQQ with amines such as methoxylamine and dimethylhydrazine, forming Schiff bases in addition to methylation of its carboxyl groups. Even with these efforts, we could not obtain any peak on the chromatograms. The only one, which successfully gave a peak, was the reaction between PQQ and PTMA (Fig. 1). The mass spectrum of the reaction product is shown in Fig. 2, and its chemical structure was estimated to be 3-[1-methyl-3,5-di(methoxycarbonyl)pyrrole-2-yl]-2,4,6-tri(methoxycarbo-nyl)pyridine by analyzing the spectrum. PTMA is usually used as an

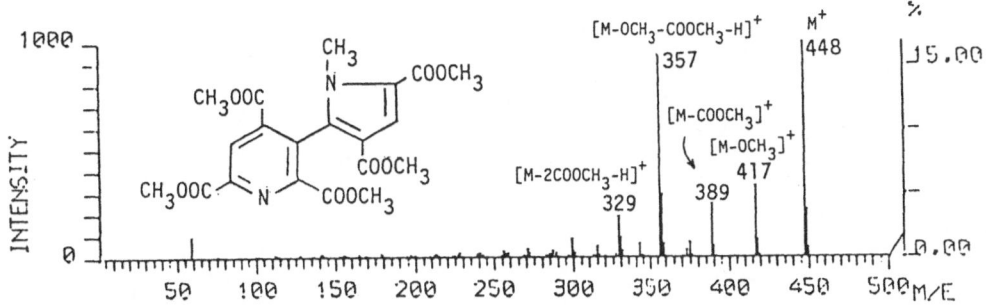

FIGURE 2. Mass sepctrum of the reaction product and its probable fragmentation mode.

on-column methylating reagent, but its reaction with PQQ is unique because it splits the bond between both carbonyl groups. The splitting seems to take place during heating at 100°C, since no peak appeared with on-column methylation without prior heating. The optimal reaction time at 100°C was found to be 15 min.

Packed column GC. Various kinds of packed columns, such as 3% OV-1, 3% OV-17, GP 2% SP-2510-DA and 5% PolyI-110, were tested for GC of PQQ. The best one was 5% PolyI-110. However, its sensitivity was rather low; the detection limit of PQQ was about 100 ng in an injected volume. In addition, the chromatograms suffered from tailing of the PQQ peak and high backgrounds.

Wide-bore capillary column GC. Wide-bore capillary GC is now gaining popularity because capillary columns can be easily attached to old-type GC instruments designed for packed columns, and compounds to be analyzed are relatively stable during passage through the column due to fast flow and thus short exposure to heat.

Figure 3 shows wide-bore capillary GC for the authentic PQQ (5 μg/0.1 ml), and extracts of 1 ml plasma with and without addition of 5 μg PQQ at the initial step. From the peak heights, the recovery of PQQ added to 1 ml plasma was about 50 %. The plasma extract without addition of PQQ gave a very small peak at the same retention time as that of PQQ (Fig. 3, right panel), but this is not due to endogenous PQQ present in plasma because the reagent blank also gave such a small impurity peak at this retention time.

The addition test was also made for human urine. The recovery of 5 μg PQQ added to 1 ml urine was 75%. Endogenous PQQ could not be detected also in urine (data not shown).

A calibration curve was made as a function of various amounts of the authentic PQQ. The curve showed linearity up to 200 ng in an injected volume. The detection limit was 10-20 ng.

Narrow-bore capillary column GC. Narrow-bore capillary column GC enjoys its high resolution ability, but sometimes shows decomposition of compounds to be analyzed due to its slow gas flow inside the column and long exposure of compounds to heat. Fortunately, the reaction product of PQQ with PTMA was relatively heat-stable and gave no problem in its assay with this type of column. High sensitivity could be also achieved with use

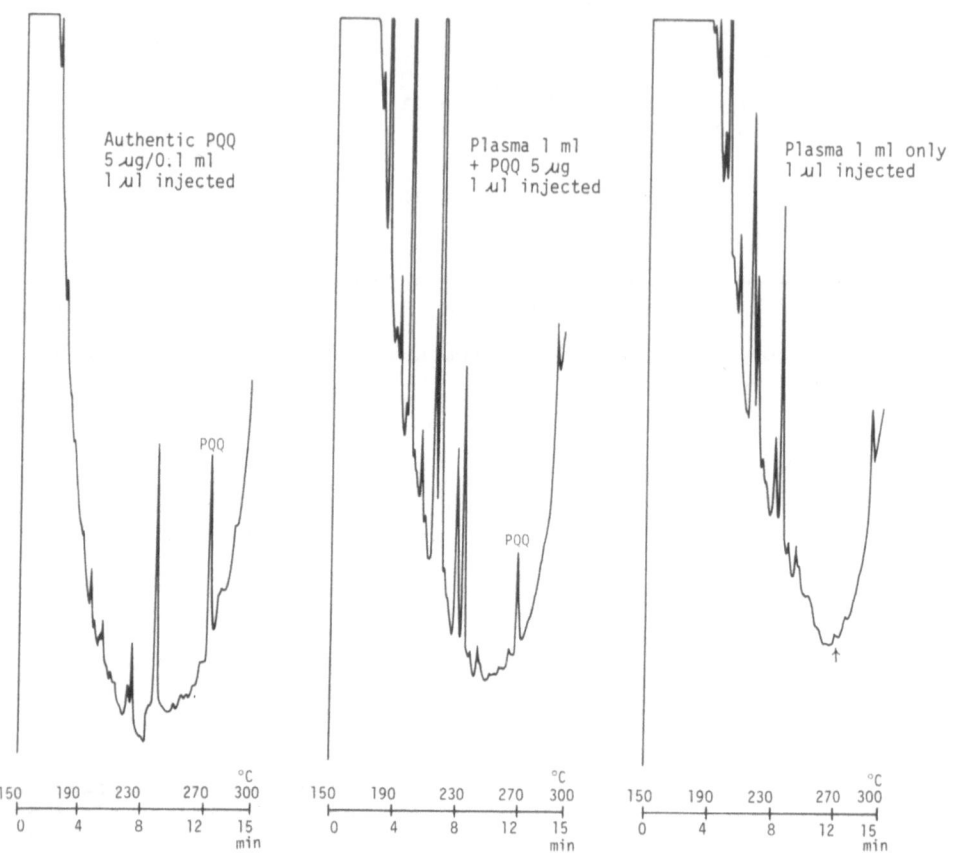

FIGURE 3. Wide-bore capillary GC for the authentic PQQ and human plasma extracts after reaction with PTMA with and without addition of PQQ. GC was carried out with a SPB-1 fused silica wide-bore capillary column (15 m x 0.53 mm i.d., film thickness 1.5 μm). Its conditions were: column temperature 150-300°C (10°C/min) and nitrogen flow rate 20 ml/min. Five micrograms of PQQ were added to 1 ml plasma (middle panel) and extracted with Sep-Pak C_{18} cartridges.

of the splitless mode upon injection of samples at a low temperature.

Figure 4 shows narrow-bore capillary GC for the authentic PQQ (100 ng injected) and extracts of 1 g human brain with and without addition of 10 μg PQQ. The peaks in the chromatograms were much sharper than those by wide-bore capillary GC; its sensitivity was also several times higher than that of the wide-bore capillary GC. The calibration curve with the authentic PQQ showed the detection limit of 5-10 ng in an injected volume. Even with this sensitive GC method, no distinct peaks due to endogenous PQQ in human brain could be detected (Fig. 4, right panel). Other samples, such as human liver, rat brain and rat liver were also tested with the

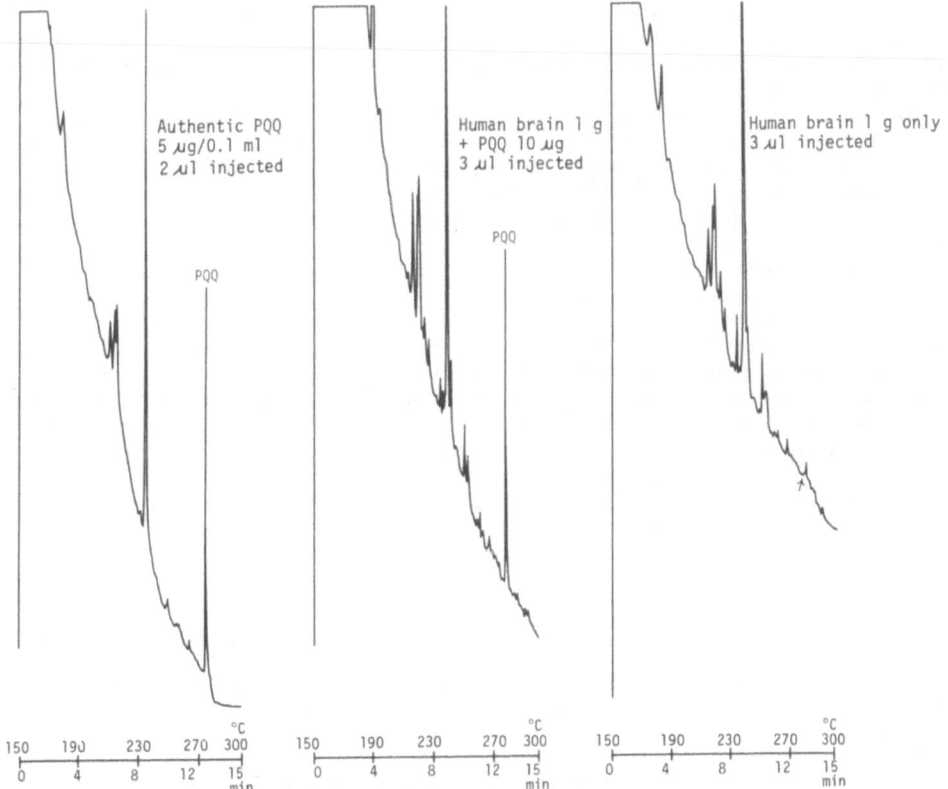

FIGURE 4. Narrow-bore capillary GC for the authentic PQQ and extracts of human brain after reaction with PTMA with and without
addition of PQQ. GC was carried out with a SPB-1 fused silica
narrow-bore capillary column (15 m x 0.32 mm i.d., film thickness
0.25 μm). Its conditions were: column temperature 150-300°C (10
°C/min) and nitrogen flow rate 3 ml/min in the split mode. At 150
°C of the column temperature, the samples were injected into GC
port in the splitless mode. After standing for 5 min, GC was
switched to the split mode for elevation of column temperature.
Ten micrograms of PQQ were added to 1 g human brain (middle panel)
and subjected to extractions.

same procedure, but their levels were almost below the detection limit
(probably less than 100 ng/g wet weight, if present).

DISCUSSION

In the present study, we have succeeded in obtaining a derivative of
PQQ, which is suitable for GC analysis (Figs. 1 and 2) and presented a
procedure for assays of PQQ in biological samples. The narrow-bore
capillary column GC in the splitless mode followed by the split mode is

most recommendable, because of its high resolution and high sensitivity. Although we have used flame ionization detection for GC in the present study, the sensitivity will probably be greatly enhanced with use of a nitrogen detector.

GC has a great advantage in that it can be easily connected to MS, which enables reliable identification and highly sensitive quantitation of a compound. Endogenous levels of PQQ in human and rat samples were all below detection limits by GC in the present experiments. Thus further studies on GC/MS detection of PQQ in human and rat tissues are now under way in our laboratory.

PQQ was identified in various dehydrogenases of bacterial origin in early studies (see review by Duine et al, 1986). Recently, mammalian enzymes, such as amine oxidases (Lobenstein-Verbeek et al, 1984; van der Meer et al, 1986; Moog et al, 1986; Knowles et al, 1987), lysyl oxidase (van der Meer and Duine, 1986; Williamson et al, 1986) and dopamine β - hydroxylase (van der Meer et al, 1988), have been reported to contain PQQ as a cofacator. However, in many of these reports, the existence of PQQ is insisted only on the bases of coincidence of retention times in high-performance liquid chromatography and similarity in fluorescence or absorption spectra. Therefore, GC/MS seems most powerful and reliable to establish the presence of PQQ as a cofactor in mammalian enzymes. The present GC method for PQQ gives a breakthrough to such GC/MS studies.

REFERENCES

Ameyama M, Nonobe M, Shinagawa E, Matsushita K and Adachi O, 1985. Method of enzymatic determination of pyrroloquinoline quinone. Anal. Biochem. 151: 263-267.

Duine JA, Frank J, Jzn and Jongejan JA, 1983. Detection and determination of pyrroloquinoline quinone, the coenzyme of quinoproteins. Anal. Biochem. 133: 239-243.

Duine JA, Frank J, Jzn and Jongejan JA, 1986. PQQ and quinoprotein enzymes in microbial oxidations. FEMS Microbiol. Rev. 32: 165-178.

Geiger O and Görisch H, 1987. Enzymatic determination of pyrroloquinoline quinone using crude membranes from Escherichia coli. Anal. Biochem. 164: 418-423.

Knowles PF, Pandeya KB, Rius FX, Spencer CM, Moog RS, McGuirl MA and Dooley DM, 1987. The organic cofactor in plasma amine oxidase: evidence for pyrroloquinoline quinone and against pyridoxal phosphate. Biochem. J. 241: 603-608.

Lobenstein-Verbeek CL, Jogejan JA, Frank J and Duine JA, 1984. Bovine serum amine oxidase: a mammalian enzyme having covalently bound PQQ as prosthetic group. FEBS Lett. 170: 305-309.

Moog RS, McGuirl MA, Cote CE and Dooley DM, 1986. Evidence for methoxatin (pyrroloquinolinequinone) as the cofactor in bovine plasma amine oxidase from resonance Raman spectroscopy. Proc. Natl. Acad. Sci. USA 83: 8435-8439.

van der Meer RA and Duine JA, 1986. Covalently bound pyrroloquinoline quinone is the organic prosthetic group in human placental lysyl oxidase. Biochem. J. 239: 789-791.

van der Meer RA, Jongejan JA and Duine JA, 1988. Dopamine β-hydroxylase from bovine adrenal medulla contains covalently-bound pyrroloquinoline quinone. FEBS Lett. 231: 303-307.

van der Meer RA, Jongejan JA, Frank J, Jzn and Duine JA, 1986. Hydrazone formation of 2,4-dinitrophenylhydrazine with pyrroloquinoline quinone in porcine kidney diamine oxidase. FEBS Lett. 206:111-114.

van Kleef MAG, Dokter P, Mulder AC and Duine JA, 1987. Detection of the cofactor pyrroloquinoline quinone. Anal. Biochem. 162:143-149.

Williamson PR, Moog RS, Dooley DM and Kagan HM, 1986. Evidence for pyrroloquinolinequinone as the carbonyl cofactor in lysyl oxidase by absorption and resonance Raman spectroscopy. J. Biol. Chem. 261: 16302-16305.

DIRECT AND AMPLIFIED REDOX-CYCLING MEASUREMENTS OF PQQ IN QUINOPROTEINS AND BIOLOGICAL FLUIDS: PQQ-PEPTIDES IN PRONASE DIGESTS OF DBH AND DAO

Paz, MA, Flückiger*, R, Henson, E and Gallop, PM: Children's Hospital and the Harvard Schools of Medicine and Dental Medicine, Boston, MA 02115 and the *Roche Research Laboratory, Basel, Switzerland.

Key words: methoxatin, pyrroloquinoline quinone, PQQ, quinoproteins, PQQ-peptides, redox-cycling, nitroblue tetrazolium, formazan, redox enzymes, superoxide, superoxide dismutase, SOD, dopamine, norepinephrine, diamine oxidase, dopamine-ß-hydroxylase, lysyl oxidase, cerebrospinal fluid, CSF, serum, urine, milk, egg-yolk.

ABSTRACT

We have found that PQQ can be directly measured with 3-methyl-2-benzothiazolinone hydrazone. PQQ can also be specifically detected with much greater sensitivity by its unique ability to support redox cycling. A reagent prepared from potassium glycinate and nitroblue tetrazolium can lead to the formation of amplified levels of formazan dye in the presence of picomoles of PQQ. Confirmation of PQQ as the coenzyme in pig kidney diamine oxidase and bovine adrenal dopamine-ß-hydroxylase is presented with isolation of PQQ-peptides from pronase digests of these quinoenzymes. PQQ is also present in trace levels in some proteins and completely absent from others. It is present in serum, CSF, urine, milk and egg-yolk. The concentration of PQQ in biological fluids may have clinical and pathophysiological significance. Reduced PQQ generates superoxide with oxygen and has the potential to become a source of oxidative stress in damaged tissues and to contribute to inflammatory processes.

INTRODUCTION

Pyrroloquinoline quinone (PQQ) or methoxatin is a highly reactive redox-responsive coenzyme. In its quinone form it can interact with carbonyl reagents [Lobenstein-Verbeek et al. 1984, Glatz et al. 1987], but it does not interact with these reagents in its reduced enediol forms (PQQ(2H)). We and others have noticed that PQQ can also oxidize certain carbonyl reagents without adduct formation, becoming reduced in the process and no longer able to react as a carbonyl compound [Dekker et al. 1982]. Thus, the reactivity of PQQ with reagents is complex and sensitive to the redox status of PQQ which is determined by the redox properties of other compounds present [Flückiger et al. 1988]. The availability of oxygen is a key parameter [van der Meer et al. 1986].

We have found that PQQ is reduced by NaBH₄ to PQQ(2H), and when excess borohydride is removed, PQQ(2H) reacts with oxygen in one electron steps with superoxide generation and reformation of PQQ. NaCNBH₃ can reduce adducts formed between primary amines and PQQ [Hartmann and Klinman, 1988]. We have noted that these reduced adducts are reoxidizable after removal of excess cyanoborohydride. Without cyanoborohydride, PQQ can interact with amines and in some cases oxidize them to products with the PQQ becoming reduced to a mixture of enediol and eneaminol products [Sleath et al. 1985, Flückiger et al., 1988, Paz et al., 1988].

The measurement of PQQ found in a variety of quinoproteins in bacteria [Salisbury et al. 1979; van der Meer et al. 1987] and mammals [Lobenstein-Verbeek et al. 1984; van der Meer et al. 1986; Ameyama et al. 1985] and in biological fluids is fraught with difficulties, even in simple cases where PQQ is in free form. In our laboratories in the last

131

J. A. Jongejan and J. A. Duine (eds.), PQQ and Quinoproteins, 131–143.
© 1989 by Kluwer Academic Publishers.

two years we have been able to develop new and sensitive methods, based on the special reactivity and redox nature of PQQ, which have enabled us to detect quinopeptides from digests of quinoproteins and to begin the task of defining the nature of the PQQ-locus in enzymes which use PQQ as a coenzyme or prosthetic group.

We first describe the reaction of PQQ and a model diketone, phenanthrenequinone with 3-methyl-2-benzothiazolinone hydrazone (MBTH) and then describe the use of MBTH in a specific method to measure nanomole levels of PQQ. Next we describe the amplified redox detection of PQQ with a special reagent made from potassium glycinate and nitroblue tetrazolium (NBT) and its application to the picomole detection of PQQ in various quino- and other proteins before and after pronase digestion. We confirm that a number of copper-dependent amine oxidases are quinoproteins and also confirm and extend the recent and important observation suggesting the quinoprotein nature of dopamine-ß-hydroxylase (DBH) [van der Meer et al.,1988]. We demonstrate for the first time with mammalian quinoenzymes, the identification of quinopeptides, detectable with the glycinate-NBT reagent, from pronase digests of pig-kidney diamine oxidase (PK-DAO) and DBH. We also measure PQQ in serum, CSF, urine, milk and egg.

MATERIALS AND METHODS

Methoxatin (PQQ) was obtained from Fluka, crude PK-DAO, bovine adrenal DBH, bovine serum albumin (BSA), chicken egg ovalbumin, Torula yeast glucose-6-phosphate dehydrogenase, pigeon breast muscle carnitine acetyl transferase, pig heart lactic dehydrogenase, human IgG, bovine milk ß-lactoglobulin, egg white lysozyme, bovine pancreas ribonuclease A and nitroblue tetrazolium (NBT) were from Sigma. Pronase (Streptomyces griseus) was obtained from Boehringer-Manheim. MBTH.HCl was from Aldrich. Bovine aorta lysyl oxidase (LO) was a generous gift from Dr. Herbert Kagan, Boston University.

Purification of PK-DAO: PK-DAO was further purified on a 3x16 cm hydroxylapatite column as described by Rinaldi et al. 1982. The fractions with diamine oxidase activity (Paz et al. 1988) were pooled and precipitated with 0.6 saturated $(NH_4)_2SO_4$. The precipitate was dissolved in 0.025 M K_2HPO_4 at pH 7.0, dialyzed and lyophilized.

Preparation of Phenanthrenequinone-MBTH Azine: A solution of 0.63g (0.003 mole) of phenanthrenequinone and 0.64g (0.0035 mole) of MBTH-free base (M.P. 145-146) in 30 ml of chloroform is heated at boiling temperature for 1 hour. The solvent is evaporated and the crude product triturated with chlorobutane to remove excess MBTH and product filtered off. Recrystalization from methylene chloride gave red crystals, 0.95g (85%), M.P. 199-203°, M.S.: M$^+$, M/e 369; calculated for C22H15N3S1, C 71.1%, H 4.1%, N 11.4%, S 8.7%; found C 70.4%, H 4.2%, N 11.5% and S 8.4%.

Reaction product of PQQ and MBTH: A solution of 0.025g of PQQ in 25 ml of methanol was added to a solution of 0.027g of MBTH-free base in 15 ml of methanol. The mixture was stirred for 2 hours and then evaporated to dryness. The dark red residue was triturated with methylene chloride and filtered. Thin layer chromatography on Merck 5735 silica sheet in methanol showed one red spot (nonfluorescent) at R_f 0.70.

Measurement of PQQ with MBTH: A typical standard curve of the reaction of PQQ with MBTH is prepared with a stock solution of 1 μmole/ml PQQ in 0.05 M sodium phosphate buffer pH 6.6. This solution is then diluted 1 to 10 with the same buffer and different concentrations are prepared in a

800 µl volume. To the PQQ is added 200 µl of MBTH solution which is prepared with 22 mg MBTH.HCl in 10 ml of distilled water. MBTH reacts with
PQQ and forms a red solution containing a mixture of a diazine and azine
with a maximum absorption at about 500 nm and a smaller absorption peak
at about 410nm. The color develops at 25 ºC and is very stable.

To measure nanomole levels of PQQ in biological samples with MBTH,
the following procedure is useful: Initially, $NaBH_4$ is added to the
samples and standards to reduce any contaminating carbonyl compounds to
alcohols and PQQ to PQQ(2H). MBTH is then added and the excess
borohydride is scavenged with glucose or destroyed with acid. As the
solution aerates, PQQ(2H) is reoxidized to PQQ which is trapped by the
MBTH before it can generate adducts or react with the various amines
and amino acids present in the biological samples. The red color is read
at 500nm and PQQ content calculated from a standard curve run at the same
time. Defined amounts of PQQ are usually added to duplicates of the
biological samples to make sure of full recovery of PQQ. The red solutions, containing the PQQ-adducts can also be extracted with isoamyl
alcohol since the PQQ material is not at all extractable from water,
whereas MBTH adducts with compound like dehydroascorbate are. Thus, if
large amounts of extraneous diketones are present and not permanently
removed by the borohydride step, a final extraction step with isoamyl alcohol will remove MBTH adducts with such compounds. Generally, this is
not necessary.

Detection of Picomole Levels of Reactive PQQ with Glycinate-NBT Reagent

In this procedure PQQ is reduced by glycinate; PQQ is not reduced by
valinate. Reduced PQQ is then reoxidized by oxygen with superoxide formation. The superoxide is then reoxidized by NBT to oxygen with irreversible production of formazan. There is also some direct oxidation of
reduced PQQ by NBT (about 20%) as can be seen by adding superoxide dismutase to the assay [Paz et al. 1988]. The rate of formazan production
depends on the content of PQQ and the extent of its exposure and accessibility to the redox reagents. Glycinate is present in large excess at 2M
and serves as both the buffer and initial reductant; valinate does not
reduce PQQ and can thus be used instead of glycinate as an additional
control. Valine is not as soluble as glycine and one can compare 0.5M
valinate with 0.5M-2M glycinate. For example, ketoamines in nanomole
amounts can be detected by both reagents (no amplification), but PQQ can
be detected in picomole amounts with the glycinate reagent.

The standard curve is run with 5-50 picomoles of PQQ, 1 mg of
treated-BSA, 1 ml of 0.24 mM NBT in 2M potassium glycinate at pH 10 in a
total volume of 1.5 ml. The BSA was pretreated with borohydride and
dialysis to reduce glycation ketoamines and lower the blank [Paz et al.
1988] and is added to the reaction to maintain the formazan in solution.
The tubes are incubated at 37º in a covered water bath protected from
direct light and the amplified formazan production is followed at 530 nm.
The detection of PQQ in quinoproteins usually require treatment with
pronase overnight at 37º to increase the exposure of PQQ. The
proteolysis is ended by boiling the samples in a water bath for 1 minute;
a pronase control treated as the samples is included in the test. The
NBT-glycinate redox reaction with pronase-treated proteins, undigested
proteins, pronase control and PQQ standards are monitored for several
hours. The reciprocal of the measured PQQ content as determined by the
formazan produced compared to standards is plotted versus reciprocal time
and extrapolated to infinite time to determine the final level of PQQ-
dependent formazan production.

Detection of PQQ-Peptides in Pronase Digests of DBH and PK-DAO Separated by High Pressure Liquid Chromatography (HPLC)

Two enzymes believed to be bonafide quinoproteins, DBH and PK-DAO, in 20 mM phosphate, pH 7.0 are digested with pronase following the procedure described above. The peptides generated are separated using a C-18 Ultrasphere IP 4.6 mm x 25 cm Beckman column with a linear gradient of buffer A (0.1% TFA in H_2O) and buffer B (0.1% TFA in H_2O containing 70% acetonitrile). Buffer B was increased from 0-30% over 20 minutes, from 30-55% over 50 minutes and from 55-100% B over an additional 20 minutes. Fractions were collected every minute (2ml/min) and the elution profile was followed by UV detection at 230 nm. The fractions were evaporated to dryness and both ninhydrin reactivity and PQQ content measured, the latter with the glycinate-NBT procedure.

RESULTS AND DISCUSSION

The Reaction of Phenanthrenequinone and PQQ with MBTH

Phenantrenequinone (I) reacts with MBTH to form a red solution from which a crude product is obtained in excellent yield. The crude product on thin layer chromatography on Merck 5735 silica gel sheet shows three spots, two red with Rf's at 0.35 and 0.45 and one yellow spot with Rf at 0.75. The yellow spot present in small amount is the MBTH adduct with both carbonyls and is the orthodiazine (II) or "osazine"product (Fig. 1).

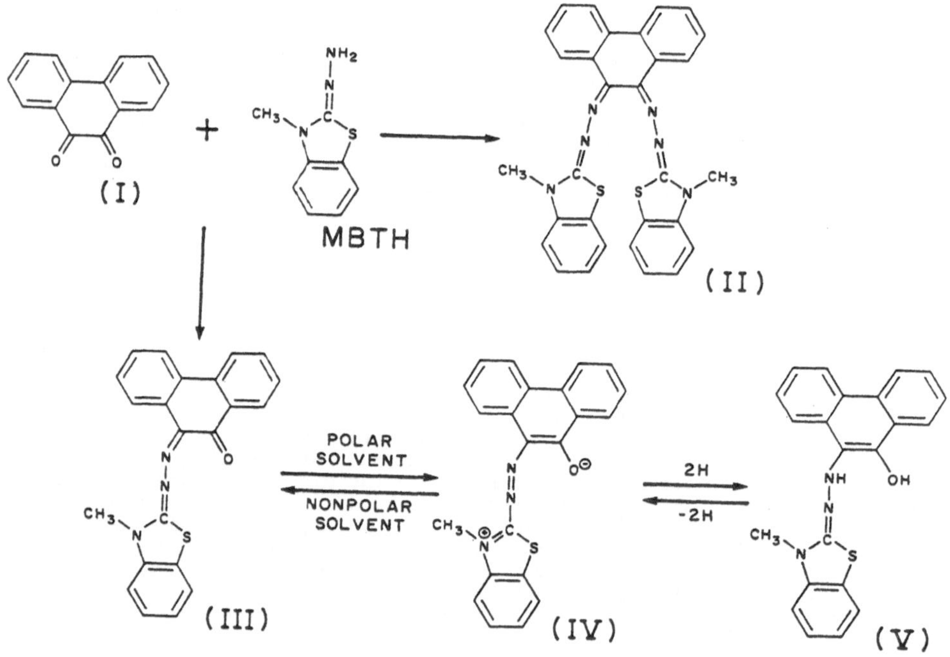

Figure 1.

Purified by TLC, it shows a mass spectra consistent with the diazine, molecular ion at 530, 26%, base ion 367, 352, 88%, 204, 83% and 208, 59%. Each red spot when isolated and rechromatographed on TLC, shows the same two spots again. The structure of each spot is the same and is that of the azine (III), consistent with the mass spectra showing a parent ion at 369, 98%, 341, 25%, 340, 48%, 207, 68 %, 206, 61%, 191, 90%, 178, 76%, 177, 65% and 164, 95%. The azine (III) is in a solvent-dependent dynamic equilibrium with the zwitter ion eneolate (IV) (Fig. 1). IV is red with absorption maximum at 482 nm (Fig. 2) and an extinction coefficent of about 17,000 and partially converts to III which is yellow in nonpolar solvents. Thus, the azine product is strongly solvatochromatic. On TLC, the azine moves as two spots the III and IV products, which always look red on the polar silica gel TLC plate.

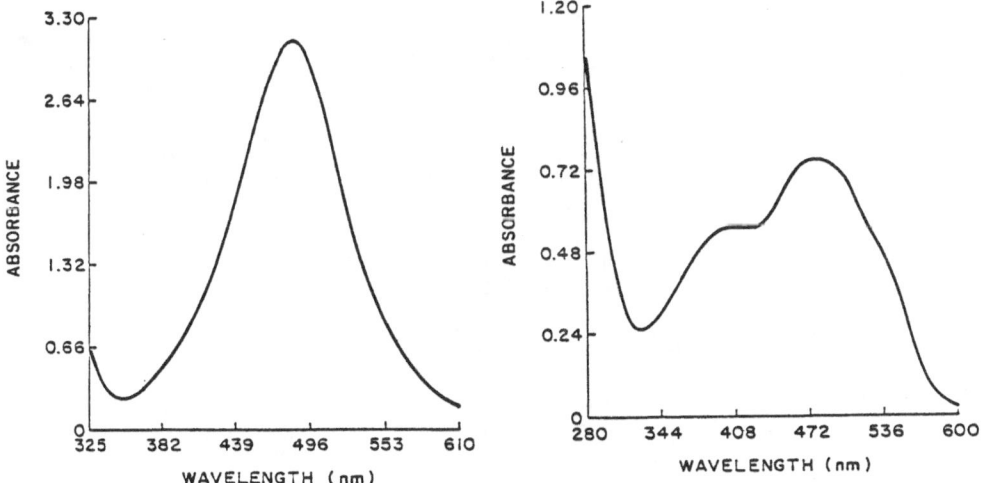

Figure 2. Absorption Spectra of
    Compound IV

Figure 3. Absortion Spectra of the
    Reaction Product of PQQ and MBTH

A red solution of IV and III in a polar solvent can be reduced with NaBH$_4$ to a yellow product V with fluorescence (Fig. 1). With excess borohydride the solution remains yellow. Swirling the yellow solution in air, a red color appears at the surface, which then rapidly changes back to yellow at rest; by removal of excess borohydride and aeration the red color will return permanently. Thus, the uncharged azine product exists in a solvent-dependent equilibrium with the zwitter ion eneolate and both are reducible with borohydride to V. V is oxidizable back to III + IV.

With PQQ and MBTH in aqueous solvents, a red solution rapidly forms which shows a spectra (Fig. 3) consistent with a mixture of PQQ-diazine (VI) (408nm) and azine<-->zwitter ion eneolate (480-500nm) VII + VIII. The structure of the various compounds is shown in Fig. 4. Crude product is refractory to further purification, since it is only soluble in polar solvents rich in water. It does not melt and decomposes and decarboxylates over a wide range of temperature. Its elemental analysis is consistent with a mixture of PQQ-diazine and PQQ-azine in about 20:80 ratio, and this ratio is also consistent with the absorption spectra. We have as yet not obtained useful mass spectral data, because of the lack of sample volatility.

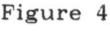

Figure 4

The red compound formed with MBTH appear to be specific for PQQ and has advantages over procedures which employ the usual hydrazine reagents. It is the base for the analytical procedure developed by us to measure free PQQ and since it is not an amplification procedure, sensitivity is limited to nanomole levels; a typical standard curve is shown in Fig. 5. Quinoproteins require proteolysis to increase exposure of PQQ to the MBTH reagent.

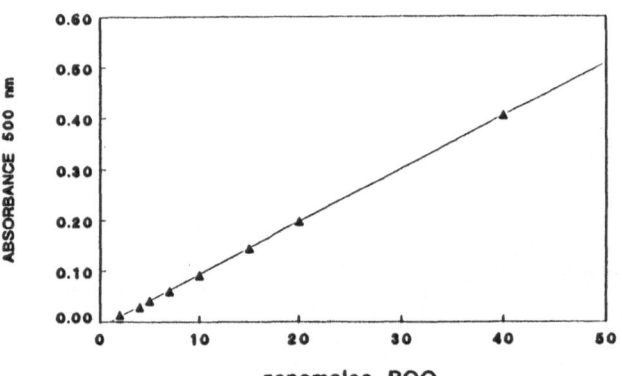

Fig. 5 Colorimetric Detection of PQQ with MBTH

The Reaction of PQQ with Glycinate–NBT Reagent: Detection of Picomole
Levels of Reactive PQQ

In order to obtain a very sensitive method for PQQ analysis
independent of the initial redox status of the PQQ, we recognized that we
could use the very special and unique redox properties of PQQ for its
amplified detection.   PQQ(2H) reduces NBT (T$^+$) to formazan (TH) both
directly, or through oxygen-PQQ(2H) generated superoxide (O$_2^-$):

$$T^+ + PQQ(2H) \rightarrow TH + H^+ + PQQ;$$

$$PQQ(2H) + 2O_2 \rightarrow PQQ + 2(O_2^-) + 2H^+; \quad T^+ + H^+ + 2O_2^- \rightarrow TH + 2O_2.$$

What one then requires to generate a PQQ-dependent redox cycle for amplified formazan production are compounds able to specifically reduce PQQ
without a concurrent reduction of NBT.   Glycinate does this most effectively and allows de-
tection of picomole
levels of PQQ (Fig.
6), ornithinate also
reduces PQQ, but it
tends to precipitate
NBT; valinate is not
able to reduce PQQ to
any significant level
and can be used as a
control.   Here one can
compare the ability of
an extract, being tes-
ted for reactive PQQ
with 0.5M valinate-pH
10, against 2M gly-
cinate, pH 10.   Valin-
ate is not sufficient-
ly soluble to use at
higher   concentration.
Compounds like keto-
amines at nanomole le-
vels will react with
and be detected by NBT
in both buffers, but

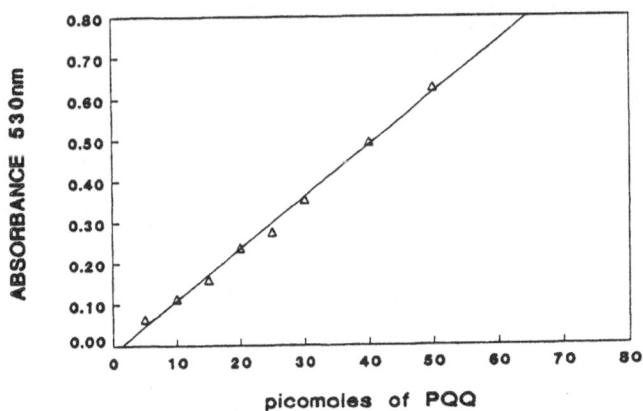

Figure 6.   Determination of PQQ by Redox-
Cycling with Glycinate-NBT

only PQQ will be detected at picomole levels and only with the glycinate-
NBT system.   Even phenanthrenequinone and amine adducts of it will not
support significant redox cycling; ascorbate-dehydroascorbate does not
undergo glycinate-NBT redox cycling.   With PQQ, one can estimate about
1000 redox cycles catalyzed by PQQ before side reactions stop the process. Cyanoborohydride + amines and NBT can also be used to generate a
useful redox system with PQQ as catalyst [Paz et al. 1988].

With   compounds   like   3-hydroxyindole   an   eneaminol<-->ketoamine,
potentially able to undergo redox cycling with cyanoborohydride and NBT,
only a very few redox cycles can be generated before crosslinking and
indigo production stop redox cycling.   With PQQ, redox cycling in the
presence of appropriate reactants is the prefered reaction since it is
likely that the multiple negative charges on PQQ and reduced PQQ inhibit
oxidative dimerization by charge repulsion, greatly increasing the
efficiency of redox cycling.

Borohydride reduces both PQQ and NBT and is not useful with assays

138

involving NBT. NaCNBH$_3$ in the presence of amines or appropriate proteins will also reduce PQQ without reduction of NBT, but also tends to precipitate some NBT, limiting the concentration of NaCNBH$_3$ that can be used. Nevertheless, it can be useful, but it is not as effective as glycinate (Paz et al. 1988). Quinoproteins usually require proteolytic digestion to expose the PQQ and the redox reaction is monitored for several hours. In Figs. 7 and 8 the apparent equivalent weight at each timepoint (pg protein/pmoles PQQ) of PK-DAO and DBH respectively, is plotted versus reciprocal time and extrapolated to infinite time (1/time -> 0). Here both the intact protein and digested protein are reacted with the glycinate-NBT reagent and the formazan color monitored over several hours and compared to a PQQ standard. For each time the apparent content of PQQ is determined and divided into the protein content and plotted against 1/time. The y-intercept gives the equivalent weight at the end of the redox reaction. It is apparent that both digested and undigested proteins react with the redox reagent but at markedly different rates. In Table I we show the equivalent weights obtained by this procedure of several intact and pronase digested proteins.

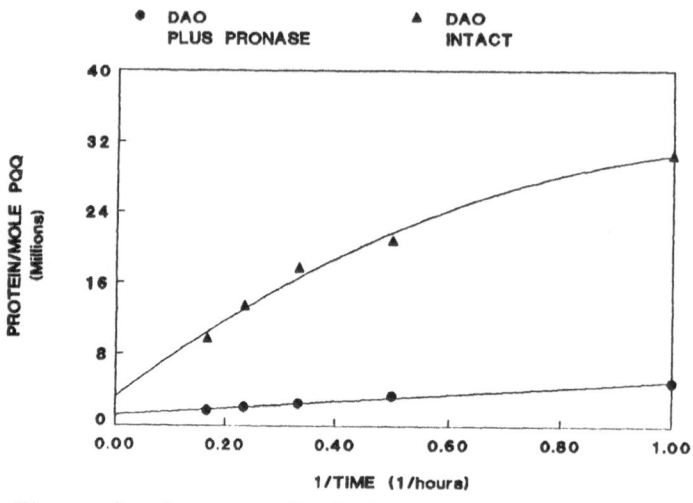

Figure 7.   Apparent Equivalent Weight (protein/ mole PQQ) of PK-DAO Extrapolated to Infinite Time

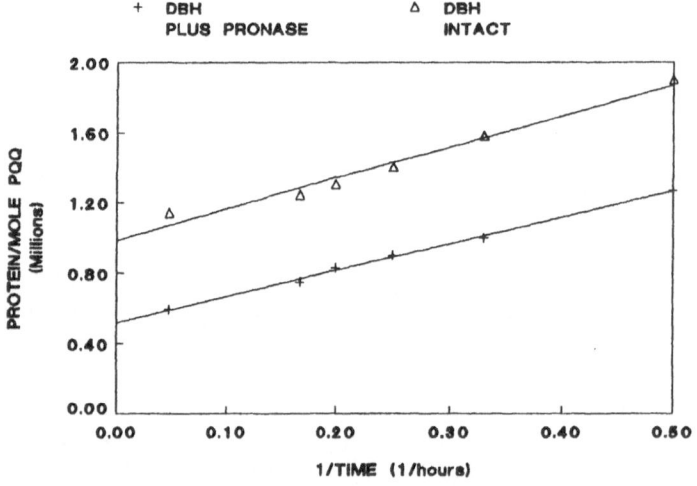

Figure 8.   Apparent Equivalent Weight (protein/ mole PQQ) of DBH Extrapolated to Infinite   Time

Of all the putative mammalian quinoproteins examined to date, DBH is the most reactive showing about one PQQ for 500,000 daltons. DBH gives a

strong reaction for PQQ even without a prior pronase digestion. Our preparation of partially purified PK-DAO, which is only about 40-50% pure
and probably a mixture of holoenzyme and apoenzyme [Paz et al. 1988],
show an equivalent weight of about 1,200,000 per PQQ molecule. Several
proteins seem to contain some PQQ after extensive pronase digestion and
give equivalent weights of 4-9 million and practically no detectable PQQ
without digestion. Other proteins seem to be entirely free of PQQ, even
after pronase treatment. It also seems likely that serum albumin is able
to carry PQQ and probably accounts for a significant portion of PQQ in
serum.

TABLE I

DETECTABLE PQQ IN INTACT AND PRONASE DIGESTED PROTEINS

PROTEIN PREPARATION	EXTRAPOLATED EQUIV. WT. per PQQ
Dopamine-ß-hydroxylase (intact enzyme)	987,000
" " " " (pronase digest)	517,000
Lysyl Oxidase (pronase digest)	680,000
Diamine Oxidase (intact enzyme)	3,200,000
" " " (pronase digest)	1,220,000
Serum Albumin (intact protein)	15,900,000
" " " (pronase digest)	1,300,000
Ovalbumin (intact protein)	no detectable PQQ
" " (pronase digest)	1,750,000
IGG (intact protein)	48,000,000
" (pronase digest)	2,700,000
Ribonuclease (intact enzyme)	no detectable PQQ
" " " (pronase digest)	4,370,000
ß-Lactoglobulin (intact protein)	no detectable PQQ
" " " (pronase digest)	4,800,000
Lysozyme (intact enzyme)	no detectable PQQ
" " (pronase digest)	9,000,000
Pyruvate kinase, glucose-6-P-dehy-drogenase, carnitine acetyl transferase (both intact enzymes and pronase digests)	no detectable PQQ

==
The Detection of PQQ Peptides in Pronase Digests of DBH and PK-DAO

Fig. 9 shows the elution profile of the peptides obtained by pronase digestion of PK-DAO. Ninhydrin reactivity and PQQ content, the latter measured with the glycinate-NBT redox reagent after overnight incubation at room temperature, were determined for each fraction. Fig. 10
shows the elution profile of the peptides obtained by pronase digestion
of DBH following a similar procedure as described for PK-DAO. It is
apparent that there are specific PQQ peptides in these digests which
chromatograph on the HPLC column. With further purification and analysis
of these PQQ-peptides it should be possible to work out the nature of the

PQQ-locus and active center of these and other quinoenzymes.

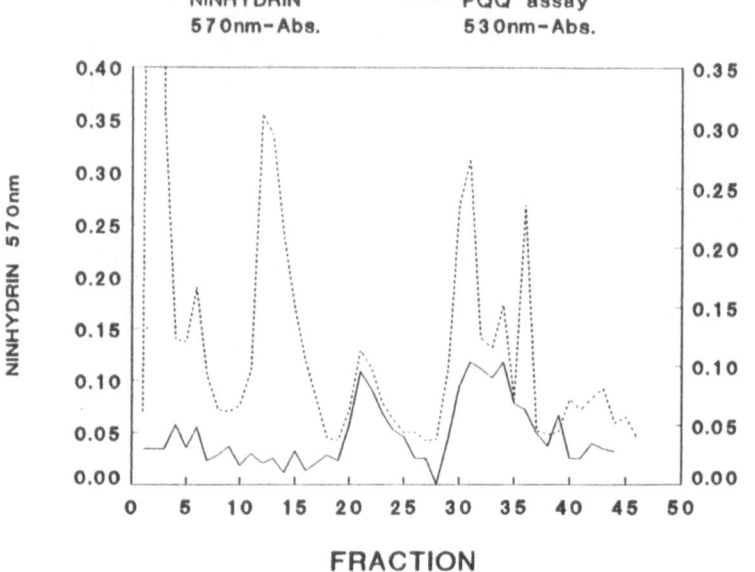

Figure 9.   Elution Profile of Pronase Digest of PK-DAO

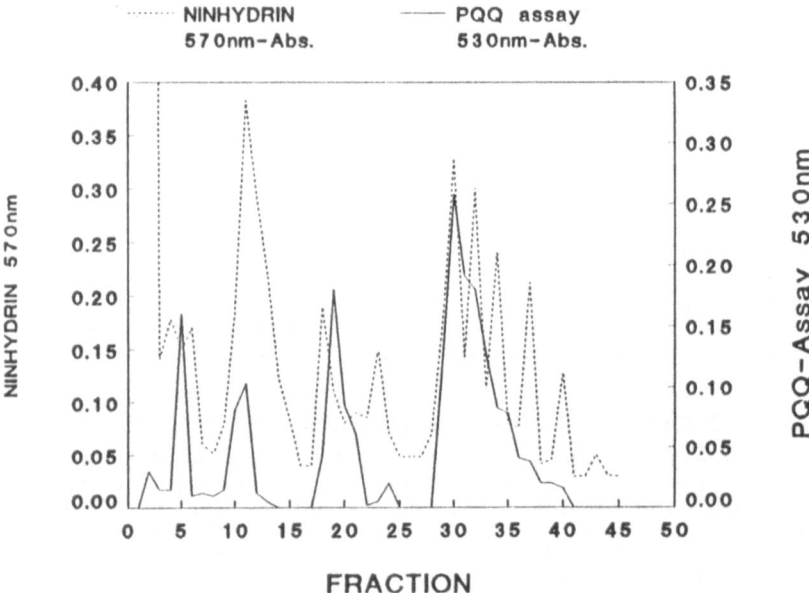

Figure 10.   Elution Profile of Pronase Digest of DBH

PQQ in Serum, Urine, CSF, Milk and Egg; PQQ as a Potential Source of Oxidative Stress

The availablity of the sensitive and specific redox procedure for PQQ analysis makes it possible to examine biological fluids and tissue extracts for free and bound PQQ content. We have found that PQQ is present in serum, urine and cerebrospinal fluid (CSF). The PQQ in human CSF and human urine appears to be dialyzable against 0.1 M glycine, pH 9. Part of the serum PQQ (45-57%) is removed by dialysis. The presence of PQQ in skim milk and egg yolk suggests that PQQ may be an essential nutrient for young and developing animals. The contents of PQQ are given in Table II.

TABLE II

PQQ CONTENT IN CERTAIN BIOLOGICAL FLUIDS

FLUID	PQQ content	
Human Serum (total)	1.6-5.0nm/ml	0.023-0.07nm/mg pro
" " " " (non-dialyzable)	0.63-2.0nm/ml	0.009-0.029nm/mg pro
CSF (total, >90% dialyzable)	1.59-3.3nm/ml	3.1-7.2nm/mg pro
Urine (all dialyzable)	30-80μmole/gr creatinine	
Skim Milk (total)	1.74nm/ml	0.05nm/mg pro
Egg Yolk (total)	50nm/ml	0.26nm/mg pro
Egg White (total)	PQQ not detectable	

The presence of PQQ, as the likely coenzyme of DBH, DAO and LO has been confirmed by the generation of a glycinate-PQQ<->PQQ(2H)-NBT redox cycle with amplified formazan production, Table 1. The LO preparation, obtained from Dr. Herbert Kagan shows a higher then expected equivalent weight per PQQ and this could be a result of poor accessibility and the presence of both apo- and holoenzyme in the preparation. With our preparation of PK-DAO the equivalent weight is also higher then expected and could be due to impurities and a mixture of apo- and holoenzyme in the enzyme preparation.

An important feature of the work reported here is the fact that we have been able to detect PQQ-peptides in digests of putative mammalian quinoenzymes in definable HPLC fractions. To our knowledge this has not been possible before because of the lack of appropriate sensitive methodology. Further analysis should now allow the characterization of these peptides with a definition of the nature of the PQQ-locus. It seems likely, from the work of Hartmann and Klinman, 1987 with cyanoborohydride and bovine serum amine oxidase and from our results with PK-DAO [Paz et al, 1988] that cyanoborohydride can reduce the PQQ attached to such amine oxidases. PQQ is probably attached to a protein amino group in such quinoproteins as a ketone-ketoimine reducible to a ketoamine <--> eneolamine. This does not eliminate the possibility of other attachments through PQQ carboxyl groups among the various possibilities [Hartmann and Klinman, 1988].

The presence of PQQ in egg-yolk and milk suggest that PQQ may be a vital nutrient for developing embryos and for newborn mammals. It remains to establish for which animals PQQ or an appropriate precursor may be a vitamin. The level of PQQ in egg-yolk is quite high, Table 2. The finding of dialysable PQQ in CSF in nanomole/ml levels is a potentially important observation. Human CSF contains about 1/200 the protein content of human serum and about the same level of total PQQ. Since PQQ is the coenzyme of DBH, the copper-containing mixed function oxidase involved in the synthesis of norepinephrine, a crucial neurotransmitter derived from the neurotransmitter, dopamine, one wonders if the levels of PQQ in various samples of patient CSF could be diagnostic and pathophysiological significance for neurological disorders such in Parkinson's and Alzheimer's diseases. The pigmented chromaffin cells and granules in adrenal and nervous tissues would be expected to be very rich in quinoenzymes and PQQ. The presence of PQQ in both dialzable and non-dialyzable form in serum is also noteworthy, suggesting both the presence of quinoproteins and PQQ-carrying proteins.

The presence of dialyzable PQQ in normal human urine in easily detectable amounts of 30–80 μmoles/g creatinine indicates that PQQ is present at levels not far removed from pyridoxamine and other vitamin B6 metabolites. It remains to be established how much of urinary PQQ is dietary, of gut-bacterial origin, or derived from de novo synthesis in the animal. If de novo synthesis is important, excretion of high levels of PQQ may turn out to be of clinical significance in situations where there are catecholamine secreting tumors.

Since free or exposed PQQ in the presence of oxygen and certain amines or other reducing compounds can generate large amounts of superoxide, exposed PQQ may become a dangerous source of toxic oxygen compounds in damaged tissues where there may be inadequate protection by SOD and catalase. Compounds like alloxan, streptozotocin, streptonigrin, paraquat cause tissue specific oxidative damage by virtue of their tendency to localize in certain tissues and generate superoxide and other oxidants [Bannister et al. 1987; Sies, 1985]. Could it be that nature has evolved appropriate protection devices against a compound as redox reactive as PQQ which is present normally as a protected coenzyme in a variety of enzyme systems? Perhaps there are special proteins designed to rapidly pick up free or exposed PQQ and shield it from further reactions that could lead to the release of unwanted superoxide. The role of PQQ as a contributor to inflammatory oxidative stress is likely to attract research attention in the future. Suffice it to say, the new methods and results presented here for PQQ and the inferences they evoke, could help advance important emerging areas of biomedical research.

ACKNOWLEDGEMENTS

We thank professor Herbert Kagan for generous samples of lysyl oxidase and helpful comments and to Dr. Anthony Lorenzo for rabbit and human CSF. We are grateful to Dr. Sam Seifter for helpful comments and advice. We are grateful to Kathleen Pearson-McIntyre and Kate Bowenkamp for expert technical assistance. We also acknowledge the help of Drs. Susan Greenspan and Julian Seifter and Mr. Oscar Rago in some of the clinically relevant studies. This work was supported by National Institute of Health Grants AG 04727, DK 34369 and AG 07723.

REFERENCES

Ameyama M, Shinagawa E, Matsushita K, Takimoto K, Nakashima K and Adachi O, 1985. Mammalian choline dehydrogenase is a quinoprotein. Agric. Biol. Chem. 49: 3623-3626.

Bannister, JV, Bannister, WH and Rotillo, G 1987. Aspects of the structure function and applications of superoxide dismutase. Critical Reviews in Biochemistry 22: 111–180.

Dekker RH, Duine JA, Frank J, Verwiel PEJ and Westerling J, 1982. Covalent addition of H_2O, enzyme, substrates and activators to pyrrolo-quinoline quinone, the coenzyme of quinoproteins. Eur. J. Biochem. 125: 69–73.

Flückiger R, Woodtli T and Gallop PM, 1988. The interaction of amino-groups with pyrroloquinoline quinone as detected by the reduction of nitroblue tetrazolium. Biochem. Biophys. Res. Commun. 153: 353–358.

Glatz Z, Kovar J, Macholan L and Pavel P, 1987. Pea (Pisum sativum) diamine oxidase contains pyrroloquinoline quinone as a cofactor. Biochem. J. 242: 603–606.

Hartmann C and Klinman JP, 1987. Reductive trapping of substrate to bovine plasma amine oxidase. J. Biol. Chem. 262: 962–965.

Hartmann C and Klinman JP, 1988. Pyrroloquinoline quinone: a new redox cofactor in eukaryotic enzymes. BioFactors 1: 41–49.

Lobenstein-Verbeek CL, Jongejan JA, Frank J and Duine JA, 1984. Bovine serum amine oxidase: a mammalian enzyme having covalently bound PQQ as prosthetic group. FEBS Lett. 170: 305–309.

Paz MA, Gallop PM, Torrelio BM and Flückiger R, 1988. The amplified detection of free and bound methoxatin (PQQ) with nitroblue tetrazolium redox reactions: insights into the PQQ-locus. Biochem. Biophys. Res. Commun. 154: 1330–1337.

Rinaldi A, Vecchini P and Floris G, 1982. Diamine oxidase from pig kidney: new purification method and aminoacid composition. Preparative Biochem. 12: 11–28.

Salisbury SA, Forrest HS, Cruse WBT and Kennard O, 1979. A novel coenzyme from bacterial primary alcohol dehydrogenases. Nature 280: 843–844.

Sies, H, editor, 1985. Oxidative Stress. Academic Press, London.

Sleath PR, Noar JB, Eberlein GA and Bruice TC, 1985. Synthesis of 7,9-didecarboxymethoxatin (4,5-dihydro-4,5-dioxo-1H-pyrrolo [2,3-f] quinoline-2-carboxylic acid) and comparison of its chemical properties with those of methoxatin and analogous o-quinones. Model studies directed toward the action of PQQ requiring bacterial oxidoreductases and mammalian plasma amine oxidase. J. Am. Chem. Soc. 107: 3328–3338.

van der Meer RA, Jongejan JA, Frank J and Duine JA, 1986. Hydrazone formation of 2,4-dinitrophenylhydrazine with pyrroloquinoline quinone in porcine kidney diamine oxidase. FEBS Lett. 206: 111–114.

van der Meer RA, Jongejan JA and Duine JA, 1987. Phenylhydrazine as probe for cofactor identification in amine oxidoreductases. Evidence for PQQ as the cofactor in methylamine dehydrogenases. FEBS Lett. 221: 229–304.

van der Meer RA, Jongejan JA and Duine JA, 1988. Dopamine ß-hydroxylase from bovine adrenal medulla contains covalently-bound pyrroloquinoline quinone. FEBS Lett. 231: 303–307.

Binding of Pyrroloquinoline Quinone to Serum Albumin

Osao ADACHI, Emiko SHINAGAWA, Kazunobu MATSUSHITA, Koji NAKASHIMA,* Koichi TAKIMOTO,** and Minoru AMEYAMA
*Laboratory of Applied Microbiology, Department of Agricultural Chemistry, Yamaguchi University, Yamaguchi 753 Japan; *St. Luke's College of Nursing, Tokyo 104 Japan; and **Radioisotopes Laboratory, Yamaguchi University, Yamaguchi 753 Japan*

Key words: PQQ binding protein/Schiff base/ serum albumin-PQQ adduct

ABSTRACT

When PQQ was mixed with serum albumin at neutral pH, a spontaneous binding of PQQ to the serum protein was observed. The resulting PQQ-albumin was fairly stable for chromatographic treatments and exhibited an absorption spectrum having a peak at 336 nm. PQQ-albumin showed a coenzyme activity to a quinoprotein glucose dehydrogenase (GDH) to almost the same level as seen with authentic PQQ. PQQ was proved to be associated via ε-amino group of lysyl residue of serum albumin forming a Schiff base. PQQ-chromophores isolated from PQQ-albumin by acid hydrolysis showed a marked reduced coenzyme activity for GDH. In a hydrolysate of native serum albumin, a similar chromophore fractions appeared and functioned as the prosthetic group for GDH. Growth stimulation for acetic acid bacteria was seen with such PQQ-chromophores as well as authentic PQQ. Thus serum albumin could be a temporary PQQ-carrier in mammals.

INTRODUCTION

In the course of detection and isolation of PQQ-chromophore from enzymes of which prosthetic groups have not been identified, a reference protein is required to show that a candidate protein yields PQQ-chromophore but the reference protein does not. In the survey for a suitable reference protein, we found that a PQQ-chromophore was generated from acid hydrolysate of bovine serum albumin (BSA) as similarly reported with other various kinds of naturally occurring substances (Ameyama et al., 1985a). The chromophore from BSA showed the least coenzyme activity for GDH (EC 1.1.99.17) but growth stimulating activity for Acetobacter aceti. It looked as if BSA were a quinoprotein.

In this report, binding of PQQ to serum albumin forming a proteinous PQQ-adduct is described. One possibility that serum albumin could be a nonspecific PQQ-carrier protein in mammals is also discussed.

145

J. A. Jongejan and J. A. Duine (eds.), PQQ and Quinoproteins, 145–147.

MATERIALS AND METHODS

Chemicals
PQQ and ^{14}C-PQQ were prepared under essentially the same methods as described previously (Ameyama et al., 1984a). Crystalline human serum albumin (Sigma) was further fractionated with DEAE-Sephadex chromatography. After two times repetition of elution of the protein at 0.1 M KCl, the protein fractions were combined, concentrated and used.

Enzymatic determination of PQQ and PQQ-chromophores
Enzymatic determination of PQQ and PQQ-chromophores was performed with Escherichia coli membranes carrying apo-GDH as mentioned elsewhere (Ameyama et al., 1985b).

Determination of growth stimulating activity of PQQ and PQQ-chromophores
Growth stimulating activity of PQQ and PQQ-chromophores for microorganisms was assayed with A. aceti IFO 3284 according to the method described previously (Ameyama et al., 1985a, 1988).

Modification of albumin
5-Phospho-pyridoxylation of serum albumin was carried out using pyridoxal 5'-phosphate (PLP) as mentioned by Dempsey and Christensen (1962). Reduction of PLP-albumin and PQQ-albumin was conducted by sodium borohydride.

RESULTS AND DISCUSSION

Binding of PQQ to serum albumin
To confirm the binding of PQQ to serum albumin, serum albumin was mixed with excess amount of PQQ (0.4 µmol of PQQ and 0.1 µmol of albumin) and subjected to gel filtration on a Sephadex G-10 column to separate the albumin fractions from unreacted PQQ. Repetition of the same experiment for several times indicated that 1 mol of serum albumin was bound by 2 mol of PQQ under these conditions. A new absorption peak at 336 nm was generated. These results potently indicated that binding of PQQ to serum albumin occurred. Such PQQ-albumin was chromatographed on a DEAE-Sephadex A-50 column. The majority of eluted protein was no longer found in 0.1 M KCl fraction but in 0.3 M and 1.0 M fractions, in spite of usage of albumin fraction which had been eluted at 0.1 M KCl. In order to confirm that the charge difference was brought about only by binding of PQQ to serum albumin, the same experiment was repeated with ^{14}C-PQQ. The resulting ^{14}C-PQQ-albumin was also no longer eluted at 0.1 M KCl as described above. These facts indicate that serum albumin bound with PQQ, a tricarboxylic acid, must be eluted at relatively higher salt concentration than authentic serum albumin.

Evidence for Schiff base formation
To check the PQQ binding site to the protein, PLP-albumin was used instead of unmodified serum albumin. As expectedly, no

additional binding of PQQ to the modified protein was observed. Since PLP is known to bind to serum albumin via ε-amino group of lysyl residue to form a Schiff base (Dempsey and Christensen, 1962), inability of binding of PQQ to PLP-albumin suggested that PQQ also binds to the same lysyl residues to which PLP is attached.

No PQQ has been found as free form in the blood serum, although occurrence of a quinoprotein, plasma amine oxidase, has been reported. This means the least possibility for PQQ to exist as a free form but a bound form. Alternatively, a PQQ-carrier protein, to which PQQ can bind specifically and from which is discharged when necessary, may exist in blood serum, though it has not been detected yet. Provided about 30-50 mg of serum albumin per ml is found in blood serum and PQQ is naturally distributed at the order of ng per ml, it can be roughly estimated that 1 mol of PQQ is surrounded by 10^4-10^5 times albumin.

Fractionation and isolation of PQQ-chromophores
In order to show the existence of PQQ to be associated to serum albumin in nature, fractionation and isolation of PQQ-chromophores were attempted. About 1 g of native serum albumin and 100 mg of PQQ-albumin were separately subjected to proteolysis and then acid hydrolysis as mentioned previously (Ameyama et al., 1984). The resulting hydrolysates were fractionated by chromatographies. Before hydrolysis, PQQ-albumin was available as the prosthetic group to GDH, while the coenzyme activity was reduced markedly after hydroysis. This indicates potently the formation of Schiff base of PQQ during handlings. However, a poor coenzyme activity was samely detected at the same position with both chromatograms of these two samples. Conversely, growth stimulating activity for microorganisms was predominated with both fractions. This indicates the existence of PQQ in serum albumin.

Encouraged by these findings, new born calf serum and whey were chromatographed. In the acid hydrolysates of both 0.3 M and 1.0 M KCl fractions from DEAE-chromatography, PQQ activity as the coenzyme for a quinoprotein GDH as well as the growth stimulating activity for microorganisms were detected. From these results, it can be concluded that serum albumin could be a temporary PQQ-carrier protein in mammals.

References
Ameyama, M., Hayashi, M., Matsushita, K., Shinagawa, E. and Adachi, O. (1984) Agric. Biol. Chem., **48**, 561-565.
Ameyama, M., Shinagawa, E., Matsushita, K. and Adachi, O. (1985a) Agric. Biol. Chem., **49**, 699-709.
Ameyama, M., Nonobe, M., Shinagawa, E., Matsushita, K. and Adachi, O. (1985b) Anal. Biochem., **153**, 263-267.
Ameyama, M. Matsushita, K., Shinagawa, E. and Adachi, O. (1988) BioFactors., **1**, 51-53.
Dempsey, W. B. and Christensen, H. N. (1962) J. Biol. Chem., **237**, 1113-1120.

GROWTH STIMULATING EFFECT OF PYRROLOQUINOLINE QUINONE FOR MICROORGANISMS

MINORU AMEYAMA
Laboratory of Applied Microbiology, Department of Agricultural Chemistry, Faculty of Agriculture, Yamaguchi University, Yamaguchi 753 JAPAN

Key words: growth stimulating activity of PQQ/reduction of lag phase by PQQ/PQQ-adducts, a potent growth stimulant

ABSTRACT

Two types of growth stimulating effect of pyrroloquinoline quinone (PQQ) for microorganisms are known. The type I PQQ effect was observed in a symbiotic polyvinyl alcohol (PVA) degradation in which one species of <u>Pseudomonas</u> excretes PQQ which in turn enables the other species to grow on PVA (Shimao et al., 1984). The latter can produce a carbon source for the former only when PQQ is supplied. Thus, PQQ is regarded as the essential growth factor of the PVA-degrading bacteria when grown on PVA but not on other usual carbon sources.

The type II growth stimulation by PQQ is characterized by a marked reduction of the lag phase in the presence of PQQ (Ameyama et al., 1984b). In this case, the subsequent growth rate at the exponential phase and the total cell yield at the stationary phase are not affected at all. Unlike the former type, PQQ is not always an essential growth factor and normal cell growth can be seen even in the absence of exogenous PQQ after a relatively prolonged lag time. This type of PQQ effect was discovered in the course of searching for a growth stimulating substance in yeast extract for acetic acid bacteria. Several lines of analytical evidence from studies with other naturally occurring substances indicated that the substance is quite probable to be PQQ or PQQ-adducts.

INTRODUCTION

It is well-known phenomenon that the growth of microorganisms is stimulated markedly by the addition of various kinds of naturally occurring substances to the culture medium. One of the reasonable popular conclusions so far is that the culture medium might be supplemented with growth nutrients such as vitamins, nucleic acids, amino acids, sugars and minerals derived from the added extract which are insufficient in chemically defined medium. In the course of investigations on the growth of acetic acid bacteria, it was observed that the growth of <u>Acetobacter</u> species, which show no appreciable nutritional requirements, was stimulated markedly

149

by the addition of yeast extract to a synthetic medium in which generally known growth factors of vitamins, amino acids or nucleic acids were fully available (Ameyama and Kondo, 1966). As to strains of the genus <u>Gluconobacter</u>, on the other hand, some species have been shown to require vitamins or others. Addition of yeast extract to a complete synthetic medium for such <u>Gluconobacter</u> strains further stimulated the growth of the organisms. A similar growth stimulating effect to that observed with yeast extract was also found with other naturally occurring substances used for culture ingredients for microorganisms (Ameyama <u>et al</u>., 1985a). They include peptone, corn steep liquor, malt extract, koji extract, casamino acids, meat extract, blood powder or blood serum, soybean cake, rumen juice and so forth. It is also generally accepted that the growth simulation observed with such naturally occurring substances is not restricted to only acetic acid bacteria but is also predominant in variety of microbe genera. The growth stimulating substance in these materials and culture broths of <u>Escherichia coli</u>, methylotrophs or pseudomonads has been isolated and identified to PQQ (Ameyama <u>et al</u>., 1984a,c, 1985a; Shimao <u>et al</u>., 1984).

Two types of growth stimulating effect by PQQ for microorganisms have been known so far (Fig. 1). One type of PQQ effect is found from a symbiotic PVA degradation in which one species of a <u>Pseudomonas</u> excretes PQQ which in turn

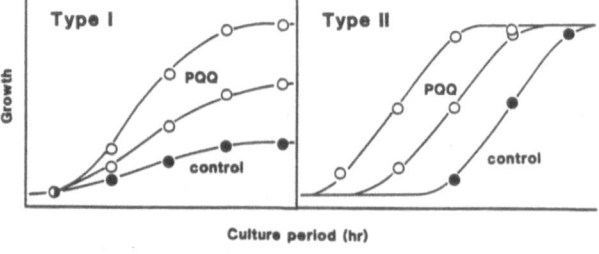

Fig. 1. Stimulating Effect of PQQ.

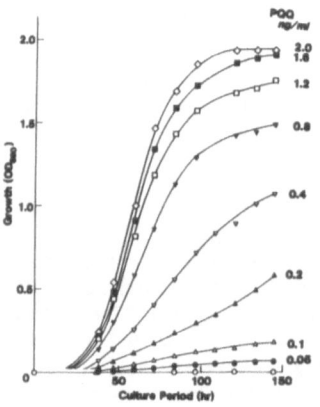

Fig. 2. Effect of PQQ on Growth of <u>Pseudomonas</u> sp. VM15C with PVA.

enables the other species to grow on PVA (Shimao <u>et al</u>., 1984). The growth extent of the PQQ acceptor is completely dependent on PQQ supplied and the growth rate as well as the cell yield are enhanced as seen in an auxotrophic growth, but an appreciable reduction of the lag phase is scarcely observed as shown in Fig. 2.

The alternative type of growth stimulation by PQQ (Type II PQQ effect) is characterized by a marked reduction of the lag phase in the presence of a trace amount of PQQ in the culture medium. In this case, the subsequent growth rate at the exponential phase are not affected at all as well as the total cell yield at the stationary phase. Unlike the type I, PQQ is not always an essential growth factor and normal cell growth, but with a relatively prolonged lag phase, can be seen even in the absence of exogenous PQQ or PQQ adduct.

In this review, the type II growth stimulating effect caused by PQQ or PQQ-adduct is mainly described.

Materials and Methods

Determination of PQQ-content

PQQ content was determined by a quinoprotein apo-glucose dehydrogenase of E. coli membranes as described elsewhere (Ameyama et al., 1985d).

Measurement of growth stimulation by PQQ

The measurement was performed under essentially the same as reported previously (Ameyama et al., 1988). Inoculum cell suspension of A. aceti (0.4 ml) having an optical density of 0.1 at 660 nm which had been prepared after rinsing several times as reported (Ameyama et al. 1985b), was inoculated under sterile conditions. Operation of a biophotometer was performed at 30°C under continuous shaking at 60 times per min. The bacterial growth was recorded automatically by plotting the absorbance at 660 nm.

Preparation of PQQ-amino acid

PQQ (100 ng/ml) (Ameyama et al., 1984a) and amino acid (10 mg/ml) was mixed and kept at 120°C for 10 min to promote Schiff base formation.

For the preparation of PQQ-serine used for isolation and further characterization as a model compound of PQQ-adduct, PQQ (2 mg) and L-serine (500 mg) were mixed in a total volume of 2.5 ml and pH of the mixture was maintained at 6.5. Formation of a Schiff base was monitored by reading optical density of the mixture at 420 nm. The mixture was kept at room temperature overnight to ensure the completion of Schiff base formation.

Isolation of PQQ-serine

An aliquot of the reaction mixture (1 ml) was subjected to DEAE-Sephadex A-25 column (1 x 10 cm), which had been equilibrated with 1 mM potassium phosphate, pH 7.0. After washed with the same buffer, separation of PQQ-serine from unreacted PQQ was performed by a linear increasing concentrations gradient of KCl to 2.0 M.

Results and Discussion

Evidence for the existence of a growth stimulating substance for acetic acid bacteria in yeast extract

The spectra of vitamin requirements were examined with acetic acid bacteria (Fig. 3). The upper parts in the figure for both strains show the effect of the single omission of each vitamin from the minimal medium containing ten vitamins. In the lower parts, the effect of the single addition of each vitamin to the minimal medium is shown. It is clear that <u>G. cerinus</u> requires nicotinic acid and pantothenic acid for growth, whereas no appreciable vitamin requirement is observed for <u>A. rancens</u>. These results are consistent with the vitamin requirements of acetic acid bacteria determined previously (Ameyama and Kondo,1966), <u>i.e.</u> the genus <u>Gluconobacter</u> requires generally pantothenic acid, except for <u>G. sphaericus</u>, and some strains to some

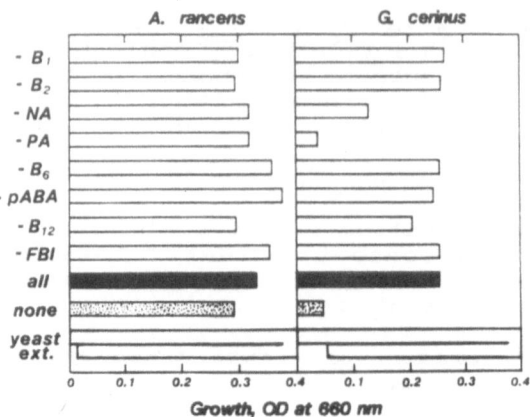

Fig. 3. Vitamin Requirement of Acetic Acid Bacteria.

extent require, moreover, thiamine, nicotinic acid and <u>p</u>-aminobenzoic acid. This is the reason why the four kinds of vitamins were included in the basal medium for assaying of the growth stimulating effect as mentioned later. No appreciable requirement for nucleic acids is observed for either strain. It is obvious what is effective, at least, excluding the amino acids, vitamins and nucleic acids used. Addition of yeast extract to the medium markedly facilitated the increase in growth of both strains. These results indicated the possibility of the existence of an unidentified growth stimulating substance in yeast extract.

Fractionation of naturally occurring substances
    The response of microbial growth to yeast extract was confirmed to be the Type II growth stimulation, when examined with various strains of microorganisms. In the tested organisms, a PQQ auxotroph was excluded. In order to determine whether such a growth stimulating effect of yeast extract is specific or universal, a concentrated solution of yeast extract was fractionated on a Sephadex gel column. The growth stimulation for <u>A. aceti</u> was specifically localized at the fraction numbers 34 to 36 and other fractions showed no appreciable growth stimulation, though the inactive fractions suggested the presence of other various kinds of materials judging from their high optical intensities. With other bacterial strains, the growth stimulating activity was similarly found around the peak fraction observed with <u>A. aceti</u>. Many kinds of naturally occurring substances were samely treated as yeast extract by gel filtration. When assayed growth stimulating effect with individual chromatograms, most of them showed the biological activity at the fraction numbers 34 to 36 (Fig. 4), indicating the presence of a similar molecular size with the same biological activity.
    It is interesting that <u>E. coli</u> broth of glucose mineral

153

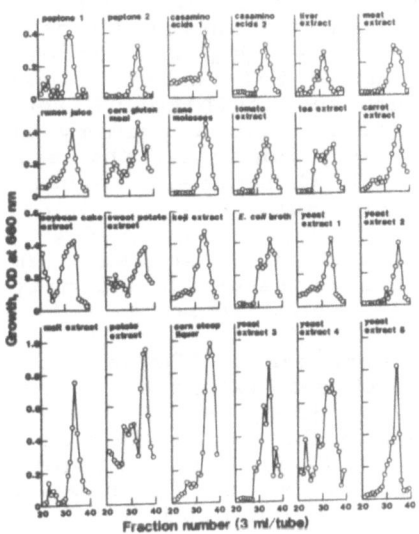

Fig. 4. Gel Filtration of Various Naturally Occurring Substances on a Sephadex Column.

medium also showed a growth stimulating activity. Thereafter, it has been clarified that E. coli as well as other microorganisms produces the growth stimulating substance into the culture medium. Isolation and characterization were made with the purified substance from E. coli broth, yeast extract, casamino acid, rumen juice and others. All preparations showed a characteristic optical properties that is identical to PQQ. They were available as the prosthetic group of quinoprotein glucose dehydrogenase, though the isolated compounds showed a reduced activity as PQQ rather than expected from their PQQ contents estimated on the basis of optical density. When isolation of PQQ-chromophore from quinoproteins was attempted in which PQQ is involved to be rather tightly bound to enzyme protein, the estimated PQQ content was usually far less than expected (Ameyama et al., 1984a, 1985b, c; Kawai et al., 1985; Nagasawa and Yamada, 1987; Shinagawa et al., 1988). On the other hand, a considerable growth stimulating activity for A. aceti was observed with such PQQ-chromophores. It can be readily presumed that when a small amount of PQQ is treated with a lot of amino compounds by means of a procedure including acid hydrolysis, extraction, concentration and so on, such PQQ must be converted to a PQQ-adduct readily and thus becomes less active as the prosthetic group.

In the followings, some important evidence is provided that PQQ becomes less active as the coenzyme for quinoprotein but rather exerts a growth stimulating activity conversely, once PQQ is treated with a compound which can make a Schiff base with PQQ.

Effect of PQQ-amino acid adducts on quinoprotein glucose dehydrogenase activity

As shown in Fig. 5, almost of L-amino acids examined were effective more or less to reduce the coenzyme

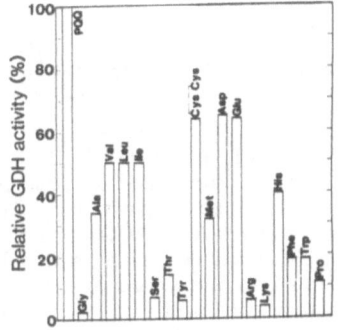

Fig. 5. Availability of PQQ-Amino Acids as the Prosthetic Group for Glucose Dehydrogenase.

154

activity. Glycine, serine, threonine, tyrosine, arginine and lysine showed a marked effect and the enzyme activity of less than 10% to that obtained with free PQQ was observed. Simultaneous addition of amino acids to the reaction mixture of glucose dehydrogenase did not show any inhibitory effect so far examined. These results indicated that a primary amino group of amino acids was selectively bound to the active carbonyl, probably 5-carbonyl of PQQ (Lobenstein-Verbeek et al., 1984; Itoh et al., 1984; van der Meer et al., 1986), to form a Schiff base. Schiff base formation with PQQ was samely observed with other amino compounds such as D-amino acids, amines, amino sugars and carbonyl reagents and gave an yellow band in absorption spectrum around 420 nm.

Isolation of PQQ-serine as a model compound of PQQ-adduct

Isolation of PQQ-adduct formed with serine was attempted by DEAE-Sephadex A-25 chromatography (Fig. 6). Fractions containing free PQQ which was probably derived either or both from unreacted PQQ and dissociated PQQ from the adduct was eluted at about 0.8 to 1.0 M of KCl as samely as previously reported (Ameyama, et al., 1984a). Unreacted serine was excluded from chromatography by washing the column before gradient elution. Two peaks of PQQ-adduct were eluted from the column somewhat later than free PQQ, since PQQ-adducts formed were expected to contain one more carboxylic group than authentic PQQ and so eluted from the column at a higher salt concentrations. PQQ activity as the prosthetic group for a quinoprotein glucose dehydrogenase was surveyed with chromatographed fractions. As expectedly, a marked coenzyme activity was found at the positions where authentic PQQ was eluted. On the other hand, a poor coenzyme activity for the enzyme, which was estimated to

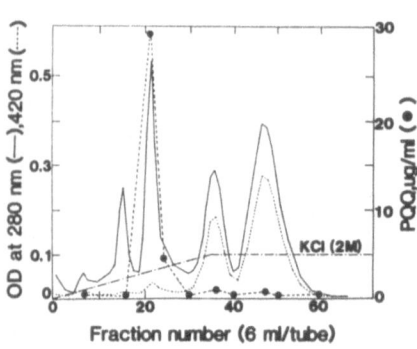

Fig. 6. Chromatographic Separation of PQQ-Serine.

be over 100 times less active than authentic PQQ, was found with both peaks of PQQ-serine, when compared each other on the basis of equimolar amount of PQQ. Conversely, when assayed the growth stimulating activity with the fractions, the two fractions of PQQ-serine equally stimulated the growth of A. aceti under the similar fashion using a biophotometer as described previously (Ameyama et al., 1988). The extent of growth stimulation by these PQQ-adducts was roughly estimated to be 10 times higher than authentic PQQ. Binding of serine to active carbonyl group of PQQ allowed it to be almost inert as the prosthetic group, whereas PQQ-adduct instead authentic PQQ looked to facilitate the incorporation of PQQ required for the initial stage of proliferation of microorganisms.

The appearance of two peaks of Schiff base having

absorption maximum at 420 nm could be explained the possibility as follows. Since PQQ has an orthoquinone structure as 4,5-dioxo- (Salisbury et al., 1979; Duine et al., 1980; Duine and Frank, 1981; Duine et al., 1987), it can be presumed that both carbonyl groups can be enrolled, but not equally, in Schiff base formation to give two species of PQQ-serine. Since the structural analysis of these compounds are now under investigations, it had better to avoid too much speculation about structural properties of PQQ-serine. Similar results were obtained with other PQQ-adducts prepared with other amino compounds (data not shown). It is interesting to recall that two peaks of PQQ-chromorphore were generated from the chromatograms of hydrolysate of quinoproteins such as amine oxidases (Ameyama et al., 1984a), methylamine dehydrogenase (Ishii et al., 1983), choline dehydrogenase (Ameyama et al., 1985c) and aliphatic amine dehydrogenase (Shinagawa et al., 1988).

Spectral properties of isolated PQQ-serine

The two isolated fractions of PQQ-serine from DEAE-Sephadex chromatography were combined separately. The first peak from the fraction numbers 32 to 39 and the second peak from 43 to 55 were adsorbed into a SepPak C_{18} cartridge. The adduct adsorbed was eluted three times with a small volume of 80% (v/v) ethanol. Each of the resulting supernatant solution was subjected to Sephadex G-10 column (1.0 x 120 cm). The PQQ-adduct was eluted from the column at 75 ml to 80 ml. Figure 7 shows the absorption spectrum of the second peak of PQQ-serine in Fig. 6. A typical absorption peak at 420 nm was characteristics in visible region. Two absorption maxima in ultra violet region suggested that the compound consists of PQQ. The spectrum was somewhat different from those obtained with 2,4-dinitrophenylhydrazine (Lobenstein-Verbeek et al., 1984; van der Meer et al., 1986). The Schiff base was fairly stable at neutral pH without an appreciable spectral changes promising no changes in biological properties. A characteristic feature was also found in its fluorescence spectra as shown in Fig. 8 by the second peak from DEAE-chromatography (Fig. 6). The same spectroscopic properties were observed with the first peak of

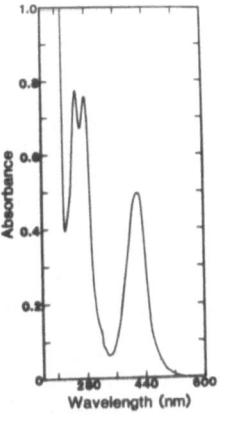

Fig. 7. Absorption Spectrum of PQQ-Serine.

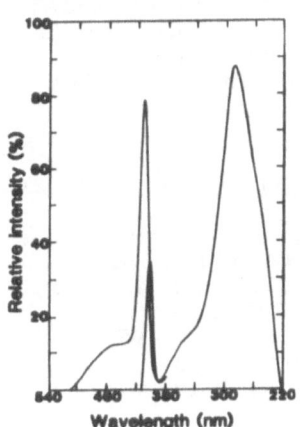

Fig. 8. Fluorescence Spectra of PQQ-Serine.

PQQ-serine in Fig. 6. An image that the compound came from PQQ was found at a broad peak around 360 nm in excitation spectrum and 460 nm in emission spectrum (Ameyama et al., 1984a). However, a sharp fluorescence peak around 420 nm which reflected the existence of a Schiff base was characteristic of all. Such characteristic peak has been obtained with chromophore fractions of acid hydrolyzate of quinoproteins and samely with the purified growth stimulating substance for microorganisms (Ameyama et al., 1984a, b, 1985a, b, c; Kawai et al., 1985; Nagasawa and Yamada, 1987; Shinagawa et al., 1988).

It is quite probable that PQQ-adduct formation readily occurs during extraction and isolation procedure for PQQ from the naturally occurring materials or PQQ-chromophore from quinoproteins. Such adduct must give a low PQQ estimation than theoretically expected. In other words, a small amount of PQQ in vivo tends to react with amino compounds to form PQQ-adducts in handlings and becomes less active as the prosthetic group.

PQQ may exist in vivo as free form or associated to a quinoprotein or a PQQ carrier protein but it seems difficult to isolate PQQ as free form, except for PQQ production by methylotrophs (Ameyama et al., 1984a, 1988). It is reasonable to consider that, in case of PQQ fermentation, since PQQ is in far excess than amino compounds, PQQ is excreted into the culture medium as free PQQ. On the other hand, in yeast cells and other natural biological materials, PQQ content is not so high to yield free PQQ and PQQ is converted spontaneously to a PQQ-adduct during handlings. So, there is no evidence that E. coli is the sole exception. The purified growth stimulating substance from culture broth of E. coli or yeast extract which have been concluded to be PQQ (Ameyama et al., 1984c, 1985a), is probably a PQQ-adduct. With such PQQ-adduct, PQQ content cannot be estimated exactly with a quinoprotein glucose dehydrogenase whereas growth stimulating activity can be estimated as samely as free PQQ. From these observations mentioned above, PQQ activity should be followed by both criteria, coenzyme activilty for quinoroteins and growth stimulating activity for microorganisms.

References

Ameyama, M. and Kondo, K. (1966) Agric. Biol. Chem., 30, 203-211.

Ameyama, M., Hayashi, M., Matsushita, K., Shinagawa, E. and Adachi, O. (1984a) Agric. Biol. Chem., 48, 561-565.

Ameyama, M., Shinagawa, E., Matsushita, K. and Adachi, O. (1984b) Agric. Biol. Chem., 48, 2909-2911.

Ameyama, M., Shinagawa, E., Matsushita, K. and Adachi, O. (1984c) Agric. Biol. Chem., 48, 3099-3107.

Ameyama, M., Shinagawa, E., Matsushita, K. and Adachi, O. (1985a) Agric. Biol. Chem., 49, 699-709.

Ameyama, M., Shinagawa, E., Matsushita, K. and Adachi, O. (1985b) Agric. Biol. Chem., 49, 853-854.

Ameyama, M., Shinagawa, E., Matsushita, K., Takimoto, K., Nakashima, K. and Adachi, O. (1985c) Agric. Biol. Chem.,

49, 3623-3626.

Ameyama, M., Nonobe, M., Shinagawa, E., Matsushita, K. and Adachi, O. (1985d) Anal. Biochem., **151**, 263-267.

Ameyama, M., Matsushita, K., Shinagawa, E. and Adachi, O. (1988) BioFactors, **1**, 51-53.

Duine, J. A., Frank, jzn, J. and Verwiel, P. E. J. (1980) Eur. J. Biochem., **108**, 187-192.

Duine, J. A. and Frank, jzn, J. (1981) Trends Biochem. Sci., **6**, 278-280.

Duine, J. A., Frank, jzn, J. and Jongejan, J. A. (1987) Adv. Enzymol., **59**, 170-212.

Ishii, Y., Hase, T., Fukumori, Y., Matsubara, H. and Tobari, J. (1983) J. Biochem., **93**, 107-119.

Itoh, S., Kato, N., Ohshiro, Y. and Agawa, T. (1984) Tetrahedron Lett., **25**, 4753-4756.

Kawai, F., Yamanaka, H., Ameyama, M., Shinagawa, E., Matsushita, K. and Adachi, O. (1985) Agric. Biol. Chem., **49**, 1071-1076.

Lobenstein-Verbeek, C. L., Jongejan, J. A., Frank, J. and Duine, J. A. (1984) FEBS Lett., **170**, 305-309.

Nagasawa, T. and Yamada, H. (1987) Biochem. Biophys. Res. Commun., **147**, 701-709.

Salisbury, S. A., Forrest, H. S., Cruse, W. B. T. and Kennard, O. (1979) Nature (London), **280**, 843-844.

Shimao, M., Yamamoto, H., Ninomiya, K., Kato, N., Adachi, O., Ameyama, M. and Sakazawa, C. (1984) Agric. Biol. Chem., **48**, 2873-2876.

Shinagawa, E., Matsushita, K., Nakashima, K., Adachi, O. and Ameyama, M. (1988) Agric. Biol. Chem., **52**, 2255-2263.

van der Meer, R. A., Jongejan, J. A., Frank, J. and Duine, J. A. (1986) FEBS Lett., **202**, 111-114.

NUTRITIONAL ESSENTIALITY OF PYRROLOQUINOLINE QUINONE

Robert Rucker, John Killgore, Lisa Duich, Nadia Romero-Chapman, Carsten Smidt and Donald Tinker, Department of Nutrition, University of California, Davis, CA 95616, U.S.A.

KEY WORDS: Nutrition, amine oxidase, lysyl oxidase, collagen and elastin crosslinking, PQQ

ABSTRACT

A chemically defined diet, based on synthetic amino acids, was fed to rats and mice. The control diet (+PQQ) was fortified with pyrroloquinoline quinone (PQQ) at 875 ug/kg. The deficient diet (-PQQ) contained no detectable PQQ. The -PQQ diet was fed to pregnant rats and, in later experiments, to mice throughout gestation. Newborn pups remained with the dams and at weaning were assigned to either the +PQQ or -PQQ diet. In each case, weight reduction occurred in the -PQQ groups (30% for mice, 10-15% for rats). However, to observe this response, an antibiotic, succinyl sulfathiazole, was added to diets at 2%. Diets were autoclaved to reduce the potential for subsequent bacterial contamination, and animals were housed within a laminar flow hood in sterile stainless steel cages. Moreover, cage bottoms were designed to prevent coprophagy. Addition of PQQ to the diets of animals in the -PQQ group consistently stimulated growth. Important physiological responses to PQQ deprivation were a 40% reduction in tendon benzylamine oxidase and an increase in the solubility of dermal collagen.

INTRODUCTION

PQQ serves as a cofactor for copper-containing amine oxidases in plants, animals, fungi and yeast. The enzymes that have been characterized in most detail are lysyl oxidase and the soluble monoamine oxidase. That PQQ serves as a cofactor for such oxidases has far-reaching consequences; one of which is the possibility that PQQ is a vitamin-like cofactor.

MATERIALS AND METHODS

Animals and diets. A chemically defined diet, void of PQQ (Table 1), was fed ad libitum to Sprague-Dawley (300 g, Exp. I) or Swiss-Webster (30-40 g, Exp. II) dams. The diet (-PQQ) was fed for 3 weeks to deplete available stores of PQQ. The rat or mouse dams were then bred and carried to term. In Exp. I, the litters of rat pups were weaned to the -PQQ diet, and at week 6 half of the rats were switched to the +PQQ diet (Fig. 1). In Exp. II, weanling mouse pups were fed a -PQQ or +PQQ diet for 6 weeks. Anatomical and morphologic changes were observed at this point.

PQQ and enzyme assays. Rat tendon soluble monoamine oxidase (MAO) was assayed as described by Chou et al. (1969) using benzylamine as substrate. Estimates of skin collagen and protein extractable into phosphate buffered saline (PBS) were made (Exp. I). PQQ in the diet and water was assayed by an HPLC method (Van Kleef et al., 1987) or by a microbiological assay (Geiger & Görisch, 1987).

159

J. A. Jongejan and J. A. Duine (eds.), PQQ and Quinoproteins, 159-161.

TABLE 1.   Composition of purified diet.

Ingredient	%
Sucrose	25.90
Cornstarch	44.40
Amino acid mixture[a]	16.00
Corn oil	5.00
Salt mix[a]	5.37
Vitamin mix[a]	0.20
Choline·Cl	0.10
NaHCO$_3$	1.00
MgSO$_4$·7H$_2$O	0.03
Succinylsulfathiazole	2.00
Total	100.00

a. Meets NRC requirements for rats
   or mice. Selected vitamins were
   added at 5-10x requirements to
   accommodate losses during auto-
   claving. (Hirakawa et al., 1982)

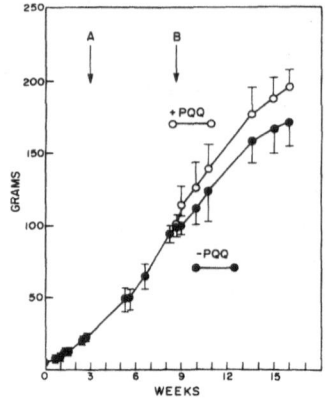

Figure 1. Growth of rats fed
a chemically-defined diet
with or without PQQ.

RESULTS

In Figure 1 is shown the growth rate for rats nursed from PQQ-deprived
dams (A) and then fed a -PQQ diet for an additional 6 weeks (B). When half
of the rats were switched to a +PQQ diet, a significant growth stimulation
was observed. Each point is the average for at least 10 to 15 rats. Moreover, the experiment was repeated (2x) with similar results.

At week 15, rats were killed and examined. Few gross lesions were observed, but the skin from -PQQ rats was more friable than skin from +PQQ
rats. Moreover, collagen of skin from -PQQ rats was 2-3 times more soluble
in PBS than that of skin from +PQQ rats. Values (units per 100 mg tissue ±
S.E.M.) for benzylamine oxidase activity extracted from tendon were: +PQQ,
3.75 ± 0.5 (n=7); -PQQ, 2.2 ± 0.2 (n=6); p<0.05.

With respect to mice, PQQ deprivation caused a 30 percent reduction in
weight by week 6 post-weaning. Many of the PQQ-deprived mice were also unthrifty in appearance (Fig. 2). One mouse (out of a group of 25) died of a
ruptured abdominal aorta aneurysm.

DISCUSSION

The possibility that PQQ may be a dietary growth factor is based on the
following:

1) When PQQ was added to a deficient diet, the rate of growth was stimu-
    lated for rats and mice housed in isolation.
2) PQQ-deprived mice and rats were often unthrifty, i.e., similar to
    lathyritic or Cu-deprived animals in appearance.
3) PQQ deprivation resulted in a decrease in MAO, measured in a functional
    assay.

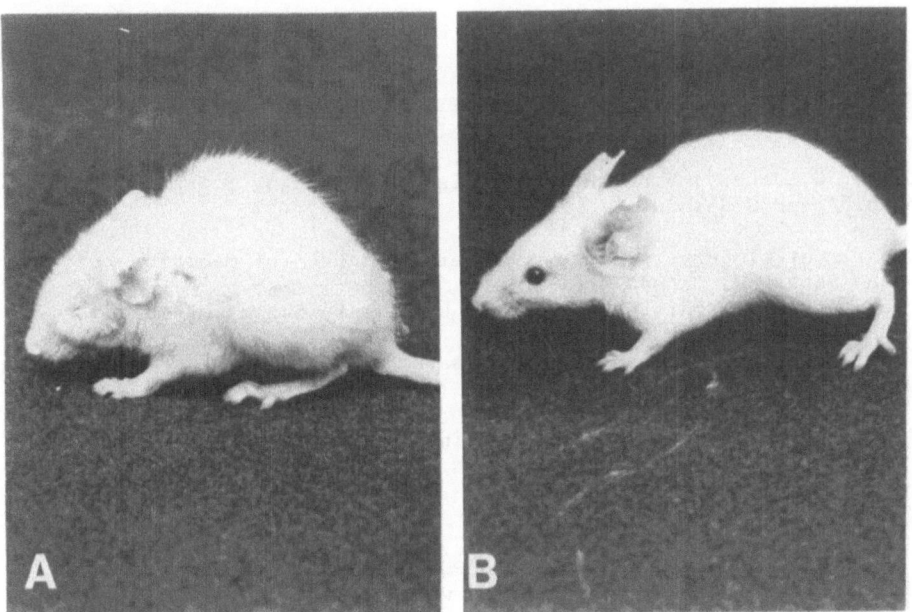

Figure 2. Mice were fed a -PQQ (A) or +PQQ diet (B). No PQQ was detected in the deficient diet. PQQ at 930 ug/kg was recovered in the supplemented diet when assayed using the glucose dehydrogenase assay.

Why PQQ has not been identified previously as a potential growth factor may relate to the potential for intestinal bacteria synthesis, or the probability that dietary ingredients used in previous studies contained some PQQ (cf. Geiger & Görisch, 1987). Moreover, some unthriftiness and slow growth are considered acceptable in germ-free studies. Consequently, it is of importance that PQQ addition improved each of the parameters that were measured herein.

REFERENCES

Chou WS, Rucker RB, Savage JE and O'Dell BL, 1970. Impairment of collagen and elastin crosslinking by an amine oxidase inhibitor. Proceedings of the Society for Experimental Biology and Medicine 134: 1078-1084.

Geiger O and Görisch H, 1987. Enzymatic determination of pyrroloquinoline quinone using crude membranes from Escherichia coli. Analytical Biochemistry 164: 418-423.

Hirakawa DA, Olson LM and Baker D, 1984. Comparative utilization of a crystalline amino acid diet and a methionine-fortified casein diet by young rats and mice. Nutrition Research 4: 891-895.

Van Kleef MAG, Dokter P, Mulder AC and Duine JA, 1987. Detection of the cofactor pyrroloquinoline quinone. Analytical Biochemistry 162: 143-149.

EFFECTS OF EXOGENOUS PQQ ON MORTALITY RATE AND SOME BIOCHEMICAL PARAMETERS DURING ENDOTOXIN SHOCK IN RATS

T.Matsumoto, O.Suzuki*, H.Hayakawa, S.Ogiso, N.Hayakawa, Y.Nimura, I.Takahashi** and S.Shionoya

First Department of Surgery, Nagoya University School of Medicine, 65 Tsuruma-cho Showa-ku, Nagoya 466, JAPAN
*Department of Legal Medicine, Hamamatsu University School of Medicine, 3600 Handa-cho, Hamamatsu 431-31, JAPAN
**Clinical Investigation Laboratory, Ekisaikai Hospital, 4-66 Syonen-cho Nakagawa-ku, Nagoya 454, JAPAN

KEY WORDS: PQQ, Endotoxin shock, Polyamine

ABSTRACT

Effects of exogenously administered PQQ on mortality rate and some biochemical and hematological parameters were studied in rats during endotoxin shock induced by a E. coli toxin. PQQ administration (6mg/kg i.p.) reduced their mortality rate from 68.8% to 28.6%. Aspartate amino-transferase(GOT), alanine aminotransferase(GPT), blood urea nitrogen(BUN), uric acid, lipoperoxide and calcium in rat sera were all improved by PQQ administration platelet counts, prothrombin time(PT), and antithrombin III (AT III) were also improved markedly.
Our results suggest the utility of PQQ for treatment or prevention of various kinds of shock in clinical medicine.

INTRODUCTION

Medical and clinical studies on PQQ are very scant at the present time. In the present study, the effect of intraperitoneal administration of PQQ on mortality rate, and some biochemical parameters were studied during endotoxin shock induced by E. coli.

MATERIALS AND METHODS

Male Donryu rats weighing 350-450g were used. All animals had free access to standard commercial pellet food and tap water. PQQ (6mg/kg b.wt.) was injected intraperitoneally 1.5h before the intraperitoneal injection of the E. coli toxin lipopolysaccharide type 026-B6 (Difco, Co. Ltd., U.S.A. 3mg/kg b.wt.), and animals were killed 5 hours later.

RESULTS

The rate of mortality induced by the endotoxin was decreased from 68.8% to 28.6% by the administration of the 6mg/kg b.wt. PQQ.

162

J. A. Jongejan and J. A. Duine (eds.), PQQ and Quinoproteins, 162–164.
© 1989 by Kluwer Academic Publishers.

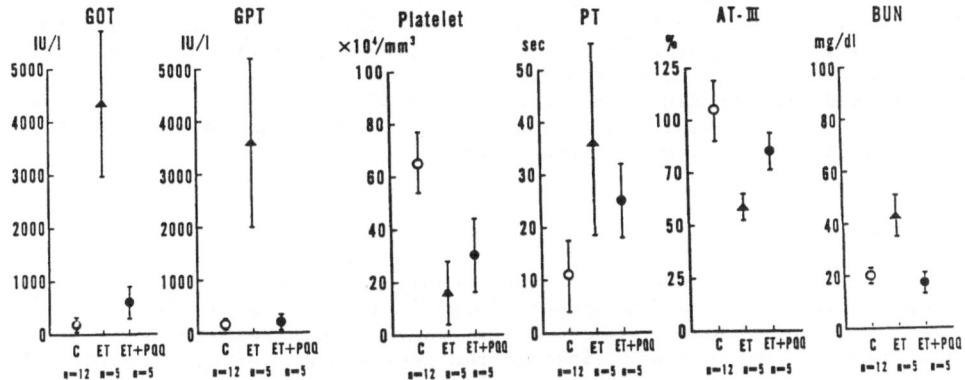

FIGURE 1. Effect of PQQ on aspartate aminotransferase(GOT), alanine
aminotransferase(GPT), blood urea nitrogen(BUN), platelet
counts, prothrombin time(PT) and antithrombin III(AT-III)
5h after endotoxin administration to rats.

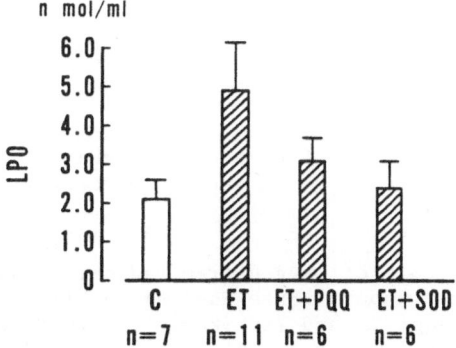

FIGURE 2. Effect of PQQ on lipoperoxide(LPO) in rat sera 5h after
endotoxin administration to rats.

Serum aspartate aminotransferase, alanine aminotransferase, BUN and uric acid were significantly improved by PQQ administration from 4367±1471 IU/1, 3609±1302 IU/1, 41.6±6.2 mg/dl, 3.9±1.8 mg/dl to 524±99.6, 176±40, 18.1±1.2, 1.13±0.12 respectively.(Fig. 1) Platelet counts, prothrombin time and antithrombin III were also improved markedly.(Fig. 1)

Increased lipoperoxide and decreased calcium in rat sera as a result of endotoxin shock were also improved by PQQ from 4.94±1.54 nmol/ml, 9.25±0.17 ng/dl to 3.2±0.86, 9.54±0.26 respectively.(Fig. 2)

DISCUSSION

The present study showed that exogenously administered PQQ markedly reduced mortality rate and improved biochemical and hematological parameters during endotoxin shock. The endotoxin was reported to cause various membrane damages and disseminated intravenous coagulopathy(DIC) through generation of superoxide anion (1, 2). The PQQ administration significantly suppressed the enhancement of serum lipid peroxide induced by the endotoxin.(Fig. 2)

Therefore, the administered PQQ seems to act as a scavenger of the superoxide anion and thus contribute to be reducing the mortality rate.

PQQ was identified in various enzymes and is now believed to be present exclusively as prosthetic group. Although the doses of PQQ administered in the present study far exceeded endogenous levels of PQQ in tissue (Suzuki et al, unpublished observation), there seems to be a possibility that free endogenous PQQ present in tissue also acts as a protecting factor against crisis of an organ like vitamin E and ubiquinoes.

From a clinical point of view, our study suggests that PQQ may be useful for treatment and prevention of severe organ damage including endotoxin shock in humans.

REFERENCES

1. Sakaguchi,S., Kanda,N., Hsu,C.C. and Sakaguchi,O. 1981. Lipid peroxide formation and membrane damage in endotoxin poisoned mice. Microbiol. Immunol. 25: 229-244.

2. Ogawa,R. 1987. The role of oxygen free radicals in the pathogenesis of tissue injury during shock. Free Radical in Clinical Medicine. Vol. II eds. Kondo,M., Oyanagi,Y. and Yoshikawa,T. NIHON-IGAKUKAN Tokyo JAPAN

Theoretical Evaluation of large scale Production of PQQ via Fermentation.

B.J. Wesselink and P. Gupta
Dept. of Chemical Engineering
Delft University of Technology
Julianalaan 136, Delft
The Netherlands.

INTRODUCTION

Large scale production of PQQ might become interesting in case its nature as a vitamin or its significance in pharmaceutics is established. For the industrial production of PQQ both fermentative and synthetic routes can be considered. However, on the basis of economics, simplicity of the process and environmental considerations, we preferred a fermentative method for further theoretical investigation. A process plant was designed (see figure 1) having an arbitrarily chosen production capacity of 100 kg of PQQ annually. A general discussion of this process design is presented in this paper.

ASSUMPTIONS AND FUNDAMENTAL ASPECTS

A few assumptions used in the process design were derived from the patent application of Ameyama [1]. Fermentation was carried out with the organism Pseudomonas Nl-1 which produced 30 mg PQQ/litre broth in fed batch using methanol as the sole source of carbon and energy. Biomass increase during the 50 hour fermentation period was assumed to be ten times the initial amount. It was assumed, further, that besides PQQ no other products were excreted into the culture medium and the substrate was completely converted into biomass, carbon dioxide and water.

Due to the lack of basic biokinetic data for the organism, reasonable assumptions had to be made. The maintenance coefficient, m_s, and the netto yield of biomass on substrate, Y_{sx}, were estimated from

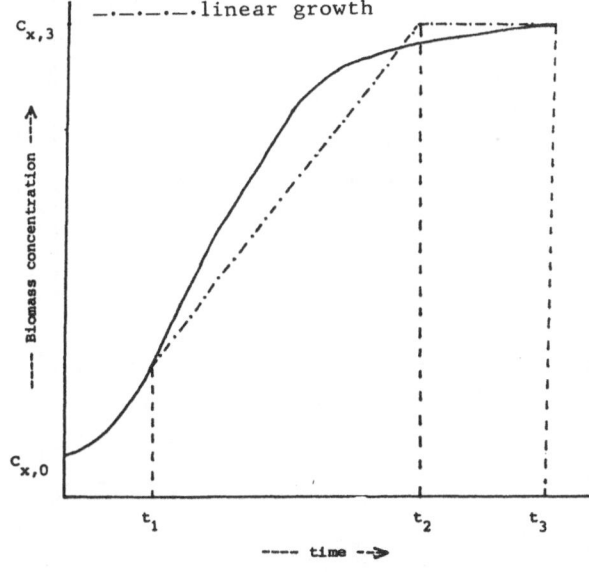

figure 2: growth curves for fermentation.

averaged parameters of selected Pseudomonas species growing on methanol [2]. The values determined were 0.05 mol substrate/mol biomass h. and 0.67 mol/mol, respectively. The specific growth rate, μ^{max}, of the organism Methylomonas methanolica [3] was assumed to be similar to that of

165

J. A. Jongejan and J. A. Duine (eds.), PQQ and Quinoproteins, 165–168.

FIGURE 1:

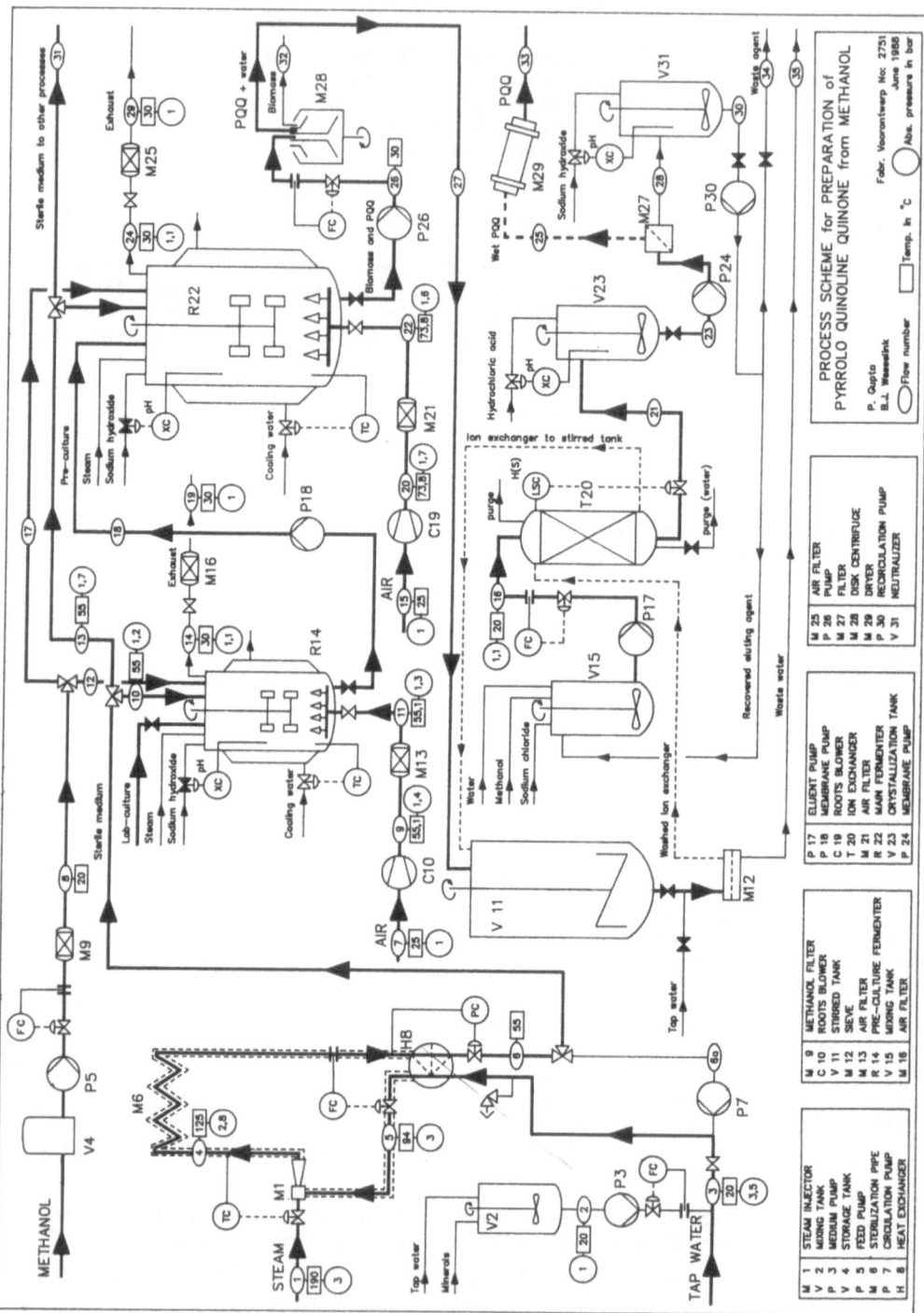

PROCESS SCHEME for PREPARATION of PYRROLO QUINOLINE QUINONE from METHANOL

Pseudomonas N1-1. The μ^{max} value determined was 0.5 h^{-1}. As evident from [3], the methanol concentration was kept below 4 g/l in order to maintain the specific growth rate.

We assumed a linear growth curve (see figure 2) for the organism as this characteristic is relatively easy to describe in the form of a model.

Fermentation was carried out at 30 °C and the pH was regulated at 7.0.

PROCESS DESCRIPTION

The technical details of the fermentors are provided in table 1. The fermentors are provided with sterile medium from a continuous sterilizer. After inoculation, the pre-fermentation is allowed for 20 hours, and the broth is then

Table 1: technical details of fermentors

	main	pre-
workvolume (m³)	50	2.5
material	AISI 316	
aeration rate (m³/hr)	593	212
purchase ($)	900,000	300,000

transported to the main fermentor. The extent of the main fermentation is fifty hours during which the organisms are allowed to grow exponentially for the initial two hours, followed by a linear growth period of forty hours and for the final eight hours the organisms are maintained at a constant level. During both fermentations growth is regulated by methanol supplementation.

After centrifugation to seperate the biomass from the culture medium, the supernatant, containing PQQ, is transported to a 60 m³ stirred tank to which 60 litres of Amberlyst A21, an ion-exchanger, is added. The ion-exchanger is subsequently separated by a sieve and packed densely in a column. PQQ is eluted with a solution of water and methanol (1:1, v/v) and 1 N sodium chloride. The eluate is acidified and the fluid mixture is left for enough time to form PQQ crystals.

Each fermentation cycle lasts for three days. Considering a 99 % recovery of PQQ from the culture medium, a total of 1.51 kg of PQQ is produced per cycle. Thus, if an annual production capacity of 100 kg of PQQ is desired, 67 cycles or approximately 200 production days are necessary. Table 2 shows the annual requirements of raw materials and utilities to produce 100 kg of PQQ.

Table 2: requirements on annual basis to produce 100 kg of PQQ.

item	amount
methanol	40200 kg
$(NH_4)_2SO_4$	6700 kg
K_2HPO_4	3350 kg
tap water	3350 m³
electricity	49000 kWh
steam	193000 kg
cooling water	56950 m³

WASTE PRODUCTS

The waste flow includes 50 m³ of spent medium, 123 kg dry weight of biomass and some eluting agent per cycle. The biomass can be treated to serve as fodder. The eluting agent contains methanol, which is burned.

COST EVALUATION

The annual production costs of PQQ on a hundred kilogram scale is estimated by applying a financial model, which requires the cost

168

factors as mentioned in table 3.
The financial model includes
overheads and indirect costs on top
of direct cost factors; examples
being administration, marketing and
personnel expenditure.

Table 3: annual direct costs

factors	$
Raw materials	50443.00
Maintenance	462876.00
Personnel	124800.00
Utilities	9698.00

An accurate cost estimate is difficult without further investigation of specific expenses. A moderately accurate costprice is predicted (although care must be taken when applied to this pre-design, because of the high investments compared to the small PQQ production capacity) if an "average" model, based on several studies, is used. The total annual costs C_t, excluding depreciation, are estimated by:

$$C_t = a \times C_p + d \times L + f \times I$$

where a, d, f constants, respectively 1.13, 2.6 and 0.13;
C_p production costs i.e. raw materials, maintenance and utilities;
L labour costs;
I investment costs.

The total annual costs approximately are 2.2 million US$, excluding depreciation. The installation is depreciated within ten years, an acceptable time for a multi purpose installation like this one. Consequently, the total annual costs including depreciation are 3.1 million US$; this gives the cost price of PQQ to be US$ 31 per gram. This price will decrease, if an organism is isolated that produces more PQQ than Pseudomonas N1-1 under similar process conditions.

CONCLUSIONS

(1) The commercialization of new products made possible through biotechnology require a coordinated network of various unit operations. In the design of the process scheme these unit operations are shown.
(2) The cost price of PQQ was estimated to be US$ 31.00 per gram. The production price of PQQ would decrease further if an organism were to be isolated that can produce more PQQ than Pseudomonas N1-1 under approximately similar conditions. From these estimates and the current price of PQQ, fermentation appears to be an attractive method for producing large amounts of PQQ.
(3) The production level of PQQ can be increased without major alterations in the process design. Although the design is in a preliminary stage, it gives a general idea of the process and offers the opportunity for project evaluation.

REFERENCES

[1]   Ameyama, M., Adachi, O. (1984) European Patent Application EP 0164943 A2.
[2]   Roels, J.A. (1983) Energetics and Kinetics in Biotecnology, Elsevier, Amsterdam.
[3]   Dostålek, M., Molin, N. (1975) Single Cell Protein II, Ed. Tannebaum and Wang, MIT Press, Mass.

GENES INVOLVED IN THE BIOSYNTHESIS OF PQQ
FROM *ACINETOBACTER CALCOACETICUS*

Nora Goosen, Harold P.A. Horsman, René G.M. Huinen, Arjan de Groot and Pieter van de Putte.
Laboratory of Molecular Genetics, University of Leiden, P.O.Box 9505, 2300 RA Leiden, the Netherlands.

(PQQ biosynthesis, *Acinetobacter calcoaceticus*, gene cloning, DNA sequence)

ABSTRACT

From a gene bank of the *Acinetobacter calcoaceticus* genome a plasmid was isolated that complements four different classes of PQQ⁻ mutants. Subclones of this plasmid revealed that the four corresponding PQQ genes are located on a fragment of 5 kilobases. The nucleotide sequence of this 5 kb fragment was determined and by means of Tn5 insertion mutants the reading frames of the PQQ genes could be identified. Three of the PQQ genes code for proteins of M_r 29700 (gene I), M_r 10800 (gene II) and M_r 43600 (gene III) respectively. In the DNA region where gene IV was mapped however the largest possible reading frame encodes for a polypeptide of only 24 amino acids. A possible role for this small polypeptide will be discussed. Finally we show that expression of the four PQQ genes in *Acinetobacter lwoffi* and *Escherichia coli* lead to the synthesis of the coenzyme in these organisms.

INTRODUCTION

Pyrrolo-quinoline quinone (PQQ) is used by a variety of organisms as a cofactor in dehydrogenase reactions. The presence of PQQ has been demonstrated in different bacterial species (for review see Duine *et al.*, 1986) but also in mammalian cells (Lobenstein-Verbeek *et al.*, 1984; van der Meer *et al.*, 1986). Some quinoprotein producing bacteria do not make PQQ. *Escherichia coli* K12 and *Acinetobacter lwoffi* synthesize an apo-glucose dehydro-genase (GDH) and are for the constitution of the holo-enzyme dependent on uptake of PQQ from the culture medium (Hommes *et al.*,1984; van Schie *et al.*, 1985).

As yet not much is known about the biosynthetic pathway of PQQ. Our approach to elucidate this problem is to clone the genes and study the gene products that are involved in PQQ synthesis. For this purpose we chose the organism *A. calcoaceticus* LMD79.41 which converts glucose to gluconic acid with the PQQ dependent enzyme GDH. First we isolated mutants of this strain that were no longer able to synthesize PQQ. Next a genomic library of the *A.calcoaceticus* genome was screened for complementing plasmids. In this way a plasmid was found which contains four genes involved in PQQ synthesis. The complete nucleotide sequence of these four genes was determined and it was shown that these four genes can direct the synthesis of PQQ in other microorganisms.

169

J. A. Jongejan and J. A. Duine (eds.), PQQ and Quinoproteins, 169–175.
© *1989 by Kluwer Academic Publishers.*

MATERIALS AND METHODS

Bacterial strains and culture conditions. The bacterial strains that were used in this study are *A.calcoaceticus* LMD79.41 (Duine *et al.*, 1980), *A.lwoffi* (van Schie *et al.*, 1985) and *E.coli* PPA41 (*ptsI*) (Hommes *et al.*, 1984). The strains were cultured in L broth (Miller JH, 1972) or defined minimal medium (Goosen *et al.*, 1987) as indicated. Transformation of *E.coli* strains (Maniatis *et al.*, 1982) and conjugation with *Acinetobacter* strains (Goosen *et al.*, 1987) were performed as described. Holo-GDH enzyme activity of *A.calcoaceticus* was tested on MacConkey agar containing 0.4% glucose. GDH activity gave rise to red colonies whereas non-acid producing mutants resulted in white colonies. Acid production by *A.lwoffi* was tested on L-plates containing 0.4% of glucose and a few drops of phenol-red. Acid production resulted in yellow plates, whereas the plates of non-acid producing strains remained red.

Cloning and DNA sequencing. All recombinant DNA techniques were essentially as described (Maniatis *et al.*, 1982). Restriction endonucleases, T4 DNA ligase and DNA polymerase (Klenow fragment) were obtained commercially. For sequencing of the PQQ genes the primer extension method (Sanger *et al.*, 1977) was used. Sequences were compiled and analyzed using the Sequence Analysis Software Package of the University of Wisconsin Genetics Computer Group (Devereux *et al.*, 1984)

RESULTS

Isolation of PQQ⁻ mutants. The GDH holoenzyme of *A. calcoaceticus* converts glucose to gluconic acid. This can be observed by the appearance of red colonies on MacConkey indicator plates containing glucose. Non-acid producing mutants of *A.calcoaceticus* LMD79.41 were isolated by selecting for white colonies on these indicator plates. These mutants were subsequently tested for their ability to regenerate gluconic acid production in the presence of purified PQQ in the culture medium. In this way 23 independent PQQ⁻ mutants were isolated.

Construction of a gene bank from *A.calcoaceticus*. As no transformation procedure for the direct introduction of plasmid DNA into *A.calcoaceticus* is available we used the vector pLV21 for the construction of a genomic bank. This vector contains a *mob* site and can therefore be transferred from *E.coli* to *A.calcoaceticus* by conjugation. The total DNA of *A.calcoaceticus* LMD79.41 was isolated and partially digested with *Sau*3A. The DNA was subsequently fractionated on a sucrose gradient and fragments of about 15 kb were ligated into the *Bam*HI site of pLV21. Finally the hybrid plasmids were transformed to *E.coli*. A total of 2700 transformants were taken to establish a gene bank of *A.calcoaceticus*. About 60% of these transformants appeared to carry a DNA insert of approximately 15 kb. Assuming that the lenght of the *A.calcoaceticus* genome is around 5000 kb the gene bank should contain at least four times the *A.calcoaceticus* genome.

Isolation of the PQQ genes. The gene bank was transferred to one of the PQQ⁻ mutants of *A.calcoaceticus* by conjugation as described (Goosen et al., 1987). Next the exconjugants were tested for acid production on MacConkey agar with glucose. In this way a plasmid could be isolated that complemented all the PQQ⁻ mutants. This plasmid (11B-35) appeared to carry an insert of 13 kb. To determine the number of PQQ genes located on plasmid 11B-35, subclones were isolated that carry only part of the 13 kb insert. Again these subclones were tested for complementation of the different PQQ⁻ mutants. The results of these experiments showed that four classes of PQQ⁻ mutants can be distinguished, each class being complemented by a different set of subclones.

One of the subclones (pSS2) has an insert of 5 kb and still complements all four classes of PQQ⁻ mutants. Therefore this 5 kb fragment is expected to carry at least four different PQQ genes.

Nucleotide sequence of the PQQ genes. The complete nucleotide sequence of the 5 kb insert of pSS2 was determined using the chain termination method described by Sanger *et al* (1977). It comprises 5087 bp which show a rather high A/T content (61%). Within the sequenced fragment different open reading frames (ORF's) can be indicated that are large enough to code for possible enzymes.

To determine whether these ORF's correspond to the different PQQ genes we isolated a series of Tn5 insertions (Goosen *et al.*, 1987). Transposon Tn5 can insert randomly into the DNA. When Tn5 is present in one of the cloned PQQ genes this can phenotypically be shown by the inability of this particular plasmid to complement the corresponding class of PQQ⁻ mutants. Furthermore the precise insertion point of the Tn5 insertion can be determined by DNA sequencing. In this way the PQQ genes could be mapped on the 5 kb DNA fragment. The result is schematically represented in Fig 1.

Fig 1

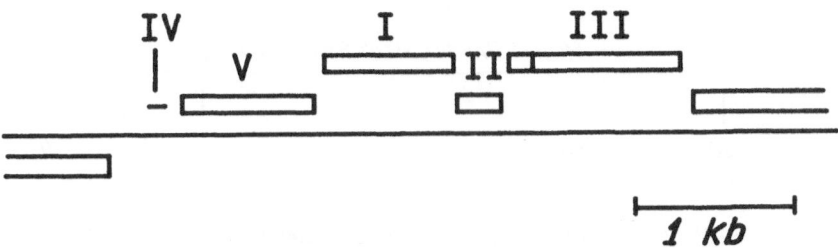

Schematic representation of the different ORF's (indicated by bars) present on the 5 kb insert of pSS2. The gene IV region is also indicated.

All ORF's in the 5 kb fragment are in the same DNA strand, which means that the genes are transcribed in the same direction (from left to right in Fig 1).

The first ORF (indicated as gene V) can code for a protein of M_r 33600. However three different Tn5 insertions that were mapped within this ORF do not affect the complementation of any of the PQQ⁻ mutants.

Apparently gene V is not involved in PQQ bio-synthesis.

Adjacent to gene V are the ORF's of gene I and gene II (Fig 1). These reading frames can code for proteins of M_r 29700 and M_r 10800 respectively. The translation start codon of gene II overlaps with the translation stop codon of gene I indicating that the expression of both genes is strongly coupled. Indeed it was found that the insertion of a Tn5 within gene I affects the complementation of two classes of PQQ⁻ mutants. Apparently both genes are involved in PQQ synthesis and most likely they are transcribed from the same promoter.

Tn5 insertions that block the complementation of the third class of PQQ⁻ mutants map in an ORF located to the right of gene II (Fig 1) which is indicated as gene III. Two putative translation starts are present from which a protein of either M_r 43600 or M_r 37400 can be synthesized. Until now we have not been able to discriminate between both possible start sites.

Finally the Tn5 insertion that affects the complementation of the fourth class of PQQ⁻ mutants was mapped to the left of gene V. Surprisingly only a very small reading frame, coding for a polypeptide of 24 amino acids, could be indicated in this region. Still by subcloning this gene IV region we could show that it has to code for a trans-acting product, as it complements class four PQQ⁻ mutants.

Fig 2

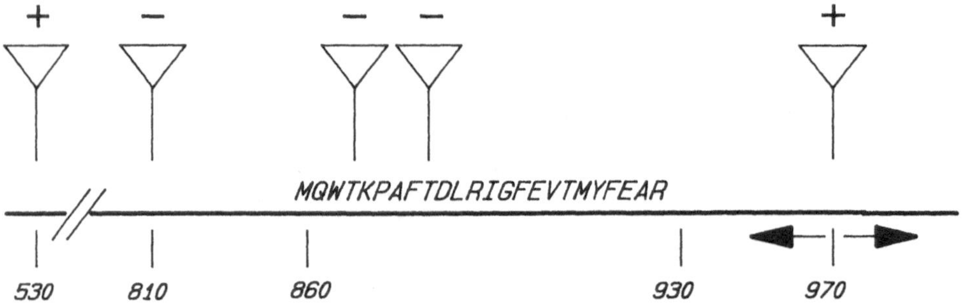

Schematic representation of the gene IV region. The sequence of the 24 amino acids polypeptide is indicated. The arrows represent the 11 bp inverted repeats. The positions of the Tn5 insertion that afect PQQ synthesis (-) and that do not affect PQQ syntheis (+) are shown.

Role of gene IV in PQQ synthesis. A closer examination of the gene IV DNA region reveals the presence of a 11 bp G-C rich hairpin structure followed by an A-T rich stretch of DNA (Fig 2). This structure is very likely to be a terminator for transcription initiated within or to the left of the gene IV region. Two possibilities for the gene IV product can now be considered. First the RNA that terminates at the hairpin structure might code for the 24 amino acids polypeptide, which in its turn has a direct or indirect function in PQQ synthesis. Secondly the RNA itself might play a role in the biosynthesis of PQQ.

To discriminate between both possibilities more Tn5 insertions in the gene IV region were isolated. The positions of these Tn5 insertions are indicated in Fig 2. Two of the Tn5 insertions that block complementation of gene IV mutants map within the 24 amino acids reading frame. Another Tn5 that maps within the hairpin structure does not affect gene IV expression. Apparently termination of transcription is not essential for the function of the gene IV region, which makes it unlikely that it is the RNA itself which is essential for PQQ synthesis. Another Tn5 insertion that blocks gene IV expression was mapped 50 bp before the translation start of the 24 amino acids polypeptide (Fig 2). Most probably transcription of the gene IV region is terminated within this Tn5. The promoter of gene IV should then be located to the left of this insertion.

Finally using site directed mutagenesis (Kunkel, 1985) we introduced a single base pair mutation into the gene IV region, thereby changing the putative ATG translation start of the 24 amino acids polypeptide to an ATA triplet. The resulting plasmid no longer complements the class IV PQQ⁻ mutants. As the mutation only changes one base in the RNA, but completely blocks the synthesis of the 24 amino acids polypeptide it is most likely that it is the polypeptide which has a role in PQQ synthesis.

Expression of the PQQ genes in other bacteria. *A.lwoffi* and *E.coli* K12 produce a PQQ dependent glucose dehydrogenase (Hommes *et al.*, 1984; van Schie *et al.*, 1985). However, these bacterial species do not synthesize PQQ. Therefore the holo-GDH can be formed only when PQQ is supplied from the culture medium. To test whether our four cloned PQQ genes contain sufficient information for the biosynthesis of PQQ, we introduced these genes into *A.lwoffi* and *E.coli* and tested whether now these bacteria produce an active GDH enzyme.

First pSS2 was introduced in *A.lwoffi* by conjugation and subsequently it was shown that the transconjugants produce acid on L-plates containing glucose (Materials and Methods). This indicates that in the presence of pSS2 a sufficient amount of PQQ is synthesized for the reconstitution of the holo-GDH enzyme. To test whether *A.lwoffi* is normally lacking all the PQQ enzymes and not just carrying a mutation in one of the PQQ genes, we also tested whether the pSS2 plasmids with the different Tn5 insertions produced PQQ. The results of these experiments showed that all four genes (I,II,III and IV) are required for the synthesis of PQQ in *A.lwoffi*.

Next we tested whether also *E.coli* K12 could synthesize PQQ upon introduction of the four genes. For this experiment an *E.coli* K12 strain carrying a *pts*I mutation (PPA41) was used. Due to the *pts*I mutation, this strain no longer grows on glucose as sole carbon source. In the presence of PQQ however the GDH enzyme can be used and growth on glucose is restored (Hommes *et al.*, 1984).

Upon introduction of pSS2 into PPA41 this strain still does not grow on minimal medium plates with glucose, indicating that no PQQ is synthesized. As this might be due to an inproper expression of the PQQ genes by the *E.coli* RNA polymerase we cloned the 5 kb DNA fragment with the four PQQ genes under control of *E.coli* *lac* promoter (pSS160). Now the resulting transformants did form colonies on minimal medium plates with glucose. As a control no growth could be observed on plates containing mannitol as carbon source indicating that the transformants were still mutated in *pts*I and that growth on glucose had to be the consequence of

PQQ synthesis. The PQQ production however was apparently low as in the presence of additional PQQ in the culture medium the growth of the transformants on glucose was increased. To test whether also in *E.coli* the gene IV product is required for PQQ synthesis, we cloned a DNA fragment containing only genes I, II and III under control of the *lac* promoter (pSQ105). PPA41 with pSQ105 does not grow on glucose meaning that also in *E.coli* the product of gene IV (which most likely is a 24 amino acids polypeptide) is essential in the biosynthetic pathway of PQQ.

DISCUSSION

Biosynthesis of the coenzyme PQQ is thought to be a multistep process involving at least five or six different enzymes. We have isolated a DNA fragment from the genome of *A.calcoaceticus* LMD79.41 which contains four genes involved in PQQ synthesis. Furthermore we have shown that expression of these four different PQQ genes in *E.coli* or *A.lwoffi* is sufficient for the synthesis of the coenzyme in these organisms. Sequence analysis of the PQQ genes showed that only three of them (genes I, II and III) code for proteins of a size (M_r 29700, M_r 10800 and M_r 43600 respectively) that can be expected for proteins with an enzymatic function. The fourth gene (gene IV) however does not seem to code for an enzyme, since the most probable reading frame comprises only 24 amino acids. An explanation for the presence of so few PQQ-specific enzymes might be that a precursor resembling the mature coenzyme structure is already present, not only in *A.calcoaceticus* but also in *A.lwoffi* and *E.coli*.

The intriguing question is what the role of gene IV in PQQ synthesis might be. Complementation studies have shown, that the gene IV region is coding for a trans-acting product. Site specific mutagenesis has shown that this product probably is the 24 amino acids polypeptide. Several functions for such a small polypeptide can be considered.

First it can be imagined that the 24 amino acids polypeptide is used as a precursor substrate. Recently it has been shown that the amino acids tyrosine and glutamate are used as precursors for PQQ biosynthesis in *Hyphomicrobium* X (van Kleef *et al.*, 1988) and *Methylobacterium* AM1 (C.J. Unkefer, pers comm). Since both amino acids are present in the 24 amino acids polypeptide it is possible that this polypeptide is used as a substrate. Through the tertiairy structure of the polypeptide both amino acids could be brought together, the enzymes could make the bonds and finally the complete (or almost complete) PQQ molecule could be cut out of the polypeptide.

As a second possibility the 24 amino acids polypeptide might also play a role in the transport of PQQ across the cytoplasmic membrane. The active site of the PQQ dependent GDH enzyme is located at the periplasmic side of the inner membrane (Duine *et al.*, 1986). Therefore eventually also the PQQ has to be present in the periplasmic space. This could be achieved in two possible ways. First the PQQ-enzymes can be transported through the inner membrane and the PQQ can be synthesized in the periplasm. The predicted amino acid sequences of the different PQQ enzymes however do not reveal the presence of N-terminal signal peptides (a stretch of hydrophobic amino acids flanked by hydrophilic residues). Alternatively the PQQ could be synthesized in the cytoplasm and subsequently translocated to the periplasm. The 24 amino acids polypeptide might be involved in this process, although it also is not highly hydrophobic.

Sandwiched between two genes that do play a role in PQQ synthesis (genes IV and I) is gene V which is not involved in this process. Gene V also does not play a role in the transport of electrons from PQQ to the electron transport chain, since deletion mutants lacking the complete PQQ region (including gene V) show normal GDH activity upon addition of PQQ (Goosen et al., 1987). A reason for the location of gene V between two PQQ genes might be that gene V codes for another PQQ dependent enzyme.

REFERENCES

Duine JA, Frank Jzn J and van Zeeland K, 1979. Glucose dehydrogenase from Acinetobacter calcoaceticus: a quinoprotein. FEBS Lett 108:186-192

Duine JA, Frank Jzn J and Jongejan JA, 1986. PQQ and quinoprotein enzymes in microbial oxidations. FEMS Microbiol Rev 32:165-178

Devereux J, Haeberli P and Smithies O, 1984. A comprehensive set of sequence analysis programs for the VAX. Nucleic Acids Res 12:387-395

Goosen N, Vermaas DAM and van de Putte P, 1987. Cloning of the genes involved in synthesis of coenzyme pyrrolo-quinoline quinone from Acinetobacter calcoaceticus. J Bacteriol 169:303-307

Hommes RWJ, Postma PW, Neyssel OM, Tempest DW, Dokter P and Duine JA, 1984. Evidence of a quinoprotein glucose dehydrogenase apoenzyme in several strains of Escherichia coli. FEMS Microbiol Lett 24:329-333

Kunkel TA, 1985. Rapid and efficient site-specific mutagenesis without phenotypic selection. Proc Natl Acad Sci USA 82:488-492

Lobenstein-Verbeek CL, Jongejan JA, Frank J and Duine JA, 1984. Bovine serum amine oxidase: a mammalian enzyme having covalently bound PQQ as prosthetic group. FEBS Lett 170:305-309

Maniatis T, Fritsch EF and Sambrook J, 1982. Molecular cloning, a laboratory manual. Cold Spring Harbor Laboratory, Cold Spring Harbor, N.Y.

Miller JH, 1972. Experiments in molecular genetics. Cold Spring Harbor Laboratory, Cold Spring Harbor, N.Y.

Sanger F, Nicklen S and Coulson AR, 1977. DNA sequencing with chain termination inhibitors. Proc Natl Acad Sci USA 74:5463-5467

van der Meer RA, Jongejan JA, Frank J and Duine JA, 1986. Hydrazone formation of 2,4-dinitrophenylhydrazine with pyrroloquinoline quinone in porcine kidney diamine oxidase. FEBS Lett 206:111-114

van Kleef MAG and Duine JA, 1988. L-tyrosine is the precursor of PQQ biosynthesis in Hyphomicrobium X. FEBS Lett in press

van Schie BJ, Hellingwerf KJ, van Dijken JP, Elferink MGL, van Dijl JM, Kuenen JG and Konings WN, 1985. Energy transduction by electron transfer via a pyrrolo-quinoline quinone dependent glucose dehydrogenase in Escherichia coli, Pseudomonas aeruginosa and Acinetobacter calcoaceticus (var lwoffi). J Bacteriol 163:493-499

PQQ: Biosynthetic Studies in *Methylobacterium* AM1 and *Hyphomicrobium* X Using Specific ^{13}C Labeling and NMR.

David R. Houck, John L. Hanners and Clifford J. Unkefer*

*Los Alamos National Laboratory, University of California, INC-4, M.S. C345 Los Alamos, NM, USA, 87545.

and

Mario A. G. van Kleef and Johannis A. Duine**

**Laboratory of Microbiology and Enzymology, Delft University of Technology, Julianalaan 67, 2628 BC Delft, The Netherlands

Abstract
Using ^{13}C labeling and NMR spectroscopy we have determined biosynthetic precursors of pyrroloquinoline quinone (PQQ) in two closely related serine-type methylotrophs, *Methylobacterium* AM1 and *Hyphomicrobium* X. Analysis of the ^{13}C-labeling data revealed that PQQ is constructed from two amino acids: the portion containing N-6,C-7,8,9 and the two carboxylic acid groups,C-7'and 9', is derived-intact -from glutamate. The remaining portion is derived from tyrosine; the phenol side chain provides the six carbons of the ring containing the orthoquinone, whereas internal cyclization of the amino acid backbone forms the pyrrole-2-carboxylic acid moiety. This is analogous to the cyclization of dopaquinone to form dopachrome. Dopaquinone is a product of the oxidation of tyrosine (via dopa) in reactions catalyzed by monophenol monooxygenase (EC 1.14.18.1). Starting with tyrosine and glutamate, we will discuss possible biosynthetic routes to PQQ .

Introduction
Pyrroloquinoline quinone (PQQ, 2,7,9-tricarboxy -1H-pyrrolo[2,3-f] quinoline-4,5-dione) was first recognized as a cofactor for the pyridine nucleotide-independent bacterial dehydrogenases (Duine and Frank, 1981 and references therein). These quinoproteins represent a novel class of dehydrogenases distinct from the well-known pyridine nucleotide and flavoprotein dehydrogenases (Duine et al., 1986 and references therein). Recent studies indicate that PQQ is a cofactor for several well-known metallo-enzymes including, bovine serum amine oxidase (EC 1.4.3.6) (Lobenstein-Verbeck et al.,1984), porcine kidney diamine oxidase (EC 1.4.3.6) (Meer et al., 1986), human placental lysyl oxidase (EC 1.4.3.13) (Meer and Duine, 1986), dopamine-β-hydroxylase (EC 1.14.17.1) (Meer et al.,1988) and soybean lipoxygenase-1 (EC 1.13.11.12) (Meer and Duine, 1988). The presence of PQQ in higher organisms raises the question of how this novel compound is biosynthesized and whether or not this compound or a related analogue is a vitamin. While the bacterial genes for the biosynthesis have been cloned from *Acinetobacter calcoaceticus* (Goosen et al.,1988) and expressed in *E. coli* (Dr N. Gossen personal communication), until recently none of the biosynthetic precursors or intermediates had been identified. This manuscript reviews two recent reports of the biosynthetic precursors of PQQ in two closely related serine-type methylotrophs. Independent studies were carried out on *Methylobacterium* AM1 by the Los Alamos group (Houck et al., 1988) and on *Hyphomicrobium* X by the Delft group (Kleef and Duine, 1988).

177

J. A. Jongejan and J. A. Duine (eds.), PQQ and Quinoproteins, 177–185.

Materials and Methods

Methylobacterium AM1 (*Pseudomonas* AM1, ATCC 14718) was cultured in a standard mineral medium (Breadsmore etal, 1982) with methanol (0.5%) or ethanol (0.5%) as the carbon source. In experiments with ^{13}C-enriched carbon sources, the organism was cultured in a stirred tank fermentor (10% inoculum) until the methanol (or ethanol) was exhausted (24-48 h). *Hyphomicrobium* X was cultivated on mineral medium (Kleef and Duine, 1988) with [^{13}C]methanol added as a carbon and energy source. This medium was supplemented with tyrosine or phenylalanine (0.27 g/l, natural abundance ^{13}C).

In both systems, PQQ was isolated from the clarified culture broth in several chromatographic steps (anion exchange followed by one or two reverse phase chromatography steps) (Duine and Frank, 1980 and Ameyama et al., 1984) and analyzed for ^{13}C enrichment by NMR spectroscopy. The yield of PQQ was typically 1 mg/l of culture broth. Mixtures of amino acids were obtained from protein hydrolysates (Putter et al., 1969) and analyzed by GC-MS (White and Rudolph, 1978) or separated for NMR analysis by ion exchange chromatography (Hirs et al., 1954).

For NMR analysis, PQQ was dissolved in 2.3 ml d_6-DMSO and placed in 10-mm NMR tubes. Spectra were obtained at 50.3 MHz on a Bruker AM200 WB spectrometer (Los Alamos) or at 100.5 MHz on a Varian VXR-400 S spectrometer (Delft). Chemical shifts are referenced to tetramethylsilane. Chemical shifts of a natural abundance sample (17.2 mg/ml in d_6-DMSO at 25C) were determined using PQQ obtained from Fluka Chemical Co. ^{13}C NMR spectra were obtained at 25 C using a 45° pulse and with the ^1H decoupler gated off for 10 s to minimize NOE effects. ^{13}C Enrichments were determined from the relative integrals of ^{13}C NMR resonances which were obtained by Lorentzian line shape analysis and normalized to the enrichment at C-8 which was determined by ^1H NMR analysis. Data were not corrected for partial T_1 saturation effects.

Table I
Chemical Shift Assignments and ^{13}C Enrichments of PQQ

		^{13}C Enrichments (atom %^{13}C) From:	
Carbon	δ, ppm	[1-^{13}C]Ethanol	[2-^{13}C]Ethanol
1a	136.7	16	54
2	127.6	23	68
2'	161.3	82	24
3	113.8	16	65
3a	123.4	16	59
4	173.4	13	59
5	179.2	27 "triplet"	46
5a	148.1	54 "triplet"	27
7	146.5	16	64
7'	165.4	80	59
8	130.3	17	61
9	142.2	n.o.	76
9'	167.2	101	n.o.
9a	126.1	35 "triplet"	37

Results and Discussion

Summarized in Table I are the one-to-one assignments for the fourteen ^{13}C NMR signals from natural abundance PQQ (Fig. 2a) which were determined from the ^{1}H-^{13}C coupling patterns ($^{1}J_{CH}$ and $^{3}J_{CH}$) and carbon-carbon correlations. These data agree with the partial assignments made by Duine and coworkers (1981). Carbon-carbon couplings were observed using a sample of [U-^{13}C]PQQ (90+ %^{13}C) isolated from cultures grown on [^{13}C]methanol (99.7%). The complete assignment was achieved by selecting for one-bond ^{13}C coupling interactions ($^{1}J_{C-C}$=55 Hz) in ^{13}C COSY experiments (Bax et al., 1981a and Bax and Freeman, 1981b).

^{13}C-Labeling Studies in Methylobacterium AM1-- During growth on methanol, *Methylobacterium* AM1 derives essentially all of its carbon from the methanol (Anthony, 1980); therefore, it is impossible to extract information pertinent to the biosynthesis of PQQ from experiments using [^{13}C]methanol as the sole carbon and energy source. A more useful labeled precursor is ethanol because, as described below, one can determine which carbons in PQQ are derived from C-1 and/or C-2 of ethanol. Ethanol is assimilated into four-carbon compounds by the action of malate synthase (Dunstan et al.,1972a and 1972b); three-carbon compounds are produced by decarboxylation of oxalacetic acid or by the action of serine transhydroxymethylase. Therefore, four-carbon compounds derived from malate should be labeled at C-1 and C-4 by [1-^{13}C]ethanol and at C-2 and C-3 by [2-^{13}C]ethanol. Three-carbon compounds such as pyruvate will be labeled only at C-1 by [1-^{13}C]ethanol and at C-2 and C-3 by [2-^{13}C]ethanol. These predicted labeling patterns were confirmed by examining the label distribution in amino acids isolated from cultures that contained [1-^{13}C]ethanol (Table II). Aspartate derived in two steps from malate was labeled predominantly at C-1 and C-4 by [1-^{13}C]ethanol; alanine derived from pyruvate was labeled only at C-1 by [1-^{13}C]ethanol. Because two-carbon units are incorporated directly into glutamate C-5 and C-4, their labeling probabilities are correlated. Thus [1-^{13}C]ethanol labels glutamate C-5 essentially without dilution; glutamate C-4 is unenriched and not subject to the background scrambling (12-17%) observed in other carbons derived ostensibly from ethanol C-2. Alternately, [2-^{13}C]ethanol yields glutamate highly enriched at C-4 (77%) and unenriched at C-5 (<3%). This is consistent with published radiolabeling data (Dunstan et al., 1972a) and is characteristic of organisms that have an incomplete TCA cycle (Walker et al., 1987). [1-^{13}C]Ethanol labels the phenol ring of tyrosine at C-3' and C-4' yielding a NMR spectrum that exhibits $^{1}J_{C-C}$ coupling; this labeling pattern is identical to that observed in tyrosine isolated from *E. coli* cultured on [1-^{13}C]lactate (LeMaster and Cronan, 1982). The adjacent labeling of C-3' and C-4' of tyrosine arises from the C-1 to C-1 joining of two trioses in gluconeogenesis and is diagnostic of compounds that arise from the shikimate pathway

Table II
Labeling of Amino Acids in *Methylobacterium* AM1 by [1-^{13}C]Ethanol

Amino Acid ^{13}C Enrichments From [1-^{13}C]Ethanol (atom %^{13}C)

	C-1	C-2	C-3	C-4	C-5	
Alanine	58	14	15			
Aspartate	68	17	17	68		
Glutamate	52	13	13	2	72	
	C-1	C-2	C-3	C-1'	C-2'(C-6')	C-3'(C-5')	C-4'
Tyrosine	73	17	17	17	18(18)	61(17)	61

The ^{13}C enrichments in PQQ biosynthesized from [1-^{13}C]ethanol based on analysis of NMR intensities are summarized in Table I. C-1 of ethanol labels predominantly the three carboxylates (C-2',7' and 9') and carbons 5, 5a and 9a. The predominantly singlet character of the carboxylates indicates that they are incorporated into positions in which their neighbors arise from C-2 of ethanol. Carbons 5, 5a and 9a each yield three resonances which are the combination of a singlet from singly labeled species and doublet ([1]J_{C-C}=60 Hz) from species labeled at C-5 and C-5a or C-9a and C-5a. The [1-^{13}C]ethanol labeling experiment coupled with the obvious structural homologies provide a working hypothesis for the biosynthetic origins of PQQ (Fig. 1). We propose that glutamate provides N-6 and carbons 7', 7, 8, 9 and 9', while the remaining nine carbons and N-1 are donated by an amino acid from the shikimate pathway, most likely tyrosine.

The precursors were identified by comparing the selective ^{13}C-labeling patterns in PQQ with those observed in amino acids. In PQQ, C-1 of ethanol significantly labels C-7' (59%) and C-9' (>99%) , but not C-9 (<2%); similarly, C-2 of ethanol labels PQQ at C-7 (64%), C-8 (61%) and C-9 (76%), but not C-9'. These labeling patterns are essentially identical to those observed in glutamate (Table II).

The incorporation of C-1 of ethanol into C-2', 5, 5a, and 9 of PQQ is equivalent to its incorporation into C-1, 3' and 4' of tyrosine. The adjacent labeling evident from the high degree of ^{13}C coupling at C4' and C-3' in tyrosine is also observed in the orthoquinone-containing ring in PQQ. While tyrosine C-3' and C-5' are chemically

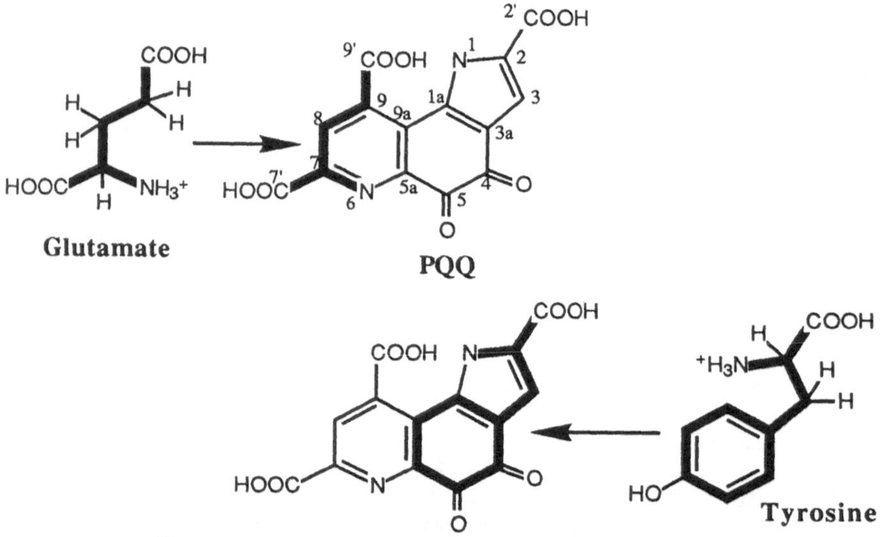

Figure 1) Proposed Biosynthetic Precursors for PQQ

equivalent, they are biosynthetically inequivalent because the aromatic ring is a product of asymmetric synthesis via the shikimate pathway (Haslam, 1974). Therefore, tyrosine C-3' arises from ethanol C-1, and C-5' from ethanol C-2. PQQ derived from [1-^{13}C]ethanol has adjacent ^{13}C labeling (doublets) at C-5a and C-5, or C-5a and C-9a. This labeling implies that the orthoquinone-containing ring arises from a symmetric compound (C$_2$ axis through C-1' and C-4') and predicts that C-5 and C-9a will be labeled equivalently and to an intermediate extent by both C-1 and C-2 of ethanol. Indeed, [2-^{13}C]ethanol labels C-5 and

C-9a, but not C-5a. This symmetric labeling pattern rules out indole as a precursor for that portion of PQQ containing the orthoquinone and pyrrole rings.

Direct evidence for the incorporation of tyrosine -- To demonstrate directly that tyrosine is a precursor of PQQ, L-[3',5'-$^{13}C_2$]tyrosine (Walker et al., 1986) was added to *Methylobacterium* AM1 cultures (0.5 mM) growing on methanol (0.5%). PQQ isolated from this culture was examined by ^1H and ^{13}C NMR spectroscopy. The ^{13}C spectrum (Fig. 2) indicates that L-[3',5'-$^{13}C_2$]tyrosine labels PQQ at C-5 and C-9a as predicted by the biosynthetic model (Fig. 1). Because C-9a is vicinally coupled to H-8, ^1H NMR analysis can be used to determine the ^{13}C enrichment at C-9a. Under these culture conditions, PQQ was labeled at C-9a (and C-5) to an enrichment of 63%. The resonances for C-5 and C-9a are doublets as a result of geminal ^{13}C-^{13}C coupling, proving that the phenol group of tyrosine is incorporated intact into the ring of PQQ containing the orthoquinone.

The results outlined above prove that tyrosine provides the six carbons of the orthoquinone-containing ring. To examine the possibility that internal cyclization of the tyrosyl backbone forms the pyrrole-2-carboxylic acid moiety, *Methylobacterium* AM1 was cultured on a mixture of [3',5'-$^{13}C_2$]tyrosine (50%) and[3-^{13}C]tyrosine (50%). The ^{13}C NMR spectrum of PQQ isolated from the culture filtrate contained only three resonances corresponding to C-5, C-9a and C-3. ^{13}C enrichments were determined by ^1H NMR. The equal incorporation of ^{13}C at C-3 and C-9a (41% and 42%, respectively) indicates that tyrosine is incorporated intact into PQQ.

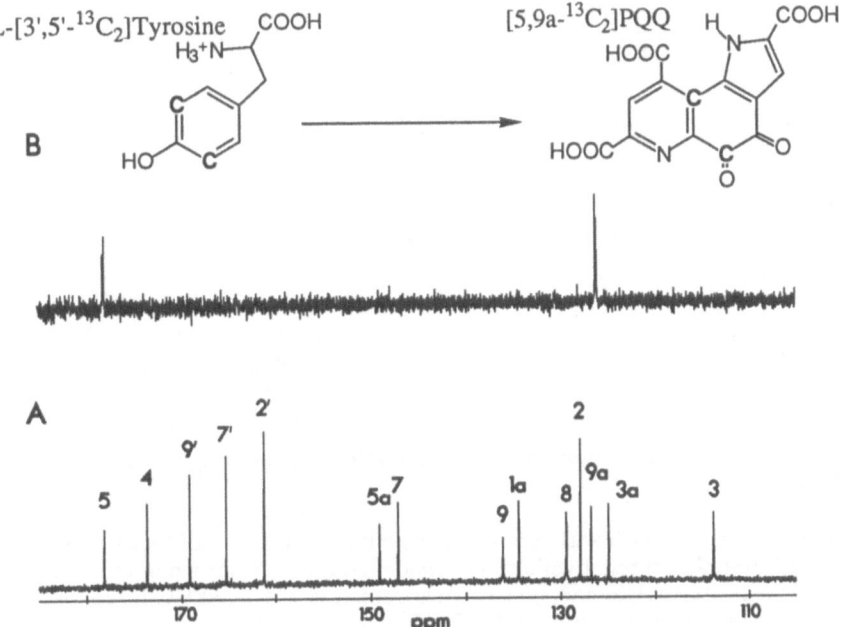

Figure 2) ^{13}C NMR spectrum of PQQ (A) and of [5,9a-$^{13}C_2$]PQQ (B) derived biosynthetically from L-[3',5'-$^{13}C_2$]tyrosine.

^{13}C Labeling studies in Hyphomicrobium X -- A simple but efficient approach to establish the intact incorporation of precursors is the replacement method. The direct incorporation of amino acids into protein of *Hyphomicrobium* X was determined by culturing the organism in a medium that contained [^{13}C]methanol as a growth substrate and was supplemented with unlabeled amino acids. Amino acids were isolated from protein hydrolysates and the dilution of the label from [^{13}C]methanol was determined by mass spectrometry. On administration of phenylalanine, 98% of this amino acid was incorporated, and no other unlabeled amino acids were found (results not shown), indicating that *Hyphomicrobium* X is not able to synthesize L-tyrosine from L-phenylalanine by the action of L-phenylalanine 4-monooxygenase (EC 1.14.16.1). Most probably, these amino acids are both synthesized from a common precursor, namely prephenic acid (Haslam, 1974 and Gorsich, 1987). On administration of L-tyrosine, this amino acid was incorporated for 94%, whereas no other amino acids were found unlabeled in the hydrolyzed protein.

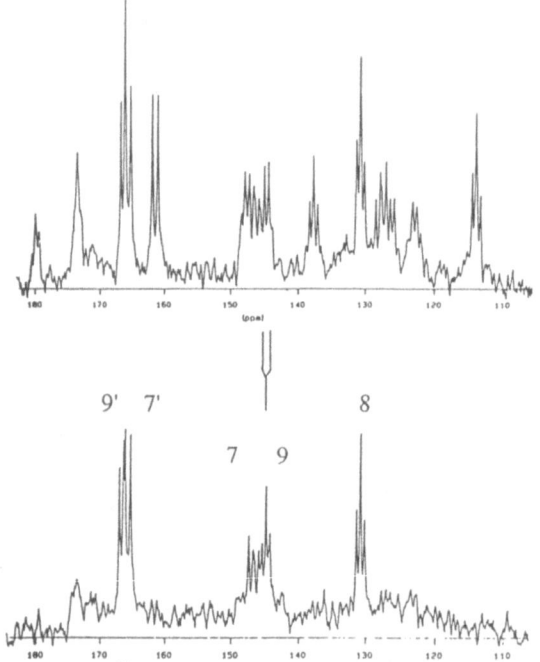

Figure 3) ^{13}C NMR spectra of [U-^{13}C]PQQ (top) and PQQ isolated from *Hyphomicrobium* X cultures grown in the presence of [^{13}C]methanol supplemented with L-tyrosine (bottom).

This replacement approach was used to examine the biosynthesis of PQQ. When PQQ was purified from the culture medium of *Hyphomicrobium* X, grown in the presence of ^{13}C-methanol plus L-phenylalanine, no significant changes in ^{1}H and ^{13}C NMR spectra were found as compared to the spectra of [U-^{13}C]PQQ, indicating that L-phenylalanine is not incorporated into the PQQ skeleton. However, PQQ isolated from cultures that were supplemented with tyrosine yielded ^{1}H and ^{13}C NMR spectra that were clearly altered. The ^{1}H-NMR resonance from H-3 collapsed to a singlet (7.15 ppm); H-8 (8.64 ppm) was a doublet (168 Hz) as a result of one-bond C-H coupling. Thus C-3 had been totally replaced with ^{12}C (<5% ^{13}C) and must have been derived from tyrosine; C-8 was enriched with ^{13}C (94%) and was derived from the labeled methanol. Inspection of the ^{13}C-NMR spectrum (Fig. 3, bottom) shows that signals only at δ 130.7 (C-8), 144.7 (C-9), 146.5 (C-7), 165.8 (C-7'), and 166.8 ppm (C-9') are present. The C-9 resonance is collapsed to a triplet due to the absence of coupling with C-9a. The absence of signals from C-2, C-2', C-3, C-3a, C-4, C-5, C-5a, C-9a, and C-1a indicates that these carbon atoms are derived from tyrosine. These results provide definitive proof that the carbon skeleton of tyrosine is incorporated intact into PQQ; the phenol side chain provides the six carbons of the ring containing the orthoquinone, whereas internal cyclization of the amino acid backbone forms the pyrrole-2-carboxylic acid moiety. Unfortunately, the replacement technique could not be used to test the potential of glutamate as a precursor for the remaining carbons (C-7,7',8,9 and 9') of PQQ in *Hyphomicrobium* X because glutamate added to the culture medium of *Hyphomicrobium* X did not significantly dilute label from [^{13}C]methanol in glutamate isolated from protein hydrolysates.

    Possible routes of PQQ biosynthesis -- As demonstrated from the results obtained in experiments on *Methylobacterium* AM1 and *Hyphomicrobium* X, PQQ arises from the condensation of glutamate and tyrosine (Fig 1). A possible route for the biosynthesis of PQQ is diagrammed in Figure 4. In this route, tyrosine or some derivative of tyrosine is oxidized to dopachrome in a reaction catalyzed by a monophenol monooxygenase-like enzyme(EC 1.14.18.1, tyrosinase). Glutamate could form a Schiff base with dopaquinone. The cyclization of the tyrosine backbone to form the pyrrole ring could occur by a Michael-type addition analogous to the known non-enzymatic cyclization of dopaquinone to form dopachrome (Canovas, 1982). Alternatively, dopachrome may be an intermediate in the biosynthesis of PQQ. Efforts to assay tyrosinase in *Methylobacterium* AM1 and have thus far failed. In addition, no sequence homology exists between tyrosinase genes from *Streptomyces glaucescens* (Huber et al., 1985) or *Neurospora crassa* (Lerch, 1982) and the PQQ biosynthesis genes of *Acinetobacter calcoaceticus* (Dr. N. Gossen, personal communication).

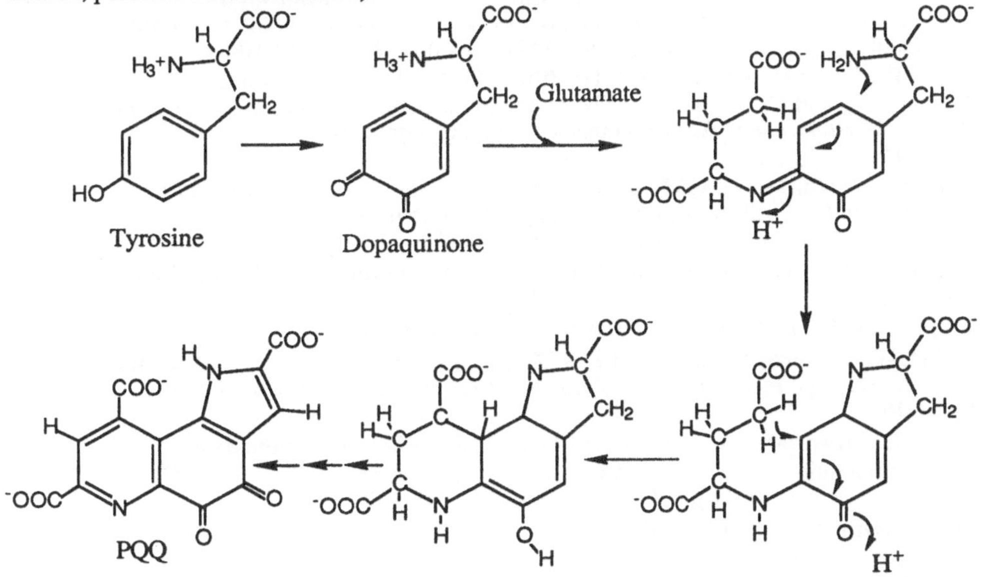

Figure 4) Possible route of PQQ biosynthesis.

References

Anthony, C., 1982. The biochemistry of methylotrophs. Academic Press, Inc., New York 1-40.

Bax, A., Freeman, R. and Morris, G., 1981a. Correlation of proton chemical shifts by two-dimensional fourier transform NMR. J. Mag. Reson. **42**:164-168.

Bax, A. and Freeman, R., 1981b. Investigation of complex networks of spin-spin coupling by two-dimensional NMR. J. Mag. Reson. **44**: 542-561.

Beardsmore, A.J., Aperghis, P.N.G. and Quayle, J.R., 1982. Characterization of the assimilatory and dissimilatory pathways of carbon metabolism during growth of *Methylophilus methylotrophus* on methanol. J. Gen Microbiol. **128**:1423-1439.

Canovas, F.G., Garcia-Carmona, F., Sanchez, J.V., Pastor, J.L.I., and Teruel, J.A.L., 1982. The role of pH in the melanin biosynthesis pathway. J. Biol. Chem. **257**:8738-8744.

Duine, J.A. and Frank Jzn. J., 1980. The prosthetic group of methanol dehydrogenase. Purification and some of its properties. Biochem.J. **187**:221-226.

Duine, J.A. and Frank Jzn. J., 1981. Quinoproteins: a novel class of dehydrogenases. Trends Biochem. Sc. **6**:278-280.

Duine, J.A. Frank, Jzn. J. and Jongejan, J.A.,1986. PQQ and quinoprotein enzymes in microbiological reviews. FEMS Microbiol Rev. **32**:165-178.

Duine, J.A. Frank, J.Jzn, and Verwiel, P.E.J., 1981. Structure and activity of the prosthetic group of methanol dehydrogenase. Eur. J. Biochem. **118**:395-399.

Dunstan, P.M., Anthony, C. and Drabble, W.T., 1972. Microbial metabolism of C_1 and C_2 compounds. The involvement of glycollate in the metabolism of ethanol and of acetate by *Pseudomonas* AM1. Biochem. J. **128**: 99-106.

Dunstan, P.M., Anthony, C. and Drabble, W.T., 1972. Microbial metabolism of C_1 and C_2 compounds. The role of glyoxylate, glycollate and acetate in the growth of *Pseudomonas* AM1 on ethanol and on C_1 compounds. Biochem. J. **128**:107-115.

Goosen, N., Vermaas, D.A.M. and Putte, P.van de, 1987. Cloning of the genes involved in the synthesis of coenzyme pyrroloquinoline quinone from *Acinetobacter calcoaceticus*. J. Bacteriol. **169**:303-307.

Gorisch. H., 1987. Chorismate Mutase from *Streptomyces aureofaciens*. Methods Enzymol. **142**:463-472.

Haslam, E., 1974. The Shikimate Pathway; Wiley, New York.

Hirs, C.H.W., Moore, S. and Stein, W.H., 1954. Chromatography of amino acids on ion exchange resins. Use of volatile acids for elution. J. Am.Chem. Soc. **76**:6063-6065.

Houck, D.R., Hanners, J.L. and Unkefer, C.J. 1988. Biosynthesis of Pyrroloquinoline quinone. 1. Identification of the biosynthetic precursors using ^{13}C labeling and NMR spectroscopy. J. Am. Chem. Soc. **110**:6920-6921.

Huber, M. Hintermann, G. and Lerch, K., 1985. Primary structure of tyrosinase from *Streptomyces glaucescens*. Biochem. **24**:6038-6044.

Kleef, M.A.G. van and Duine, J.A., 1988. L-Tyrosine is the precursor of PQQ biosynthesis in *Hyphomicrobium* X. FEBS Lett. **237**:91-97.

LeMaster, D.M. and Cronan, J.E., Jr., 1982. Biosynthetic production of ^{13}C-labeled amino acids with site-specific enrichment. J. Biol. Chem. **257**: 1224-1230.

Lerch, K., 1982. Primary structure of tyrosinase from Neurospora crassa II. Complete amino acid sequence and chemical structure of a tripeptide containing an unusual thioether. J. Biol. Chem. **257**:6414-6419.

Lobenstein-Verbeek, C.L., Jongejan, J.A., Frank, Jzn. J. and Duine, J.A., 1984. Bovine serum amine oxidase: a mammalian enzyme having covalently-bound PQQ as a prosthetic group. FEBS Lett. **170**:305-309.

Meer, R.A. van der and Duine, J.A., 1986. Covalently-bound Pyrroloquinoline quinone is the organic prosthetic group in human placental lysyl oxidase. Biochem. J. **239**: 789-791.

Meer, R.A. van der and Duine J.A., 1988. Pyrroloquinoline quinone is the organic cofactor in soybean lipoxygenase-1. FEBS Lett. **235**:194-200.

Meer, R.A. van der, Jongejan, J.A., and Duine, J.A., 1988. Dopamine-β-hydroxylase from bovine medulla contains covalently-bound pyrroloquinoline quinone. FEBS Lett. **231**:303-307.

Meer, R.A. van der, Jongejan, J.A., Frank, Jzn. J., and Duine, J.A., 1986. Hydrazone formation of 2,4-dinitrophenylhydrazine with PQQ in porcine kidney diamine oxidase. FEBS Lett. **206**:111-114.

Putter, I., Barreto, A., Markley, J.L. and Jardetzkey, O., 1969. Nuclear magnetic resonance studies of the structure and binding sites of enzymes. X. Preparation of selectively deuterated analogs of staphylococcal nuclease. Proc. Nat. Acad. Sci. USA **64**:1396-1403.

Walker, T.E., Matheny, C., Storm, C.B. and Hayden, H., 1986. An efficient chemomicrobiological synthesis of stable Isotope-labeled L-tyrosine and phenylalanine. J. Org. Chem. **51**:1175-1179.

Walker, T.E. and London, R.E., 1987. Biosynthetic preparation of L-[13C]- and [15N]glutamate by Brevibacterium flavum. Appl. Env. Microbiol. **53**:92-98.

White, R.H. and Rudolf, F.B., 1978. The Origin of the nitrogen in the thiazole ring of thiamine in Escherichia coli. Biochim. Biophys. Acta. **542**:340-347.

THE ROLE OF PQQ IN K.aerogenes AND CLONING OF pqq GENES

J.J.M. MEULENBERG, W.A.M. LOENEN, E. SELLINK AND P.W. POSTMA.
Laboratory of Biochemistry, Biotechnology Centre, University of Amsterdam, P.O.Box 20151, 1000 HD Amsterdam (The Netherlands).

INTRODUCTION

Enterobacteriaceae transport glucose into the cell by means of the PEP-dependent carbohydrate phosphotransferase system (PTS). In Klebsiella aerogenes an alternative pathway exists involving a glucose dehydrogenase (GLD, facing the periplasm) which uses pyrrolo-quinoline quinone (PQQ) as cofactor and is linked to the respiratory chain. It probably plays a role in the overflow metabolism of the cell, being induced under conditions of energetic stress. Glucose is converted to gluconate and can subsequently be used as a carbon source.

Escherichia coli and Salmonella typhimurium cannot synthesize PQQ, but do contain the GLD-apoenzyme which can be activated by the addition of PQQ. Growth defects in galactose (gal) and mannose (ptsI) metabolism can also be restored by PQQ (galactonate and mannonate being the oxidation products, respectively), possibly via the same apoenzyme.

Our aim is to study the expression and regulation of the genes involved in the synthesis of GLD and PQQ. We will clone the genes and make lacZ fusions in order to solve the following questions:
- How many genes are involved in PQQ biosynthesis in K. aerogenes and which (if any) of these genes are present in E. coli and S. typhimurium?
- Is the regulation of the GLD-apoenzyme and PQQ coordinate?
- Is the regulation at the transcriptional or translational level, or both?
- Is the observed PQQ-dependent oxidation of various carbohydrates in E. coli and
  S. typhimurium all due to the activity of GLD, or are different apoenzymes involved?

MATERIALS AND METHODS

The following K. aerogenes strains were utilized: NCTC418, parent; KA56, ptsI NCTC418; KA11, ptsI pqq NCTC418. E. coli strains: PPA186, ptsI; PPA42, ΔptsHI. Cosmid pJRD215 (J. Davison et al., 1987) was used for the construction of a K. aerogenes genomic library. A lamB encoding plasmid (A. Harkki et al., 1985) was introduced to make K. aeroge-nes sensitive for λ. λ Transposon mutagenesis was performed according to Way et al 1984., Glucose dehydrogenase activity was measured as described by Duine et al., 1979.

J. A. Jongejan and J. A. Duine (eds.), PQQ and Quinoproteins, 187–189.

RESULTS

We set out to isolate mutants in PQQ biosynthesis and construct a cosmid bank of K. aerogenes to identify the genes by complementation studies.

pqq Mutants. KA56, a ptsI derivative of K. aerogenes NCTC418 was used for transposon mutagenesis with λNK1098, a phage carrying a Tn10 deletion transposon. Mutants unable to grow on minimal medium with glucose as sole source of carbon (MM glc) were isolated. 8 of these, KA126-KA133, still grew on MM with gluconate (table 1), indicating a defect some-where in the GLD/PQQ pathway or in the respiratory chain. Since the GLD activity in cell-free extracts of the mutants and parent strain (KA56) was similar after preincubation with PQQ, they evidently had an active GLD-apoenzyme. The values of NADH and glucose oxida-tion in membranes of KA126-KA133 (preincubated with PQQ) showed that these mutants were neither defective in the respiratory chain. They were stimulated by PQQ on McConkey glucose plates (McC glc, table 1). However, unlike a previously isolated K. aerogenes pqq point mutant, KA11, KA126-KA133 failed to grow on MM glc after addition of PQQ. Finally, these mutants grew rather poorly on a number of carbon sources and even rich media compared to KA56 and KA11.

Table 1. Growth of K.aerogenes mutants on various carbon sources.

Strain	Relevant Genotype	Growth on:[1]				
		MM gluc	MM glc	MM glc + PQQ	McC glc	McC glc + PQQ
KA56	ptsI	++	+	+	red	red
KA126-KA133	ptsI pqq::Tn10?	+	-	-	white	red
KA11	ptsI pqq	+	-	+	white	red

[1]Abbreviations: MM, Minimal Medium; gluc, gluconate; glc, glucose; McC, McConkey.

Cosmid bank. Partial Sau3A fragments of K. aerogenes NCTC418 DNA were ligated into the BamH1 site of vector pJRD215. After in vitro packaging, the cosmid bank was tested for complementation of KA11, the pqq point mutant. One clone, COS4 complemented KA11 both on MM glc and McC glc plates (table 2). COS4 contained a 4.2 kb insert (figure 1). Slight stimulation of several E. coli pts mutants by COS4 was observed on MM glc and McC glc plates. Though COS4 stimulated KA126-KA133 on McC glc, growth was not restored on MM glc, in line with our previous results.

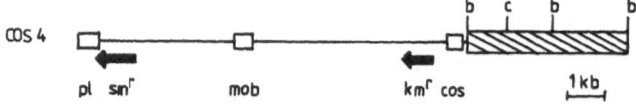

Figure 1. Map of COS4. —— = pJRD215 segment, ▨▨ = 4.2 kb K. aerogenes DNA. Abbreviations: sm[r], streptomycin resistance; mob, plasmid mobilization functions; km[r], kana-mycin resistance; cos, phage cohesive ends; pl, polylinker.

Table 2. Complementation of <u>K.aerogenes</u> and <u>E.coli</u> glucose⁻ mutants by COS4.

Strain	Relevant Genotype	Growth on:[1]			
		MM glc	MM glc +PQQ	McC glc	McC glc +PQQ
<u>K. aerogenes</u>					
KA11	<u>ptsI pqq</u>	–	+	white	red
KA11/COS4	<u>ptsI pqq</u>	+	+	red	red
KA126-KA133	<u>ptsI pqq</u>::Tn<u>10</u>?	–	–	white	red
KA126-KA133/COS4	<u>ptsI pqq</u>::Tn<u>10</u>?/COS4	–	–	red	red
<u>E. coli</u>					
PPA186	<u>ptsI</u>	–	+	white	red
PPA186/COS4	<u>ptsI</u>/COS4	±	+	pink	red
PPA42	Δ<u>ptsI</u>	–	+	white	white
PPA42/COS4	Δ<u>ptsI</u>/COS4	±	+	white	white

[1]Abbreviations: MM, Minimal medium; glc, glucose; McC, McConkey.

<u>Southern blots</u>. In order to establish whether the Tn10 insertions in KA126-KA133 were within the same <u>pqq</u> locus as present on COS4, we performed Southern blot analysis with two of these mutants using the 4.2 kb insert of COS4 as probe. Since the same size fragments hybridized to the probe in both mutant and wild type strains, we concluded that the Tn10 insertions are at a different locus, at least 20 kb away. The 4.2 kb fragment hybridized only weakly to <u>E. coli</u> and <u>S. typhimurium</u> chromosomal DNA, indicating that this locus shares little homology.

DISCUSSION

We have cloned a <u>pqq</u> locus, which complements a <u>K. aerogenes</u> <u>pqq</u> point mutant strongly and <u>E. coli</u> <u>pts</u> mutants weakly. Since we expect at least 5-10 genes to be involved in the biosynthesis of a rather complicated molecule such as PQQ, several of these genes may be present in <u>E. coli</u>, the cloned locus being only 4.2 kb in length.

We have isolated mutants which display a partial <u>pqq</u> phenotype, using Tn10 transposon mutagenesis. Southern blots showed the Tn10 insertions to be outside the cloned <u>pqq</u> locus. At present it is not clear whether these mutants are true <u>pqq</u> mutants or whether the phenotype is due to some pleiotropic mutation or transcriptional polarity caused by the Tn10 insertion.

REFERENCES

Davison J., Heusterspreute M., Chevalier N., Ha-Thi V. and Brunel F., 1987. Gene 51: 275-280.
Duine J.A., Frank Jzn J., and van Zeeland J.K., 1979. FEBS Letters 108: 443-446.
Harkki A. and Palva E.T., 1985. FEMS Microbiology Letters 27: 183-187.
Way C., Davis M.A., Morisato D., Roberts D.E. and Kleckner N., 1984. Gene 32: 369-379.

MUTANTS OF <u>Methylobacterium organophilum</u>
UNABLE TO SYNTHESIZE PQQ

F. Biville*, P. Mazodier**, E. Turlin* and F. Gasser*, Unité de Régu-
lation de l'Expression Génétique* et Unité de Génie Microbiologique**,
Institut Pasteur, F-75724 Paris Cedex, France.

Key words : Pyrroloquinoline-quinone, Methanol dehydrogenase, PQQ mutants
<u>Methylobacterium organophilum</u>, Plasmid pSUP 106

ABSTRACT
The phenotype of mutants unable to synthesize PQQ is analyzed for
different categories of methylotrophic bacteria. The advantages offered
by strains dissimilating methylamine through methylated amino-acids are
discussed. In <u>M.organophilum</u>, 40% of the mutants unable to grow in
methanol medium but with normal methylamine utilization, were affected in
PQQ metabolism. The genetic properties of <u>M.organophilum</u> useful to study
PQQ mutants are discussed, mainly the use of pSUP106 to create insertion
mutations in the bacterial chromosome and to replace wild-type genes by
modified genes. An example is given of the possibility to create R´
plasmids containing large fragments of <u>M.organophilum</u> DNA. Some
physiological properties of a PQQ mutant are described, regarding growth
kinetics, PQQ uptake and accumulation.

INTRODUCTION
Enzymes of PQQ biosynthesis have not yet been characterized.
Identification of the corresponding genes might give useful information
on the number and nature of the enzymatic steps involved in PQQ
biosynthesis and on the regulation of these steps. Goosen et al., (1987)
have isolated a plasmid containing 5 kb of <u>Acinetobacter calcoaceticus</u>
DNA that could complement four classes of mutants unable to synthesize
PQQ (PQQ⁻ mutants). In methylotrophic bacteria only one study on a single
PQQ⁻ mutant has been published (Biville et al., 1988). We present in this
paper the advantages of isolating PQQ⁻ mutants in certain categories of
methylotrophic bacteria, including the Gram-negative facultative
methylotroph <u>Methylobacterium organophilum</u>. Some of the genetic
properties of <u>M.organophilum</u> useful to study these mutants, and their
main physiological characteristics are discussed.

PHENOTYPE OF PQQ⁻ MUTANTS.
In <u>A.calcoaceticus</u>, PQQ is associated with glucose dehydrogenase and
mutants deficient in PQQ biosynthesis are revealed by a deficience of
acid production in the presence of several aldoses that glucose
dehydrogenase can use as substrate (Goosen et al., 1987). In
methylotrophic bacteria, methanol dehydrogenase (MDH) and methylamine
dehydrogenase are two important quinoproteins directly associated with
the utilization of methanol and methylamine respectively. Methanol
oxidation by MDH is an inescapable step of methanol catabolism and PQQ⁻
mutants will always be found among the cells unable to grow on methanol.
In methylotrophs, methylamine oxidation can be achieved in two different
ways :

J. A. Jongejan and J. A. Duine (eds.), PQQ and Quinoproteins, 190–194.
© 1989 by Kluwer Academic Publishers.

(i) **by methylamine dehydrogenase,** a PQQ dependent enzyme which is present for example in Methylobacterium extorquens strain AM1 or methylophilus methylotrophus. In these organisms, PQQ mutants should be unable to grow on methanol **and** on methylamine. But this particular phenotype is also found in the mutants of formaldehyde assimilation pathways, serine or ribulose monophosphate (RuMP). This has two consequences: first, the finding of PQQ⁻ mutants would suppose the screening of a large collection of independent mutants of the above mentioned phenotype and this might be difficult to obtain; allyl alcohol is a substrate of MDH which gives toxic products (acrolein), and it can be used as a suicide substrate for positive selection of mutants with a MDH defect, but to our knowledge, no such suicide substrate has been described for methylamine dehydrogenase. Secondly, a satisfactory growth substrate other than methanol or methylamine has to be found. This may be difficult with Gram-negative organisms using RuMP pathway because most of them are obligate or at best restricted facultative methylotrophs.

(ii) **Through methylated amino–acids** such as methylglutamate which is oxidized by methylglutamate dehydrogenase, a flavoprotein enzyme which is present for example in Methylobacterium organophilum DSM 760 (Biville et al., 1988) or in Methylophaga marina 92b (Janvier et al., 1985). In these organisms, PQQ⁻ mutants can be found among cells impaired only in methanol utilization and growing normally in methylamine medium since PQQ is not involved in methylamine oxidation. This is a great advantage because the numerous mutants of formaldehyde assimilation pathway cannot be confused with the PQQ⁻ mutants.

Thus, PQQ⁻ mutants of M.organophilum DSM 760 were obtained after EMS treatment, followed by a growth of 6-7 generations in succinate medium and positive selection on succinate medium containing 0.05% allyl alcohol. Fourty per cent of the mutants unable to grow on methanol medium were affected in PQQ biosynthesis.

GENETIC PROPERTIES OF M.organophilum USEFUL IN THE STUDY OF PQQ MUTANTS
1. Directed mutagenesis and gene exchange with pSUP 106 hybrids
The conjugative plasmid pSUP 106 was initially designed to allow the transfer of large DNA fragments to a wide range of Gram-negative bacteria (Priefer et al., 1985). In the Gram-negative organism M.organophilum, pSUP106 cannot replicate and behave as a suicide plasmid (Mazodier et al., 1988). With E.coli as donor, the pSUP106 hybrids containing cloned DNA are transferred at a high frequency into M.organophilum where they can produce a transient complementation under some conditions. Beside the transient expression of plasmidic genes, a recombination occured between homologous, cloned and chromosomal, DNA. This procedure of DNA exchange allows the possibility to introduce modifications in defined regions of the genome. Thus, insertion mutations can be made inside a defined gene or in the vicinity of a defined gene. In the latter case wild type genes can be replaced by modified genes previously cloned in pSUP 106.

For example: Precise localization of the gene pqqA by directed mutagenesis in the chromosomal pqqA region (Mazodier et al., 1988).
The plasmid pMO512 - an hybrid of pSUP106 containing 2.5 kb of M. organophilum DNA - complemented MTM1 a mutant of M.organophilum unable to synthesize PQQ. A Kan^r cartridge (pUC4K, Pharmacia, cat. n° 24-4987-00) was cloned into two different sites of pMO512, BamHI and SalI, yielding plasmids pMO5121 and pMO5122 respectively. After transformation in E.coli S17-1 and subsequent transfer in M.organophilum DSM 760, kanamycin

resistant <u>M.organophilum</u> exconjugants were selected : MDQ1 which received pMO5121 and MDQ2, pMO5122.

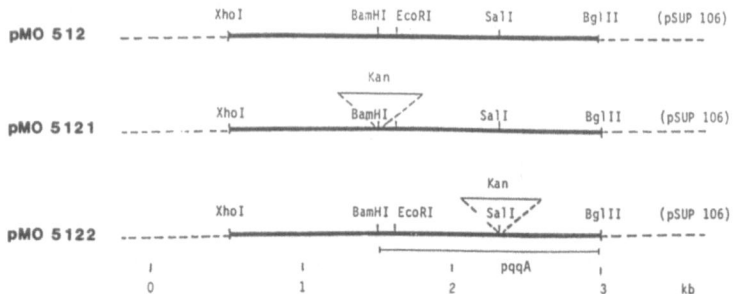

In strains MDQ1 and MDQ2, the plasmids pMO5121 and pMO5122 did not replicate and were lost. In the selected kan^r exconjugants, a double-crossing-over occured between the homologous DNA, chromosomal and plasmidic, the latter on either side of the resistance marker which was eventually inserted by this mechanism into the bacterial chromosome.

The growth abilities of MDQ1 and MDQ2 reported on table 1 showed that the mutation affecting MTM1 was located in a 1.5 kb <u>BglII</u>-<u>BamI</u> fragment of <u>M. organophilum</u> chromosome (gene <u>pqqA</u>). By this procedure, all possible interference of multiple mutations, frequently found after chemical mutagenesis, were excluded. Moreover since no plasmid remained present in the recipient strain, problems involving the interpretation of the results of conjugation-complementation <u>viz</u> recombination are excluded.

2. <u>Preparation and use of R´ plasmids</u>

Taking advantage of the introduction of precisely located chromosomal markers, prime plasmids can be selected including large fragments of chromosomal DNA in the vicinity$_s$ of these markers. The chromosome mobilizing vector was pJB3J1, a Kan^s derivative of R68-45, a plasmid with high mobilizing ability (Haas & Holloway, 1976).

For example: Selection of R´ plasmids which could complement both moxF and pqqA genes.

MD5 is a strain of <u>M.organophilum</u> DSM 760 containing the transposon Tn5 inserted in a region close to <u>moxF</u>, the gene encoding apo-MDH. Regarding methanol utilization, MD5 has a wild-type phenotype (table 1). In a first step MD5 received pJB3J1 from <u>E.coli</u>. In a second step MD5(pJB3J1) was crossed with <u>E.coli</u> HB101, a <u>recAB</u> deficient strain of <u>E.coli</u>. Several Kan^r <u>E.coli</u> exconjugants clones were selected. In one of these clones, a plasmid R´51 complemented both <u>moxF</u> and <u>pqqA</u> mutants. The size of <u>M. organophilum</u> DNA in R´51 was estimated to be ≤ 30 kb. This gives a first approach of the respective location of <u>pqqA</u> and <u>moxF</u> genes on the chromosome of <u>M.organophilum</u>

GROUPS OF PQQ MUTANTS

A preliminary analysis showed that the mutants unable to synthesize PQQ belong to three distinct groups. Group I includes all of the mutants complemented by R´51, that is the mutants affected in <u>pqqA</u> and other mutants affected in gene(s) close to <u>pqqA</u>. From a library of chromosomal DNA cloned in the cosmid pLA 2917 (Allen and Hanson, 1985), two plasmids pMO 200 and pMO 600 were isolated which complemented two other groups of mutants the groups II and III respectively. A detailed analysis of these groups is in progress.

M.organophilum strains and relevant characteristics	MDH (1) Activity (2)	PAGE	PQQ(1)(3) concn. in crude extracts	culture supernat.
DSM 760 wild type	0.51	++	1.15	0.33
MD5 Tn5 near moxF (4)				
MD14 Tn5 in moxF (4)	0.04	+w	0.09	2.0
MDQ1 Kan^r insertion near pqqA	0.54	++	0.85	0.33
MDQ2 Kan^r insertion in pqqA	0.0	+w	0.02	0
MTM1 EMS mutant pqqA	0.0	+w	0.04	0

TABLE 1.   Biochemical characteristics relevant to methanol oxidation in M.organophilum DSM 760 and in some mutants (After Mazodier et al., 1988).

(1) Crude extracts of succinate grown cells, centrifuged in early stationary phase. Methanol-grown cells presented similar characteristics ; (2) umol DCIP reduced per min, per mg protein ; (3) nmol PQQ per mg protein or per ml culture supernatant ; (4) moxF encodes apo-MDH.

PHYSIOLOGICAL PROPERTIES OF PQQ MUTANTS OF M.organophilum DSM 760

PQQ was not detected in crude extracts of the mutants nor in their culture supernatants. PQQ was assayed with the apo-alcohol dehydrogenase from Pseudomonas testosteroni (Groen et al., 1986).
An other important common feature of all the PQQ mutants of M. organophilum was the absence or the low level of apo-MDH in crude extracts. Apo-MDH was visualized after PAGE or by immunodetection.
Three main characteristics of the mutant strain MTM1 affected in pqqA were studied in more detail (Biville et al. 1988) :

1. Growth kinetics with limiting PQQ concentrations.
In methanol medium, with PQQ added at concentrations below 1 uM, the exponential growth shifted progressively toward a linear growth rate (fig. 1A). Without added PQQ, a linear growth rate was obtained (fig. 1B).

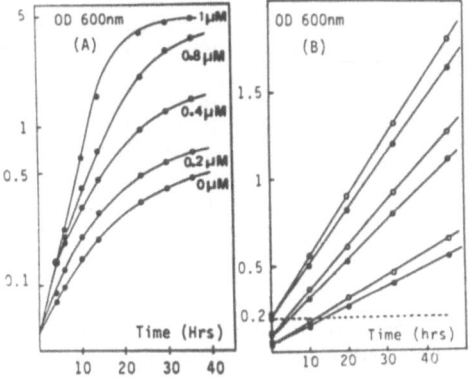

Fig. 1. (A) Growth kinetics of MTM1 in methanol medium with different concentrations of PQQ. Inoculum was a 24 h culture in methanol medium + PQQ 2 μM, washed twice in the initial volume of minimal medium without PQQ. Growth curve of the wild type (not shown) was superimposable on the growth curve with 1 μM PQQ. (B) Residual growth in methanol medium without PQQ: effect of the inoculum size and of PQQ concentration in the cultures used as inoculum. These concentrations were 1 μM (●———●) and 5 μM (○———○). Inocula were prepared as above. No residual growth was observed without methanol (------).

2. PQQ cannot accumulate in the cells of M.organophilum DSM 760

As shown on fig. 1B, a preculture of MTM1 grown in methanol medium with
an excess of PQQ (5 uM) did not produce subsequently, in methanol medium
without PQQ, a significantly higher residual growth than a preculture
grown with a minimal amount of PQQ. This experiment demonstrates that
externally added PQQ cannot accumulate in mutant cells. Neither can free
PQQ accumulate in cells not affected in PQQ synthesis. This is shown by
the results of PQQ assays (table 1) in the moxF insertion mutant MD14
(Mazodier et al., 1988). The synthesis of a non-functional apo-MDH by
this mutant results in a depletion of intracellular PQQ and an excess of
PQQ in the culture supernatant.

3. PQQ uptake by the PQQA mutant MTM1

The sharp initial decrease of PQQ concentration in the culture medium was
unspecific since it could also be observed with heat killed cells. After
this initial fall, the decrease of PQQ concentration was logarithmic as
the cell mass increased exponentialy.

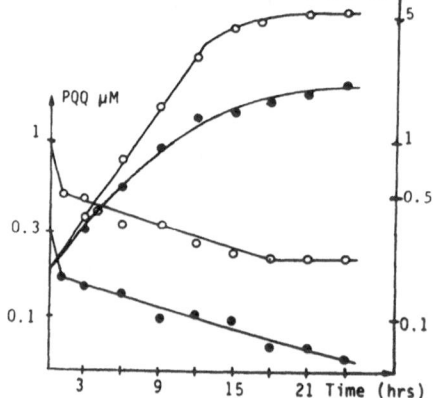

Fig. 2. Kinetics of PQQ uptake during growth of MTM1 with
initial concentrations of 0.3 (●————●) and 1 μM
(○————○) PQQ. Inocula were prepared as in experiments of
Fig. 1. Samples for assays at time 0 were taken before inoculation of medium.

REFERENCES

Allen L and Hanson R, 1985. Construction of broad host range cloning
vectors : identification of genes necessary for growth of M.organophilum
on methanol. J. Bacteriol. 161: 955-962.
Biville F, Mazodier P, Gasser F, Van Kleef MAG and Duine JA, 1988.
Physiological properties of a pyrroloquinoline quinone mutant of
Methylobacterium organophilum. FEMS Microbiol. Lett. 52: 53-58.
Goosen N, Vermaas D and Van de Putte P, 1987. Cloning of the genes
involved in synthesis of coenzyme pyrroloquinoline quinone from
Acinetobacter calcoaceticus. J. Bacteriol. 169: 303-307.
Groen W, Van Kleef MAG and Duine JA, 1986. Quinohaemoprotein alcohol
dehydrogenase apoenzyme from Pseudomonas testosteroni. Biochem. J. 234:
611-615.
Haas D and Holloway B, 1976. R-factor variant with enhanced sex factor
activity in Pseudomonas aeruginosa. Mol. Gen. Genet. 144: 243-251.
Janvier M, Frehel C, Grimont F and Gasser F, 1985. Methylophaga marina
gen. nov., sp. nov. and Methylophaga thalassica sp. nov. marine
Methylotrophs. Intern. J. of System. Bacteriol. 35-2: 131-139.
Mazodier P, Biville F, Turlin E and Gasser F, 1988. Localization of a
pyrroloquinoline quinone biosynthesis gene near the methanol dehydrogenase structural gene in Methylobacterium organophilum DSM 760. J. Gen.
Microbiol., in the press.
Priefer UB, Simon R and Pühler A, 1985. Extension of the host range of E.
coli vectors by incorporation of RSF 1010 replication and mobilization
functions. J. Bacteriol. 163: 324-330.

THE CHEMISTRY AND BIOMIMETICS OF PQQ

Yoshiki OHSHIRO and Shinobu ITOH
Department of Applied Chemistry, Osaka University
Yamadaoka 2-1, Suita, Osaka 565, Japan

Key Words: PQQ, quinoprotein, dehydrogenase (oxidase) model, redox reaction catalytic oxidation, amine, amino acid, alcohol, glucose, nitroalkane

Abstract: Chemical simulation of redox functions of PQQ-containing enzymes was studied to clarify the roles of coenzyme PQQ. Thus catalytic oxidations of amines, amino acids, glucose, alcohols, thiols, and nitroalkanes with PQQ were investigated from the stand point of organic chemistry and effective redox system was found as oxidase models.

NADP-dependent and flavin-dependent enzymes have been widely interested in the field of coenzyme chemistry. In addition to these enzymes, it has become clear that there is a third class of dehydrogenases, that is quinoproteins. In these enzymes, the compound PQQ, pyrroloquinoline quinone, plays important roles as a novel coenzyme. Recently many oxidoreductases have been found to be PQQ-containing enzymes and have been under exciting discussions.[1]

The structure of a novel coenzyme PQQ was first identified by Salisbury in 1979.[2] Since then we have been interested in its organic chemistry, particularly in redox reactions, of coenzyme PQQ because PQQ has the unique structure of a fused heteroaromatic o-quinone.

Thus we have been studying chemical simulations of redox functions of PQQ-containing enzymes using various substrates to get fundamental knowledge on the roles of coenzyme PQQ in dehydrogenases.

Chemical Simulation of the PQQ-Containing Amine Oxidases Applying Micellar Reaction.

Amine oxidases are known to catalyze oxidative deamination of amines in the presence of molecular oxygen to produce the corresponding carbonyl compounds, ammonia, and hydrogen peroxide.[3]

$$RR'CHNH_2 + O_2 + H_2O \xrightarrow{PQQ} RR'C=O + NH_3 + H_2O_2$$

We first tried aerobic oxidation of α-phenethylamine in the presence of a catalytic amount of PQQ. As shown in the table, phenethylamine was oxidized in a neutral aqueous solution containing no surfactant to give acetophenone as a sole product in 828% yield based on PQQ. This result clearly shows PQQ acts as a catalyst in the nonenzymatic oxidation of the amine and the reaction constructs a model system of the PQQ-containing amine oxidases.[4]

It is well known that micelles provide the major driving forces for binding and activation of substrates.

Surfactant	PQQ/Surfactant/H₂O 30°C, under air, 48h pH		Yield(%)
	Initial	Final	
None	8.20	8.75	828
CTAB	8.20	8.80	3395
Brij-35	8.20	8.75	1443
SDS	8.30	8.80	1214

CTAB : Cetyltrimethylammonium Bromide
Brij-35 : Polyoxyethylene Lauryl Alcohol Ether
SDS : Sodium Dodecyl Sulfate
[Surfactant]=1.67×10^{-2}M
Yields were determined by GLC based on PQQ.

195

J. A. Jongejan and J. A. Duine (eds.), PQQ and Quinoproteins, 195–204.
© 1989 by Kluwer Academic Publishers.

The above table shows that a nonionic surfactant, polyoxyethylene lauryl ether (Brij-35), and an anionic surfactant, sodium dodecyl sulfate, have little catalytic ability. On the other hand, a cationic surfactant, cetyltrimethylammonium bromide, provided a good environment for the oxidation, and higher turnover number of PQQ was achieved.[5]

In the following figures is shown kinetical study on the reaction of benzylamine with PQQ under anaerobic conditions. The reaction was performed under pseudo-first-order conditions using excessive amount of the amine in the presence of CTAB in an alkaline buffer solution. The figure on the left shows the UV absorption of the the reduced PQQ (max 320 nm) increases quickly along the progress of the reaction. As shown in the figure on the right, introduction of oxygen into the final reaction mixture (spectrum 1) led to the spectrum 2, which is consistent with quantitative regeneraton of the starting PQQ.

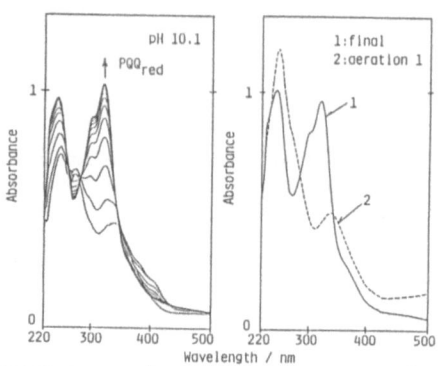

$[PQQ]=4 \times 10^{-5}M$, $[PhCH_2NH_2]=8 \times 10^{-3}M$ $[CTAB]=2 \times 10^{-3}M$, 0.05M carbonate buffer (pH 10.1), $\mu=0.2$ with KCl, 35°C anearobic condition (N_2)

Kinetic studies on the oxidation of several amines with PQQ are summarized in the table. The reaction was studied under psuedo-first order conditions in the presence of micelle at pH 10.1.

The reaction was first-order in PQQ and the amine concentrations, respectively. Since a free amine is considered to be an active species, the second-order rate constant is calculated from the pKa value of the amine. It is clear that the oxidation actually proceeds also in the case of α-phenethyl-amine, although the expected oxidation product was not isolated. The secondary and the tertiary amines showed much lower reactivity than those of primary amines.

amine	$10^4 k_{obsd}$ (s^{-1})	pKa	10^3 [amine]$_f^{a)}$ (M)	$k_{obsd}/$[amine]$_f$ (s^{-1} M^{-1})
⬡–NH$_2$ [b)]	17.5	9.45	3.26	0.536
⬡(Me)–NH$_2$ [b)]	7.69	9.33	3.42	0.225
⬡–NH$_2$ [b)]	45.1	9.78	2.71	1.67
⬡–NH$_2$ [b)]	9.69	10.33	1.49	0.650
⬡–N(H)(Me) [c)]	2.44	9.57	9.29	0.026
⬡–N(Me)(Me) [c)]	0	9.02	11.09	0

a) Free amine concentration at pH 10.1 calculated from pKa values. b) [amine]$_T$=4 $\times$ 10^{-3}M, c) [amine]$_T$=12 $\times$ 10^{-3}M

Oxidation of benzylamine with PQQ at lower pH was monitored by UV spectrometer. The reaction at pH 7.1 proceeded much slower than that at pH 10 and three absorptions at 335, 390, and 410 nm slowly increased for about 1 h. Then these absorptions started to decrease and the absorption at 320 nm based on reduced PQQ gradually increased. These phenomena imply existence of intermediates other than reduced PQQ.

The isotope effect on the oxidation of benzylamine with PQQ was also studied. The observed large kinetic isotope effect indicates that the C-H bond cleavage is the rate determining step of this oxidation reaction. From the results of kinetic studies and product analysis, we suggest the reaction mechanism of the oxidation as follows.

⬡–CD$_2$NH$_2$

$k_H/k_D= 7.24$

$k_H= 5.16 \times 10^{-3}$ s^{-1}
$k_D= 7.13 \times 10^{-4}$ s^{-1}

The oxidation of amine proceeds mainly via covalent addition of the amine to the quinone to form the carbinolamine intermediate. It is followed by rate determining α-proton removal providing a quinol shown by

an upward arrow. Alternative path is formation of the imine intermediate shown by a downward arrow. Successive rearrangement and α-proton removal gives the aminophenol product. These two paths are competing, but the latter path may be the minor process. Both types of reduced PQQ, quinol and aminophenol, are found to be reoxidized rapidly to the original PQQ by oxygen, and thus constitutes catalytic cycles.[6]

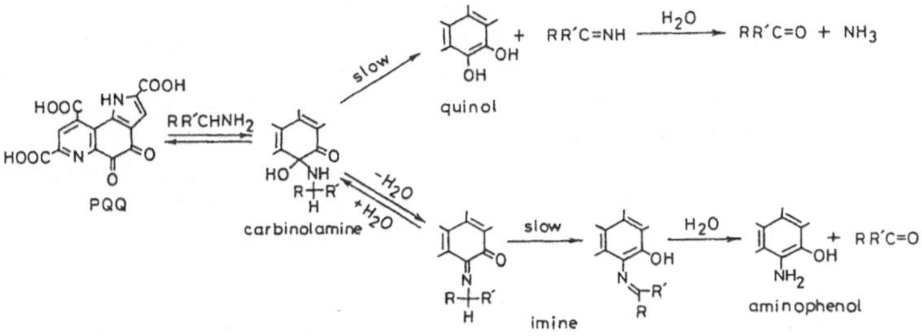

In the following table, the second-order rate constants of the oxidation of several amines under anaerobic conditions are listed to know effects of structure of aliphatic amines on hydrophobic interactions with micellar system. Since none of these aliphatic amines are oxidized at all in the absence of the CTAB micelle, it is clear that the reaction is drastically enhanced by the micelle. While methylamine was oxidized very slowly, reactivity of the amines increased markedly with an increase in the number of carbons. These results show that hydrophobic binding or association of the substrate with the micelle is the major

Oxidation of Aliphatic Amines by Coenzyme PQQ in the Presence of CTAB Micelle.

Substrate	pKa [a]	pH [b]	$k_2 (M^{-1}s^{-1})$
CH_3NH_2	10.62	9.88	0.15 [a]
$\sim\sim NH_2$	10.62	9.92	5.6
$\sim\sim\sim NH_2$	10.64	9.91	23.6
$\sim\sim\sim\sim NH_2$	10.65	9.90	80.8
$\sim\sim\sim\sim\sim NH_2$	10.64	9.91	111
$\sim\sim\sim\sim\sim\sim NH_2$	10.63	9.90	113
$\sim\sim\sim\sim\sim\sim\sim NH_2$	10.62	9.87	114
$\sim\sim\sim NH_2$	10.28 [c]	9.88	3.5
$\sim\sim\sim NH_2$	10.13 [c]	9.89	32.4
$\sim\sim NH_2$	10.35 [c]	9.88	7.1

[PQQ] = $4.0 \times 10^{-5}M$, [Amine] = $3.83-4.45 \times 10^{-4}M$, [CTAB] = $2.0 \times 10^{-3}M$, 0.05M carbonate buffer ($\mu = 0.2$ with KCl), 35°C, under aerobic conditions (N_2)
a) authentic pKa value   b) pH value of the reaction mixutre   c) determined by volumetric titration
d) $[CH_3NH_2] = 8.00 \times 10^{-3}M$

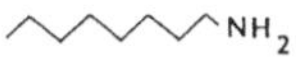

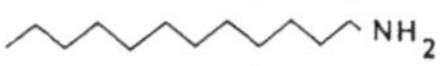

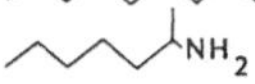

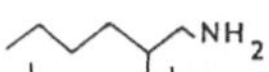

factor for the rate acceleration. Reactivities of the amines having 10 to 14 carbon chains were almost equal because of good incorporation of the substrate into the micelle. An amino group attached to a secondary carbon gave smaller rate constants.[7]

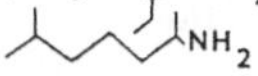

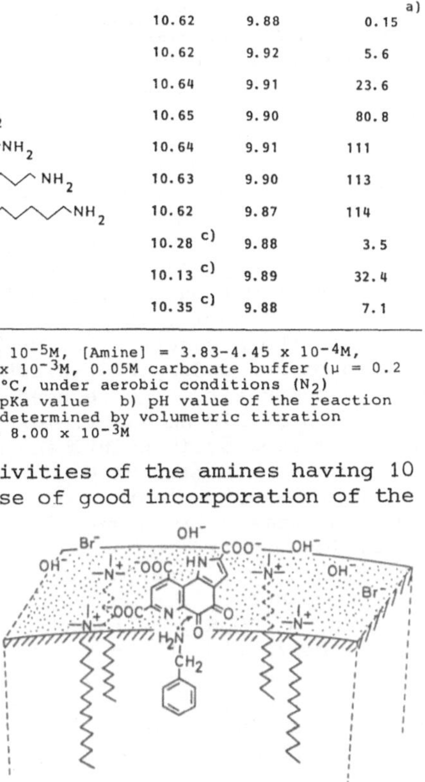

From the results so far obtained, the binding and reaction sites in the micellar system are suggested to be such as shown on the right. PQQ is bound onto the micellar surface with face to face via electrostatic interaction between the carboxylate groups of PQQ and the charged surface of the micelle. On the other hand, a non-polar alkyl or aryl group of the substrate is inserted into the hydrophobic core and the polar amino group is located in the interfacial layer of the micelle to approach to the quinone group.

Thus the chemical simulation of the PQQ-containing amine oxidases could be constructed by using CTAB-micelle system and amines were catalytically oxidzied to the corresponding ketones with high turnover numbers. We believe modification of the surfactants by introducing various functional groups will lead to more effective enzyme models.

The Reaction of Amino Acids with Coenzyme PQQ.

Oxidation of amino acids is very important process in celullar metabolism and flavin and pyridoxal coenzymes are known to play important roles in such systems. In these oxidation systems, the different oxidation products have been reported. Hence reactions of amino acids with the novel coenzyme PQQ are quite interesting from the view point of biomimetic chemistry.

When phenylglycine was reacted with PQQ under the same conditions employed in the reactions with amines, benzaldehyde and its air-oxidized product, benzoic acid, were obtained in good yields. However, α-keto acid was not detected at all. Thus we could show Strecker degradation of α-amino acids is also catalyzed by PQQ in the non-enzymatic systems.[8]

$$\underset{\text{NH}_2}{\text{C}_6\text{H}_5\text{CH–COOH}} \xrightarrow[\substack{\text{pH 7.13} \rightarrow 8.82 \\ 30°C, \text{ under air, 24h}}]{\text{PQQ / CTAB / H}_2\text{O}} \text{C}_6\text{H}_5\text{CHO} + \text{C}_6\text{H}_5\text{COOH}$$

3265%   1718

Preparation and Reoxidation of Reduced PQQ (PQQH$_2$).

Formation of the reduced PQQ was clearly observed in the reaction with amines under anaerobic conditions. Thus the reduced PQQ was prepared to investigate its reoxidation. In the reactions of PQQ with the reductants listed in the table under anaerobic conditions, PQQH$_2$ was obtained as a sole product in good yields.

The oxidation of the reduced PQQ is expexted to be responsible for catalytic cycle of the oxidation reaction. In the figure under the table, the UV spectral change along the progress of oxidation of PQQH$_2$ with oxygen underpseudo-first order conditions under air. The reduced PQQ having an absorption at 320 nm was rapidly oxidized to PQQ with the isosbestic points and an intermediate such as a semiquinone was not detected. On the other hand, we confirmed the formation of hydrogen peroxide in about 50% yield based on PQQ by iodometric titration.

The following figure on the next page shows the pH dependence of the initial rate constant of oxidation of PQQH$_2$. Although the plots are somewhat complicated, the rate generally increases with an increase in the pH from 4 to 8, and almost no change was observed when the pH went beyond 8 and thus we obtained the curve showing the existence of the optimum pKa

Preparation of PQQH$_2$

Reductant[a]	pH[b]	Time(h)	Yield(%)
C$_6$H$_5$SH	6.8	2	84
BNAH[c]	6.8	3	77
Na$_2$S$_2$O$_4$	11.4	5	99
NaBH$_4$[d]	7.2	4	99
C$_6$H$_5$NHNH$_2$	3.1	3	82
H$_2$/PtO$_2$	7.0	5	65

a) 10-fold excess over PQQ b) The pH of PQQ aqueous solution (10ml). c) in the dark d) See Ref. 9

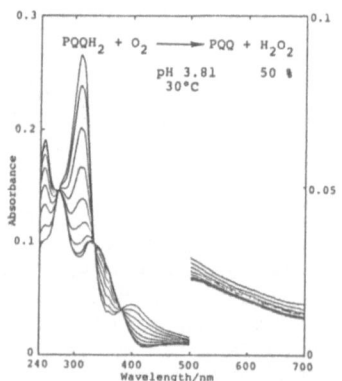

value around here.

In the oxidation reactions with PQQ, we may expect possibility of existence of semiquinone radical which is formed by the reaction with the reduced PQQ and PQQ. Thus PQQH$_2$ was treated with an equimolar amount of PQQ at neutral pH under anaerobic conditions, and the reaction was monitored by UV spectrometer. However, no absorption characteristic to the semiquinone radical was observed.

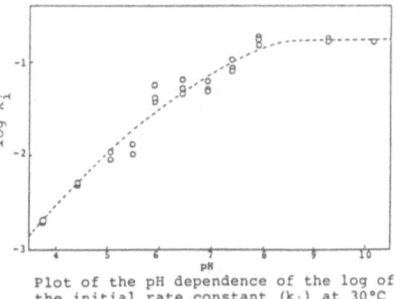

Plot of the pH dependence of the log of the initial rate constant (k_i) at 30°C

From the results of the kinetic study on the oxidation of the reduced PQQ, we consider the reaction path as follows.

The active species in the oxidation of PQQH$_2$ with oxygen is suggested to be PQQH$^-$ ion and the reaction proceeds mainly via covalent addition of oxygen to the quinol carbon of the quinolate anion followed by elimination of hydrogen peroxide.[10] In the reaction of the reduced flavin with oxygen, it is generally considered that the siginificant intermediate is the 4a-adduct,[11] and this may be the similar case to the oxidation of the reduced PQQ.

Oxidation of Alcohols by PQQ.

PQQ is known as a coenzyme of many types of alcohol dehydrogenases,[1] but our attempt to oxidize alcohols with PQQ in aqueous solutions gave unfavorable results. We examined the oxidation in nonaqueous media by using trimethyl ester of PQQ.

Oxidation of benzylalcohol with trimethyl ester of PQQ in acetonitrile under anaerobic conditions gave no benzaldehyde even after a long period or with such an additive as a protonic acid or an organic strong base or under irradiation.

$$\text{\textcircled{}-CH}_2\text{OH} \xrightarrow[\substack{168h,\ under\ N_2 \\ additive:\ H^+,\ DBU,\ h\nu}]{PQQTME/CH_3CN} \nrightarrow \text{\textcircled{}-CHO}$$

PQQ trimethyl ester alone was not effective enough to oxidize the alcohol on the contrary to the oxidation of amines. It is probably because of relatively low oxidation potential of PQQ.[12]

Alcohols could be oxidized by trimethyl ester of PQQ in THF or acetonitrile in the presence of a Lewis acid under anaerobic conditions. The results of Lewis acid-promoted oxidation of several alcohols are listed in the table. AlCl$_3$ was the most effective among the Lewis acids studied, TiCl$_4$ was less effective, and the other acids shown here have no

$$\text{RR'CHOH} \xrightarrow[\substack{under\ N_2}]{PQQTME/Lewis\ acid/Solvent} \text{RR'C=O}$$

Substrate	Lewis acid	Solvent	Time(h)	Temp(°C)	Yield(%)
⬡-CH$_2$OH	AlCl$_3$	CH$_3$CN	24	rt	76
"	"	"	9	52	99
"	"	THF	24	rt	10
"	"	CH$_2$Cl$_2$	"	"	trace
(CHOH-CH$_3$)	TiCl$_4$	CH$_3$CN	"	"	25
"	AlCl$_3$	"	9	52	64
(Ph$_2$CHOH)	"	"	"	"	13
◯-OH	"	"	24	rt	44
(CHOH)	"	"	"	"	72
"	"	"	9	52	92

PQQTME(0.025 mmol), Lewis acid(0.083 mmol), Substrate (0.25 mmol), Solvent(10 ml), Yields were determined by GLC based on PQQTME.
BF$_3$, Mg(ClO$_4$)$_2$, CuCl$_2$, ZnCl$_2$ --- no reaction

catalytic activities. As a solvent, acetonitrile was more favorable, and the yield of benzaldehyde from benzylalcohol went up to 99% at 52°C. The other alcohols were also converted into the corresponding carbonyl compounds, but benzhydrol was not so reactive because of steric hindrance.

Thus alkoxides are expected to be good substrates of oxidation by PQQ. Although lithium or potassium benzyloxide was not oxidized by PQQTME, benzyloxymagnesium bromide was oxidized to benzaldehyde in 51% yield. In the case of aluminum alkoxide the reaction proceeded more smoothly (yield 77%). From these results, we propose that the reaction mechanism is an Oppenauer type one.

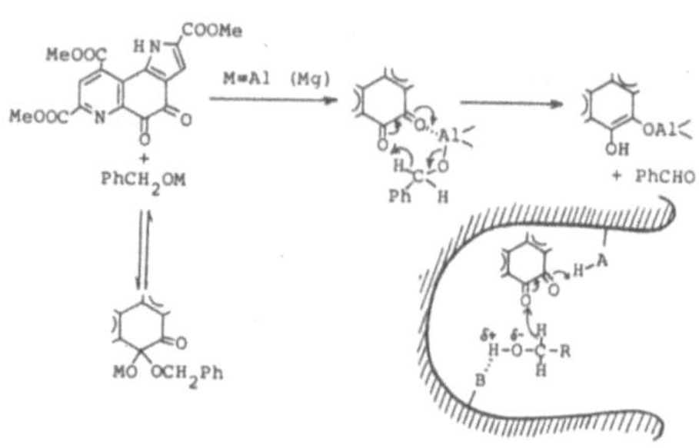

An aluminum alkoxide formed in situ coordinates to carbonyl oxygen of PQQ trimethyl ester and aluminum as a Lewis acid withdraws electrons to accelerate abstraction of hydride from benzyl position of the quinone. A lithium or potassium alkoxide is considered to add to C-5 carbonyl carbon of the quinone, which will not give rise to oxidatioin. Prof. Duine reported the formation of C-5 adduct of methanol with PQQ.[12]

The mechanism of oxidation of alcohols by PQQ in enzymatic system is not known, but the present result is the first example of nonenzymatic oxidation. Our results suggest that enzymatic oxidation of alcohols by PQQ proceeds with the aid of concerted general acid-base catalysis in the active site postulated as the above figure.[13]

Oxidation of Glucose by PQQ.

In the last few years, one of the most exciting enzymatic chemistries has been developed in the field of glucose dehydrogenases.[14] To clarify the chemical function of quinoproteins for D-glucose and some other sugars, our attention was focused on the reaction between PQQ and sugars.

At first the oxidation of D-glucose with PQQ was elucidated under anaerobic conditions by UV spectrometry. The reaction was smoothly proceeded at pH 10.4 and produced reduced PQQ and gluconic acid.

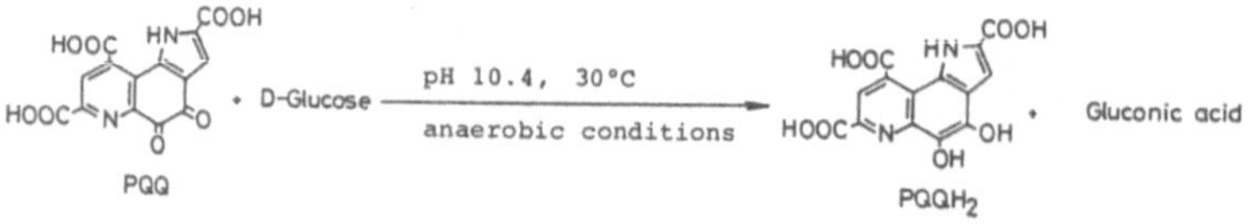

The absorption spectra of reduced PQQ was increased with isosbestic points along the progess of the reaction. After the reaction, aeration of the reaction system caused to generate original PQQ. This phenomena show PQQ can act as a turnover catalyst.

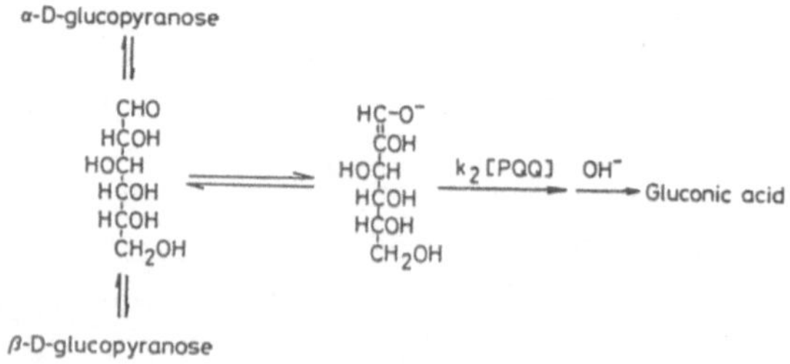

The reaction path was assumed as shown in the scheme on the the basis of kinetic studies. Glucose is in equilibrium between cyclic and acyclic structures. An acyclic glucose gives an endiolate intermediate by deprotonation with a base.

The endiolate intermediate reacts fast with PQQ followed by hydrolysis to give gluconic acid. It is clarified that the rate determining step of this reaction is the formation of the endiolate intermediate.

It is easily recognized that the oxidation of a sugar is affected significantly by the concentration of an acyclic form. Isbell and his coworkers reported the relative rates of 1,2-endiol formation of various sugars by hydrogen tritium exchang rate at around pH 11.[15] These values are shown in parentheses in the table. The relative rates of oxidation of sugars are comparably equal to the relative rates of tritium uptake reaction. This result concluded that concentration of an acyclic sugar is very important in the oxidation with PQQ.

Sugar	pH	k_{2app} $(M^{-1}s^{-1})$	Relative rate of the oxidation
D-glucose	11.21	3.6×10^{-2}	1.0 (1.0)[a]
D-mannose	11.13	3.7×10^{-2}	1.0 (0.5)
D-arabinose	11.23	1.5×10^{-1}	4.2 (4.1)
D-fructose	11.15	3.0×10^{-1}	8.3 (10.7)
DL-glyceraldehyde	11.20	11.8	328 (—)

$[PQQ]=4.0 \times 10^{-5}$M, $[Sugar]=8.0 \times 10^{-2}$M, 0.2M Na_2CO_3, 25°C, Under anaerobic conditions. a) Relative rate of tritium uptake.

Catalytic turnover reaction was monitored by UV spectrometry. After the reaction finished, the reaction vessel was opened to air. It was observed that the reduced PQQ immediately changed to oxidized form, PQQ. The reaction vessel was then closed again. After the consumption of contained oxygen, the increase in the absorption of $PQQH_2$ was observed. The recycle was smoothly repeated without deactivation of PQQ. Thus PQQ is established as an effective turnover catalyst in the oxidation of sugars (See the figure on the left).

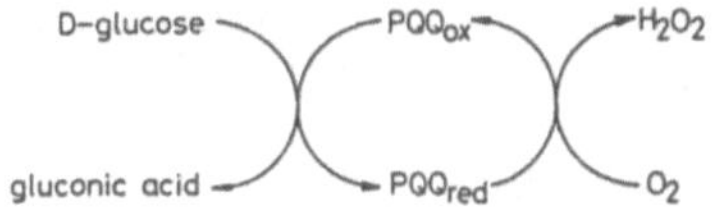

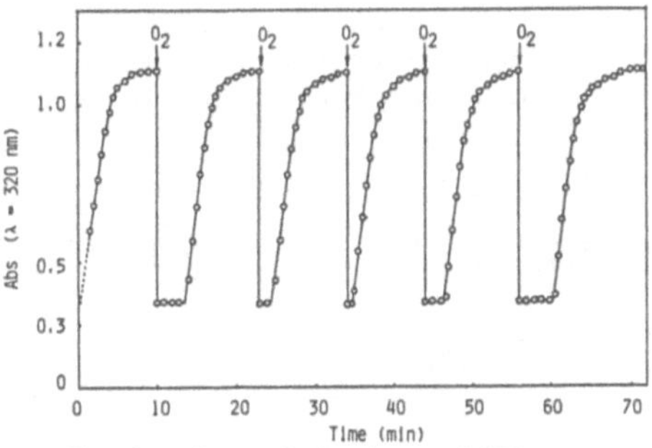

Time-dependence of absorbance of PQQ
$[PQQ] = 4.0 \times 10^{-5}$ M, $[D-glucose] = 8.0 \times 10^{-1}$ M
pH 9.70 with 0.05 M carbonate, 30 °C

As shown above, an endiolate intermediate in the oxidation with PQQ is very important. There are some substrates which give an enediolate form such as aldehydes and ascorbic acid. Deprotonation of benzoin gives an endiolate anion and an aldehyde also gives an endiolate equivalent intermediate in the presence of some nucleophiles such as cyanide ion and an N-substituted thiazolium anion.[16]

Benzoin was oxidized with PQQ at pH 10.6 to give benzil in 83% yield. p-Chlorobenzaldehyde was also oxidized with PQQ to give p-chlorobenzoic acid in the presence of the nucleophiles. In the case of the aldehyde the presence of CTAB is required essentially. The oxidation reaction did not occur without micellar system. These reaction may proceed via similar manner to the amine oxidation.

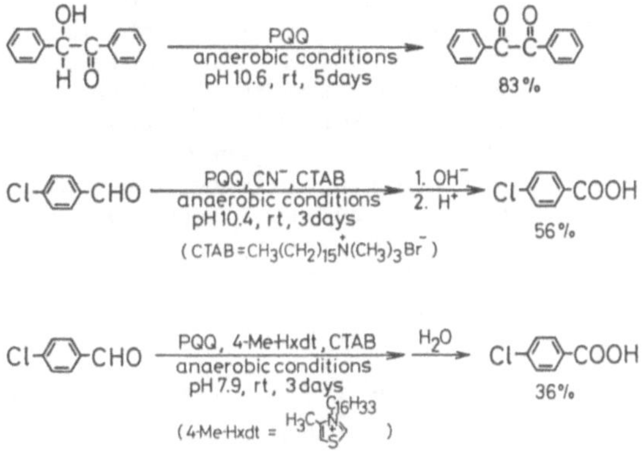

The second-order rate constants for glyceraldehyde,

(CTAB = $CH_3(CH_2)_{15}\overset{+}{N}(CH_3)_3 Br^-$)

benzoin, p-chlorobenzaldehyde, and L-ascorbic acid compared to that of D-glucose are listed in the table. The reactions of the aldehydes and benzoin are zero-order in PQQ, as same as that of D-glucose. In general, reaction rates of aldehydes are greater than that of D-glucose. In the case of ascorbic acid, the reaction was first-order in PQQ. Ascorbic acid has an endiolate skelton in nature and, therefore, oxidation reaction proceeds very fast.[17)

Substrate	pH	order	k_2 $(M^{-1}s^{-1})$
D-Glucose	9.9	0	8.25×10^{-3}
Glyceraldehyde	9.9	0	2.15
Benzoin	9.9	0	5.13×10^{-2}
p-Cl-C_6H_4CHO + KCN + CTAB [a)]	10.4	0	1.20
" + 4-Me-Hxdt + CTAB[b)]	7.9	0	1.88
L-Ascorbic acid	4.5	1	59.25

a) [KCN] = 1.0×10^{-1}M, [CTAB] = 2.0×10^{-3}M

b) [4-Me-HxdT] = 4.0×10^{-4}M, [CTAB] = 2.0×10^{-3}M

4-Me-HxdT :

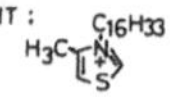

Reaction of Nitroalkanes with PQQ.

Recently Professor Soda and his coworkers proposed that nitroalkane oxidase of Fusarium Oxysporum contains PQQ and FAD as prosthetic groups and the nitroalkane is metabolized to the corresponding aldehyde.[18)

Nitroalkane Oxidase of

Fusarium Oxysporum

$CH_3CH_2CH_2NO_2 + O_2 + H_2O$

$\longrightarrow CH_3CH_2CHO + HNO_2 + H_2O_2$

Coenzyme : PQQ , FAD

In a vitro system, the reaction between PQQ and nitroethane under anaerobic condition at pH 10.1 gave two unexpected products. The spectral change was also abnormal as compared to the preceding oxidations. No absorption based on $PQQH_2$ was observed in the course of the reaction. The reaction was irreversible and was not affected by the presence of air. The structures of the two products are not yet elucidated because they are rather complicated. Therefore, we tried to clarify the reaction mode between heteroaromatic o-quinones and nitroalkanes using model systems.

The models of PQQ such as shown in the table on the right gave abnormal addition products having a dioxole structure in relatively high yields. These adducts indicate that, in the case of nitroalkanes, the reaction proceeded in different manner from those of the other substrates. Formation of the dioxole structure was observed only for o-quinones. Comparison of the spectral data between the adducts of PQQ and those of models suggests the similarity of the structures of the both adducts.

Quinone	Nitroalkane	Time(h)	Product	Yield(%)
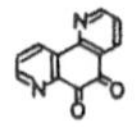	$CH_3CH_2NO_2$	5	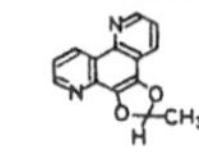	88
"	CH_3NO_2	5	many products	
"	$CH_3CH_2CH_2NO_2$	5	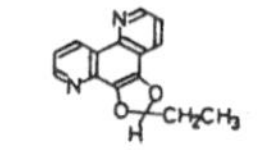	70
"	$(CH_3)_2CHNO_2$	5	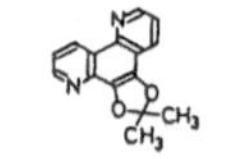	67
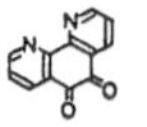	$CH_3CH_2NO_2$	5	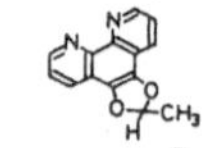	67
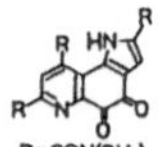 R=CONHCH$_3$	$CH_3CH_2NO_2$	5	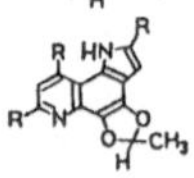	53

[Quinone] = 4.8 mM, [Nitroalkane] = 0.48 mM
[Carbonate buffer] = 0.1 M (containing 50%
acetonitirle, pH 9.45, μ = 0.2 with KCl)
55°C, anaerobic conditions (N₂)

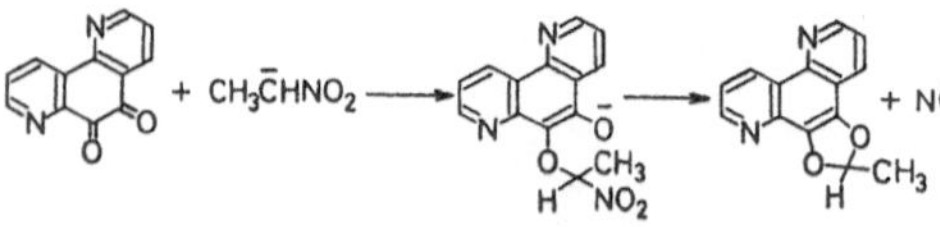

It is wellknown that a nitroalkane reacts with a carbonyl compound at carbonyl carbon under basic conditions.[19)] But the reaction of o-phenanthraquinone with vinyl Grignard

reagent gives a dioxole type adduct.[20] Our reaction may proceed similar manner to this Grignard reaction. The details of reaction mechanism are under investigation.

The abnormal adduct of 1,7-phenanthrolinequinone and nitro-ethane was easily hydrolyzed to reduced orthoquinone and the corresponding aldehyde under acidic condition. This reduced o-quinone is readily oxidized to the starting quinone under basic conditions in the presence of oxygen. These results can be estimated as one of a model of catalytic oxidation of a nitroalkane with PQQ.[21]

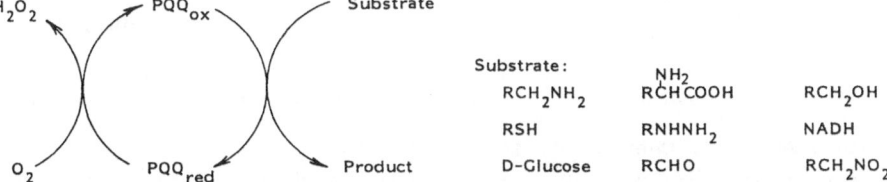

Reactions of Thiols, NADH-Models, and Other Substrates.

PQQ easily oxidizes thiols under anaerobic conditions to disulfides and serves as a good oxidation catalyst under aerobic conditions. The details are already reported.[22]

Redox reaction between NADH and PQQ is of great interest and the reactions of PQQ with NADH models as well as their application are reported by us at this Symposium and the abstract is printed elsewhere in this book.

The results of the reactions of PQQ with diamines, guanidines, and aminoguanidines, which are known as inhibitors against PQQ-containing enzymes, are also mentioned in the other pages.

So far we have studied on the oxidation reaction system with PQQ for various substrates and were successful in simulation of the function of PQQ in vitro systems.

H_2O_2 ⟶ PQQ_{ox} ⟶ Substrate

O_2 ⟶ PQQ_{red} ⟶ Product

Substrate:
RCH_2NH_2 $RCHCOOH$ (with NH_2) RCH_2OH
RSH $RNHNH_2$ $NADH$
D-Glucose $RCHO$ RCH_2NO_2

Most of the substrates except for alcohols and nitroalkanes are able to construct an effective redox system of coenzyme PQQ. In the cases of alcohols and nitroalkanes, an effective redox system was not found, but we believe the possibility should be realized in near futrue.

Syntheses and applications of models of PQQ directed toward artificial enzymes are of our current interest and molecular design of such compounds are under investigation.

References

1) J. A. Duine, J. Frank Jzn, and J. A. Jongejan, "Advances in Enzymology and Related Areas of Molecular Biology", Ed. by A. Meister, John Wiley, New York, 1987, p169.
2) S. A. Salisbury, H. S. Forrest, W. B. T. Cruse, and O. Kennard, Nature (London), 280, 843 (1979).
3) H. Yamada and K. Yasunobu, J. Biol. Chem., 237, 1511 (1962); O. Oda, T. Manabe, and T. Okuyama, J. Biochem., 89, 1317 (1981).
4) Y. Ohshiro, S. Itoh, K. Kurokawa, J. Kato, T. Hirao, and T. Agawa, Tetrahedron Lett., 24, 3465 (1983).
5) S. Itoh, Y. Kitamura, and Y. Ohshiro, J. Jpn. Oil Chem. Soc. (Yukagaku), 35, 91 (1986).
6) S. Itoh, Y. Kitamura, Y. Ohshiro, and T. Agawa, Bull. Chem. Soc. Jpn., 59, 1907 (1986).
7) S. Itoh, M. Mure, and Y. Ohshiro, J. Jpn. Oil. Chem. Soc. (Yukagaku), 36, 90 (1987).
8) S. Itoh, N. Kato, Y. Ohshiro, and T. Agawa, Tetrahedron Lett., 25, 4753 (1984).
9) J. A. Duine, J. Frank Jzn, P. Eugene, and J. Verwiel, Eur. J. Biochem., 108, 187 (1980).
10) S. Itoh, Y. Ohshiro, and T. Agawa, Tetrahedron Lett., 25, 4753 (1984).
11) T. C. Bruice, Acc. Chem. Res., 13, 256 (1980).
12) J. A. Duine, J. Frank Jzn, P. Eugene, and J. Verwiel, Eur. J. Chem., 118, 395 (1981).
13) S. Itoh, M. Mure, Y. Ohshiro, and T. Agawa, Tetrahedron Lett., 26, 4255 (1985).
14) J. A. Duine, J. Frank Jzn, and J. K. Van Zeeland, FEBS Lett., 108, 443 (1979).
15) H. S. Isbell, H. L. Frush, C. W. R. Wade, and C. E. Hunter, Carbohydr. Res., 9, 163 (1969).
16) S. Shinkai, T. Kunitake, and T. C. Bruice, J. Am. Chem. Soc., 96, 7140 (1974); S. Shinkai, T. Yamashita, Y. Kusano, T. Ide, and O. Manabe, ibid., 102, 2335 (1980); S. Shinkai, T. Yamashita, Y. Kusano, and O. Manabe, J. Org. Chem., 45, 4974 (1980).
17) S. Itoh, M. Mure, and Y. Ohshiro, J. Chem. Soc. Chem. Commun., 1580 (1987).
18) T. Kido and K. Soda, Seikagaku, 57, 1065 (1985).
19) S. Patai, "The Chemistry of the Nitro and Nitroso Groups", John Wiley, New York, 1969.
20) D. Wege, Aust. J. Chem., 24, 1531 (1971).
21) S. Ito, K. Nii, M. Mure,and Y. Ohshiro, Tetrahedron Lett., 28, 3975 (1987).
22) S. Itoh, N. Kato, Y. Ohshiro, and T. Agawa, Chem. Lett., 135 (1985); S. Itoh, N. Kato, M. Mure, and Y. Ohshiro, Bull. Chem. Soc. Jpn., 60, 420 (1987).

STRUCTURAL PROPERTIES OF PQQ INVOLVED IN THE ACTIVITY OF QUINOHAEMOPROTEIN ALCOHOL DEHYDROGENASE FROM *PSEUDOMONAS TESTOSTERONI*

J.A. Jongejan, B.W. Groen and J.A. Duine
Department of Microbiology and Enzymology, Delft University of Technology
Julianalaan 67, 2628 BC, The Netherlands

Abstract: *Quinohaemoprotein alcohol dehydrogenase apoenzyme from Pseudomonas testosteroni was used in reconstitution experiments with PQQ and PQQ analogues to determine relevant factors contributing to binding and activity of PQQ. Preliminary results suggest the electrophilic properties of the C_5-carbonyl group to be particularly important. For N_1- and C_8-alkyl-substituted PQQ analogues a correlation of enzymatic activity and covalent hydration at C_5 can be accounted for in terms of different conformations resulting from rotation of the C_9-carboxyl group. Specific release of acetone from one enantiomer of racemic PQQ-acetone adduct upon incubation with apo-enzyme, suggests the presence of a nucleophilic group in the close vicinity of the PQQ-C_5-carbonyl site in the enzyme. A mechanism for the dehydrogenation of aliphatic alcohols by quinohaemoprotein alcohol dehydrogenase from Pseudomonas testosteroni is presented.*

INTRODUCTION

Following the detection and structure elucidation of tricarboxy-pyrroloquinoline quinone, PQQ, as the prosthetic group of methanol dehydrogenase from methylotrophic bacteria (Duine *et al.*, 1978, Westerling *et al.*, 1979, De Beer *et al.*, 1979, Salisbury *et al.*, 1979, Duine and Frank, 1980a, Duine *et al.*, 1980b), PQQ has been shown to function as a cofactor in a variety of bacterial dehydrogenases and oxidases (reviewed by Duine, Frank and Jongejan, 1985, 1986, 1987). Recently, the presence of PQQ has been established in several enzymes from higher organisms as well (reviewed by Duine and Jongejan, 1989a, 1989b).

From the structural and chemical properties of PQQ, some clues regarding its function in various enzymes can be readily deduced: *i*, the redox properties of PQQ are clearly related to its function as a cofactor in dehydrogenases; *ii*, facile reaction of $PQQH_2$ with molecular oxygen likewise explains its presence in oxidases; *iii*, complex formation of PQQ with metal ions (Jongejan *et al.*, 1987, Suzuki *et al.*, 1988) can be an important factor in metal containing quinoproteins, e.g lipoxygenase (van der Meer and Duine, 1988); *iv*, reactivity of PQQ with amino acids (Itoh *et al.*, 1984, van Kleef *et al.*, 1989) may well explain its presence in e.g. dopa decarboxylase (Groen *et al.*, 1988) and related enzymes.

J. A. Jongejan and J. A. Duine (eds.), PQQ and Quinoproteins, 205–216.

A basic feature of PQQ, most probably involved in the catalytic mechanism of the majority of quinoproteins that have been described up till now, is the ready formation of adducts with a variety of nucleophiles (Dekker *et al.*, 1982). Starting with the formation of an enzyme bound C_5-PQQ-substrate adduct subsequent reactions can, in a number of cases, be deduced from straightforward chemical principles. However, formation of a C_5-adduct does not, by itself, provide a clue as to how dehydrogenation of simple aliphatic alcohols takes place in the case of quinoprotein alcohol dehydrogenases. In fact, the inability to show reduction of free PQQ by aliphatic alcohols under a variety of experimental conditions, strongly emphasizes the importance of the protein environment in these enzymes. In order to gain insight into the interaction of cofactor and apoenzyme, we investigated binding and activity of PQQ and PQQ analogues with quinohaemoprotein alcohol dehydrogenase apoenzyme from *Pseudomonas testosteroni*.

MATERIALS AND METHODS

Quinohaemoprotein alcohol dehydrogenase apoenzyme from *Pseudomonas testosteroni* was prepared as described previously (Groen *et al.*, 1986). Binding of analogues was performed by incubating apoenzyme (200-400 pmol) with appropriate amounts of analogue in 0.1 M Tris/ HCl, pH 7.5, containing 2 mM $CaCl_2$, for 10 min. at room temperature.

To check the identity of the bound compound, the reconstitution mixture was applied to a PD-10 gelfiltration column to remove traces of non-bound material. Fractions containing holoenzyme were pooled and denatured by addition of SDS (0.2 %). The detached analogue was separated from the protein by gelfiltration (HPLC, 0.1 M sodiumphosphate, pH 6.5, containing 0.2 % SDS). Photo-diode array detection (Hewlett Packard HP-1040 A) allowed identification and quantification of the detached analogue by comparing its spectral- and chromatographic properties with those of authentic samples.

Activity measurements were performed by addition of 50 µl of 6 mM n-butanol to a mixture of 750 µl of 0.1 M Tris/ HCl, pH 7.5, containing 2 mM $CaCl_2$ and 1.5 mM $K_3Fe(CN)_6$, and 200 µl of reconstitution mixture in a 1 ml cuvette at 25 °C. Initial reaction rates were determined by measuring the decrease in absorbance at 420 nm using a molar absorption coefficient for ferricyanide of $1.02 \ mM^{-1} cm^{-1}$.

PQQ was prepared by the method of Corey (Corey and Tramontano, 1981). 3-Methyl- and 3-n-propyl-PQQ were obtained by a modification of this procedure (Jongejan *et al.*, unpublished). N_1-alkylated derivatives were prepared as described by Itoh (Itoh *et al.*, 1987). 8-Methyl-PQQ was synthesized by a modification of the procedure developed for the preparation of C_8-deuterated PQQ (Jongejan *et al.*, 1988). PQQH$_4$ was prepared as described previously (Duine *et al.*, 1979). 5-Hydroxy-pyrroloquinoline-tricarboxylic acid was obtained by dehydration of PQQH$_4$ (Frank, 1988b). 4-Hydroxy-pyrroloquinoline-tricarboxylic acid was a kind gift from Prof. Rees (MacKenzie, Moody and Rees, 1986), while pyrroloquinoline- tricarboxylic acid was kindly donated by Dr. de Vries (Hendricksen and de Vries, 1982). PQQ-acetone adduct was prepared by a modification of the method described by Salisbury (Salisbury *et al.*, 1979).

Purification of PQQ and analogues was performed on a C_{18} Radial PAK HPLC column using gradient elution (10 - 90 % MeOH adjusted to pH 2 with 85 % H_3PO_4).

RESULTS

Quinohaemoprotein alcohol dehydrogenase apoenzyme from *Pseudomonas testosteroni* appears to be particularly suited for reconstitution experiments. It can be readily prepared and purified, no back-ground activity is observed, several electron acceptors can be used and no activator (e.g. ammonia) is required. Moreover, it shows a high affinity for PQQ, while bound cofactor can eventually be removed by SDS extraction.

The presence of a c-type haem forms an intriguing aspect of the enzyme. From the observation that almost complete reduction of the haem takes place when PQQ is added to apoenzyme in the presence of (endogenous) substrate (Fig. 1), one may conclude that the haem is somehow involved in the electron transfer process. This might be either as the principal dehydrogenation site or as a secondary electron acceptor. In the former case, the function of PQQ could be restricted to that of a (powerful) modulator. It was thus highly relevant for the present investigation to ascertain the role of PQQ in reconstituted enzyme.

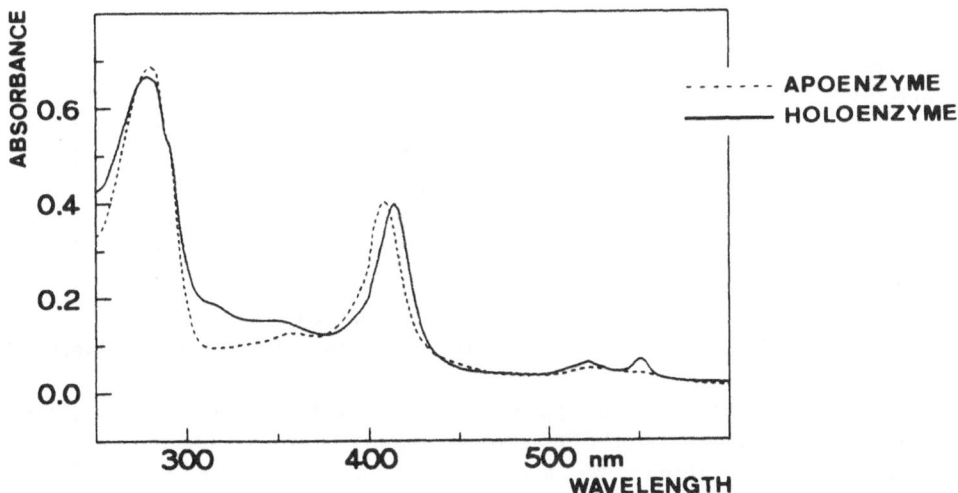

Figure 1. Absorption spectra of quinohaemoprotein alcohol dehydrogenase from *Pseudomonas testosteroni*

As it is generally accepted that the redox properties of PQQ reside in the ortho-quinone moiety, several analogues that are modified in this position were investigated for their ability to reconstitute holoenzyme activity. Initial results, obtained with redox-inactive PQQ analogues 2-5 (structures given in Table 1 A), were highly confusing. Compounds 3, 4 and 5, carrying oxygen containing substituents at either C_4 or C_5 showed activity to a greater or lesser extent.

Table 1 A.
Activity of PQQ and PQQ analogues

PQQ and PQQ-analogues	Activity		
	(%)*	(%)**	(%)***
1	100	----	----
2	0	100	89
3	0	100	37
4	0	N.D.	N.D.
5	0	100	18

* Activity after incubation of equimolar amounts of apoenzyme and analogue

** Activity after incubation of equimolar amounts of apoenzyme, analogue and PQQ

*** Activity after incubation of equimolar amounts of apoenzyme and analogue, followed by incubation for further 5 min. with one equivalent of PQQ

N.D., not determined
(Conditions as under Materials and Methods)

Upon further investigation it appeared that the activity measurements were biased by the fact that Wurster's Blue, the commonly employed electron acceptor, was able to oxidize analogues 3 - 5 to PQQ in a non-enzymatic reaction during (pre-)incubation. By performing initial rate measurements with ferricyanide as an electron acceptor, we were able to circumvent this problem. Special care was taken to ascertain whether analogues were converted to PQQ during the activity test by detaching the factor from the reconstituted enzyme and establishing the identity of the recovered compound.

As shown in Table 1 A, none of the analogues **2-5** is able to reconstitute holo-enzyme activity. Since all analogues bind strongly to the apo-enzyme (as judged from titration experiments), it must be concluded that PQQ functions as a redox-cofactor for this enzyme. It should be noted that binding of PQQ (and perhaps of analogues possessing the ortho-quinone structure), may involve complex kinetics, as suggested by different readings for simultaneous and sequential addition of equimolar quantities of analogue(s) and PQQ.

Table 1 B
Activity of PQQ analogues

PQQ-analogues	Activity (%)[*]	(%)[**]	(%)[***]
6	1 2	9 0	6 1
7	1	9 6	1 0 0
8	3	8 6	2 8
9	1	9 0	1 0 0
10	5	1 0 0	1 0 0
11	1 0 0	- - - -	- - - -

*,**,*** As described for Table 1 A

Initial experiments with N_1-methyl-PQQ, routinely prepared by the method of Itoh, indicated rather high activities (up to 80 % of the value observed for PQQ) of enzyme reconstituted in the presence of 100-fold excess of **8**. However, on closer examination it turned out that these preparations contained minor amounts of PQQ, originating from incomplete separation of methylated and unmethylated interme-diates during the 'flash chromatography' procedure (Clark Still *et al.*, 1978) used by the authors (Itoh *et al.*, 1987). By detaching and identifying the bound factor from reconstituted enzyme, we were able to prove that preferential binding of contaminating PQQ was responsible for the high activities resulting from these preparations.

Incubation of apoenzyme with extensively purified **8** resulted in the low value shown in Table 1 B. Upon addition of a large excess of this preparation no further raise in activity could be effected. In separate experiments we determined the activity when glucose dehydrogenase apoenzyme from *Escherichia coli* was reconstituted with either the original or the purified preparation of **8**. It was found that addition of the original preparation gave rise to high activities, as described by Shinagawa in a similar experiment (Shinagawa *et al.*, 1986). The purified preparation showed only minor activity (approx. 3–5 % of PQQ-reconstituted holoenzyme activity at saturating concentrations of **8**).

Incubation of alcohol dehydrogenase apoenzyme with alkyl-substituted PQQ derivatives **6–10** resulted in low activities of reconstituted enzymes (Table 1 B). Although both 3-methyl-PQQ (**6**) and 3-n-propyl-PQQ (**7**) appear to be completely bound, as judged from titration experiments (results not shown), both compounds are easily displaced by PQQ. N_1-methyl-PQQ (**8**) appears to be more strongly bound, while 8-methyl-PQQ (**10**) is again readily displaced by PQQ.
Spectral properties of **8** and **10** are shown in Fig. 2. A close resemblance is seen for the spectra of **8** and PQQ-hydrate on the one hand and **10** and unhydrated PQQ on the other hand.

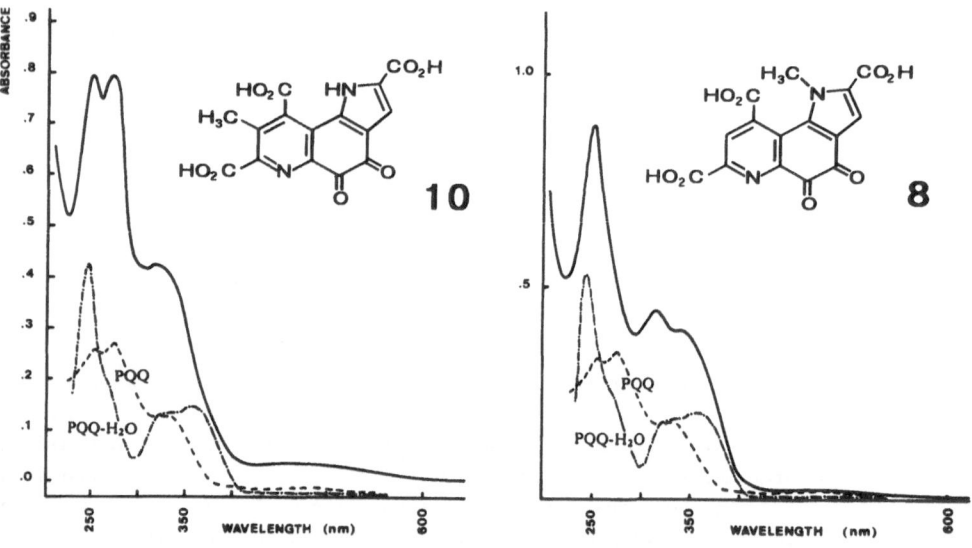

Figure 2. Spectral properties of C_8- and N_1-alkylated PQQ derivatives

Figure 3. Titration of apoenzyme (120 pmol) with (racemic) PQQ-acetone adduct

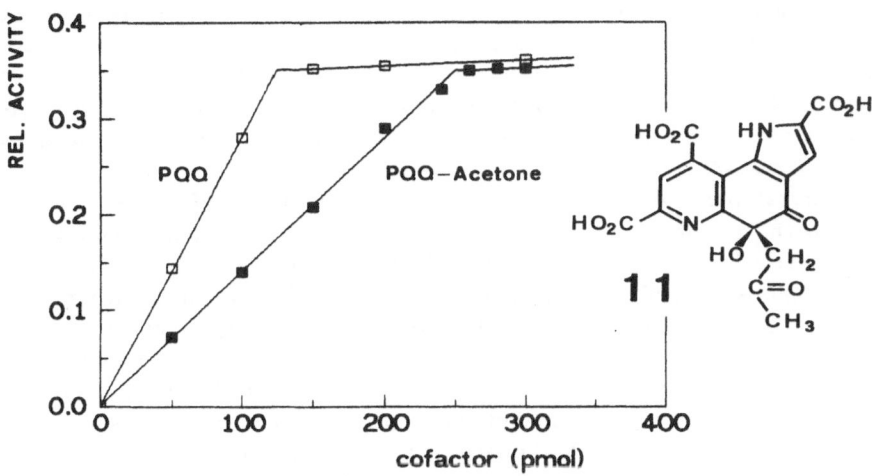

Upon titration of apoenzyme with PQQ-acetone adduct (11) a remarkable pheno-menon was observed (Fig. 3). Not only could a linear increase in activity be noted, it was also found that the maximal activity, corresponding to the activity that can be obtained with PQQ, was reached at 1 : 2 stoichiometry. As a possible explanation for this behaviour we suspected a highly selective affinity of the apo-enzyme for one of the enantiomers of the racemic adduct. The mixture obtained from incubation of apoenzyme with two equivalents of PQQ-acetone adduct was separated by gel-filtration. Low-molecular weight fractions were pooled and analyzed. UV/Vis-spectroscopy revealed the presence of one equivalent of PQQ-acetone adduct. Enzyme-containing fractions were treated with SDS, revealing the presence of the second equivalent as PQQ. Incubation of apoenzyme with the PQQ-acetone adduct preparation recovered from the former incubation mixture did not lead to detectable activity. Incubation of PQQ-acetone adduct in the absence of (apo-)enzyme did not lead to a significant conversion to PQQ.
Prolongued heating (80°C) of the adduct in solutions containing relatively high concentrations of phosphate buffer (pH 7) resulted in slow release of acetone.

To determine whether a nucleophilic group present in the active site might be responsible for the displacement of acetone from the bound adduct, incubations with iodoacetamide were performed. Preliminary results show that incubation of apoenzyme with (relatively high concentrations of) iodoacetamide completely inhibits the formation of catalytically active enzyme upon addition of PQQ. Treatment of holoenzyme under the same conditions does not lead to an appreciable loss of activity.

DISCUSSION

The role of PQQ in quinoprotein alcohol dehydrogenation can be addressed in several ways. As for the mode of hydrogen and electron abstraction, conflicting evidence with regard to possible one-electron versus two-electron transfer has been presented (Duine and Frank, 1981, Mincey et al., 1981). Reaction of cyclopropanol as a suicide substrate has been explained in terms of a radical recombination (Dijkstra et al., 1984), probably unrelated to the catalytic events during substrate dehydrogenation. Recently, evidence in favor of an anionic (two-electron transfer) reaction has been reported (Frank et al., 1989).

From a chemical point of view, addition of alcohol at C_5 of PQQ can hardly contribute to the required activation of hydrogen. It may be argued, though, that addition invokes a favorable geometry for subsequent internal hydrogen transfer in the adduct. Reaction of substrate alcohol at C_4 of PQQ, resulting in a vinylogous ester, has been proposed at one time (Duine et al., 1987) to draw attention to the possibility of a different type of PQQ-alcohol adduct in which a (para-)quinoid structure is still resident. Present results, showing a low but definite activity of N_1-alkylated PQQ derivatives, contradict such a proposal. Finally, from the fact that reduction of free PQQ by simple aliphatic alcohols has not been found to occur under various conditions, an important contribution of the (apo-)enzyme environment must be considered.

In view of these considerations, we argued that the reconstitution of quinoprotein alcohol dehydrogenase apoenzyme with suitably modified PQQ analogues, might be an appropriate approach to gain insight in the catalytic events during enzymatic alcohol dehydrogenation. Three reports concerning the reconstitution of apoquinoprotein glucose dehydrogenases with PQQ analogues have appeared in the literature. In an initial communication by Duine (Duine et al., 1979), absolute requirement for an intact ortho-quinone moiety was reported. The results of Conlin (Conlin et al., 1985) confirm a high selectivity of glucose dehydrogenase apoenzyme from Acinetobacter calcoaceticus for PQQ, albeit that a rather limited set of analogues was screened. Shinagawa (Shinagawa et al., 1986) investigated the importance of the 7- and 9-carboxylgroups of PQQ in glucose dehydrogenase catalysis by reconstituted apoenzyme from Escherichia coli. They concluded that the presence of a carboxylgroup at C_9 of PQQ is indispensible for proper functioning of reconstituted enzyme.

Inspection of the data presented in Table 1 B and Fig. 2 allows a more specific conclusion as far as the role of C_9-COOH is concerned. Apparently, the contribution to productive binding of factor and apoenzyme is not its only function. In addition, a major effect on the electrophilic character of C_5 seems obvious.

Molecular modelling shows that the bulky methyl group present at C_8 in **10** will force the carboxyl group to rotate away from coplanarity with the quinoline ring system (Fig. 4). Alkyl–substitution at N_1 in **8** and **9** can be accomodated in a different way. In these cases a relatively strain–less configuration can be realized by combined displacement of N_1–alkyl and covalent hydration of C_5–carbonyl: the over–all shape of the cyclohexadiene–(hydrated)–dione system assuming the form of a 'puckered' ring. Mesomeric effects of the C_9–carboxyl group will likewise be stronger in case a more planar orientation can be adopted, thus contributing to the electron deficiency at C_5 in **8**, favoring covalent hydration. Spectral properties of free and hydrated PQQ have been deduced from absorption and NMR spectroscopy (Dekker *et al.*, 1982). Recently, we were able to obtain a metastable solution of unhydrated PQQ in acetonitrile. The spectrum of this preparation was in complete agreement with that reported for unhydrated PQQ. Close correlation of the spectra of N_1–alkyl–PQQ and PQQ–hydrate on the one hand, and C_8–methyl–PQQ and unhydrated PQQ on the other hand suggests the respective chromophores to be highly similar. From these considerations, it seems likely that the relative displacement of N_1 and C_5 in **8** favors formation of C_5–hydrate, while rotation of C_9–COOH as in **10** leads to the opposite effect. As both **8**, **9** and **10** show low activities in the reconstitution test, it must be assumed that the electrophilic character of C_5 in PQQ is finely tuned to the catalytic need for addition and release of substrate and product.

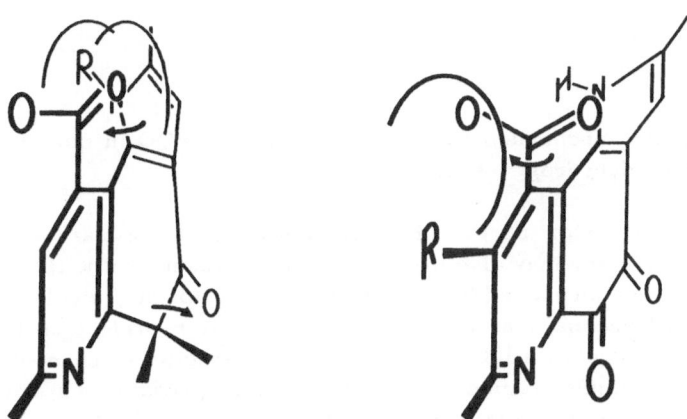

Fig. 4. Steric effects on rotation of C_9–COOH

Preferential binding of one enantiomer of PQQ–acetone adduct reflects the high selectivity of the apoenzyme cofactor site. Specific release of acetone from bound adduct suggests the presence of a nucleophilic group positioned in the vicinity of the addition site. Reactivity of apo–, but not of holoenzyme, with iodoacetamide strongly supports the involvement of such a group. In view of the relatively high concentrations of iodoacetamide that are required to inhibit the apoenzyme (approx. 1 mM), this group may well be an amine. From the evidence that is discussed above, it is clear that a working hypothesis of alcohol dehydrogenation catalyzed by this enzyme should credit the following observations:

i, optimal performance is strongly dependent on a proper fit of the cofactor in the apoenzyme active site *ii*, the electrophilic character of the C_5-carbonyl group must be finely tuned *iii*, functional presence of a nucleophilic residue positioned in the close vicinity of this group in reconstituted enzyme is required. A possible mechanism, accounting for these arguments is given in Scheme 1.

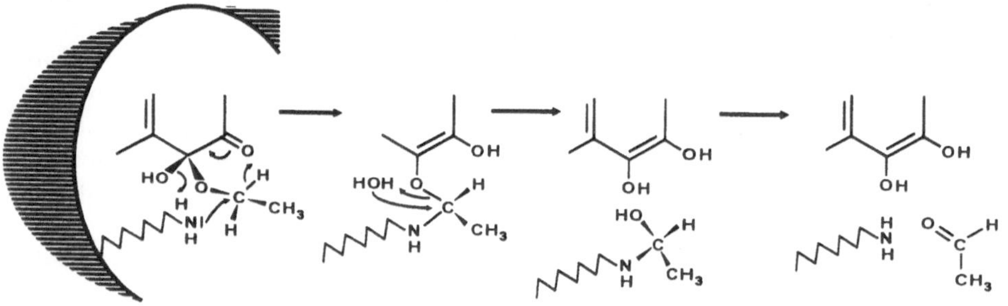

Scheme 1. Proposed mechanism for the involvement of PQQ in quinohaemoprotein alcohol dehydrogenation

As a logical extension of the ideas discussed above, primary association of apoenzyme and cofactor may also involve C_5-adduct formation with the presumed amino group. In such a case, addition of substrate alcohol may displace the amino-group, which will become free to promote subsequent hydride transfer from the α-position of the alcohol to C_4-carbonyl by a push-pull mechanism. From the alkylidene-bridged amine-(red)cofactor adduct resulting from this hydride shift, the aldehyde product can be released by hydrolysis. It should be noted that the release of aldehyde is pictured as a two-stage process, involving a Schiff-base intermediate. Hydrolysis of such an intermediate might well constitute a limiting step in the dehydrogenation reaction catalyzed by e.g. methanol dehydrogenase from methylotrophic bacteria. Stripping the aldehyde by additional factors (factor X ? (Dijkstra et al., 1988)) could form a physiological save-guard against formation of (highly toxic) free formaldehyde in these reactions. Ammonia as an artificial activator for this enzyme might serve the same purpose in vitro.

CONCLUSIONS

From the evidence presented above, it is clear that the presence of a redox-active ortho-quinone moiety is indispensible for activity of all analogues that have been investigated sofar. This allows us to conclude that PQQ formes the actual site of alcohol dehydrogenation in the haemoprotein alcohol dehydrogenase from *Pseudomonas testosteroni*. By virtue of the high affinity of apoenzyme for PQQ and PQQ analogues and the possibility to quantify the amount of bound factor by subsequent detachment, we were able to determine the catalytic activity of PQQ analogues under conditions where full binding was ensured. Exept for racemic PQQ-acetone adduct, in which case two equivalents were needed, full saturation of apoenzyme was reached by addition of a slight molar excess of the analogues mentioned in this report.

Low activities of PQQ derivatives carrying alkyl substituents at position 3, suggest that a proper fit of the cofactor is of utmost importance. For N_1- and C_8-substituted derivatives the same reasoning may apply, although, from the comparison of spectral data still another conclusion can be drawn. Thus it appears that compared to PQQ, the C_5-carbonyl group in N_1-alkyl-PQQ shows a more pronounced electrophilic character, while the opposite applies to C_8-methyl-PQQ. As both types of analogue show low but definite activities in the reconstitution experiments, the importance of a well-balanced electrophilic character of this carbonyl group is evident.

Displacement of acetone from one enantiomer of PQQ-acetone adduct upon binding to apoenzyme could result from the presence of a nucleophilic group. Inhibition of apoenzyme with iodoacetamide suggests this group to be an amine. A probable course of enzymatic alcohol dehydrogenation by this enzyme may involve formation of an intermediate alkylidene-bridged amine-cofactor adduct, from which aldehyde is released by subsequent hydrolysis. From the steric demands set by this mechanism a highly enantioselective conversion of alcohols containing a chiral α-carbon must be expected.

REFERENCES

de Beer, R., van Ormondt, D., van Ast, M.A., Banen, R., Duine, J.A. and Frank, J. (1979) *J. Chem. Phys. 70*, 4491-4495
Clark Still, W., Kahn, M. and Mitra, A. (1978) *J. Org. Chem.*, *43*, 2923-2925
Conlin, M., Forrest, H.S. and Bruice, T.C. (1985) *Biochem. Biophys. Res. Commun.*, *131*, 564-566
Corey, E.J. and Tramontano, A. (1981) *J. Am. Chem. Soc. 103*, 5599-5600
Dekker, R.H., Duine, J.A., Frank, J., Jzn. Verwiel, P.E.J. and Westerling, J. (1982) *Eur. J. Biochem. 125*, 69-73

Dijkstra, M., Frank, J., Jzn, Jongejan, J.A. and Duine, J.A. (1984) *Eur. J. Biochem.*, *140*, 369-373

Dijkstra, M., Frank, J., Jzn. and Duine, J.A. (1988) *FEBS Lett.*, *227*, 198-202

Duine, J.A., Frank, J. and Westerling, J. (1978) *Biochim. Biophys. Acta.* *524*, 277-287

Duine, J.A., Frank, J., Jzn. and van Zeeland, J.K. (1979) *FEBS Lett.* *108*, 443-446

Duine, J.A. and Frank, J., Jzn. (1980a) *Biochem. J. 187*, 221-226

Duine, J.A., Frank, J., Jzn. and Verwiel, P.E.J. (1980b) *Eur. J. Biochem. 108*, 187-192

Duine, J.A. and Frank, J., Jzn (1981) in *Microbial Growth on C_1 Compounds* (Dalton, H.. ed.) pp. 31-41, Heyden, London

Duine, J.A., Frank, J., Jzn. and Jongejan, J.A. in *Proceedings of the 16th FEBS Congress*, part A, Yu. Ovchinikov, Ed., VNU Science Press, Utrecht 1985 pp. 79-88

Duine, J.A., Frank, J., Jzn. and Jongejan, J.A. (1986) *FEMS Microbiol. Rev.*, *32*, 165-178

Duine, J.A., Frank, J., Jzn. and Jongejan, J.A. (1987) *Adv. Enzymol. 59*, 169-212

Duine, J.A. and Jongejan, J.A. (1989a) *Vitamins and Hormones.* In press

Duine, J.A. and Jongejan, J.A. (1989b) *Annu. Rev. Biochem.* In press

Frank, J., Jzn., Dijkstra, M., Duine, J.A. and Balny, C. (1988a) *Eur. J. Biochem.*, *174*, 331-338

Frank, J., Jzn, Thesis (1988b), Delft, Krips Repro Meppel, pp. 83-91

Frank, J., Jzn., Dijkstra, M., Balny, C., Verwiel, P.E.J. and Duine, J.A. (1989) *These Proceedings*

Groen, B.W., van Kleef, M.A.G. and Duine, J.A. (1986) *Biochem. J. 234*, 611-615

Groen, B.W., van der Meer, R.A. and Duine, J.A. (1988) *FEBS Lett.*, *237*, 98-102

Hendricksen, J.B. and de Vries, J.G. (1982) *J. Org. Chem.*, *47*, 1148-1150

Itoh, S., Kato. N., Ohshiro, Y. and Agawa, T. (1984) *Tetrahedron Lett. 25*, 4753-4756

Itoh, S., Kato, J., Inoue, I., Kitamura, Y., Komatsu, M. and Ohshiro, Y (1987) *Synthesis*, 1067-1071

Jongejan, J.A., van der Meer, R.A., van Zuylen, G.A. and Duine, J.A. (1987) *Recl. Trav. Chim. Pays-Bas, 106*, 365

Jongejan, J.A., Bezemer, R.P. and Duine, J.A. (1988) *Tetrahedron Lett.*, *29*, 3709-3712

van Kleef, M.A.G., Jongejan, J.A. and Duine, J.A. (1989) *These proceedings*

MacKenzie, A.R., Moody, C.J. and Rees, C.W. (1986) *Tetrahedron*, *42*, 3259-3268

van der Meer, R.A. and Duine, J.A. (1988) *FEBS Lett.*, *235*, 194-200

Mincey, T., Bell, J.A., Mildvan, A.S. and Abeles, R.H. (1981) *Biochemistry 20*, 7502-7509

Salisbury, S.A., Forrest, H.S., Cruse, W.B.T. and Kennard, O. (1979) *Nature (London) 280*, 843-844

Shinagawa, E., Matsushita, K., Nonobe, M., Adachi. O., Ameyama, M., Ohshiro, Y., Itoh, S. and Kitamura, Y. (1986) *Biochem. Biophys. Res. Commun.*, *139*, 1279-1284

Suzuki, S., Sauray, T., Itoh, S. and Ohshiro, Y. (1988) *Inorg. Chem. 27*, 591-592

Westerling, J., Frank, J. and Duine, J.A. (1979) *Biochem. Biophys. Res. Commun.*, *87*, 719-724

FACTORS RELEVANT IN THE REACTION OF PQQ WITH AMINO ACIDS.
Analytical and mechanistic implications.

Mario A.G. van Kleef, Jaap A. Jongejan, and Johannis A. Duine

Department of Microbiology and Enzymology, Delft University of Technology,
Julianalaan 67, 2628 BC Delft, The Netherlands

ABSTRACT

In order to reveal the stability of PQQ in complex samples, its reaction on incubation with amino acids was followed spectrophotometrically, by monitoring oxygen consumption, and with a biological assay. For several α-amino acids, the formation of a yellow coloured compound (λ_{max} = 420 nm) was accompanied by oxygen uptake and disappearance of biological activity from the reaction mixture. The yellow product appeared to be an oxazole of PQQ, the exact structure depending on the amino acid used. Besides the condensation reaction, there is also a catalytic cycle in which an aldimine adduct of PQQ and the amino acid is converted into the aminophenol form of the cofactor and an aldehyde resulting from oxidative decarboxylation of the amino acid. Addition of NH_4^+-salts, and sometimes as well as that of certain divalent cations greatly stimulated both reactions. With basic amino acids, oxazole formation scarcely occurred. However, as oxygen consumption was observed (provided that certain divalent cations were present), conversion of these compounds took place. Oxazole formation also occurred under anaerobic conditions with concomitant formation of $PQQH_2$, suggesting that PQQ is able to oxidize the presumed oxazoline to an oxazole. A reaction scheme is proposed for the aerobic oxidation of glycine with PQQ. Although acid hydrolysis of PQQ-oxazoles was feasible, hydrolysis in the presence of amino acids did not lead to PQQ, since tryptophan appeared to have a deleterious effect on the cofactor under the conditions of acid hydrolysis. The results here described explain therefore why analysis methods for free PQQ can fail in case the samples contain amino acids.

INTRODUCTION

Being a quinone, PQQ is very reactive towards nucleophiles. In view of the nucleophilic character and the omnipresence of amino acids in nature, problems in the detection and quantitation of PQQ in (complex) samples can be expected (van der Meer and Duine, these Proceedings). Therefore, in order to establish the presence of PQQ and PQQ-adducts, characterization of the adducts formed from PQQ with nucleophiles and determination of factors relevant for their formation are very important. As it is highly improbable that PQQ is available in vivo in its free form, it is also necessary to find out - especially if PQQ turns out to be a vitamin - in what form it is taken up, transported and converted by the cells into free PQQ. For the latter aspect, research on the conversion of condensation products into PQQ is also relevant.

Some information on the reaction of PQQ with amino acids has been provided recently (Itoh et al., 1984; Sleath et al., 1985; Fluckiger et al., 1988).

J. A. Jongejan and J. A. Duine (eds.), PQQ and Quinoproteins, 217–226.

218

Oshiro and coworkers (1984) have reported on the oxidative decarboxylation of α-amino acids with coenzyme PQQ [1]:

$$
\underset{\substack{| \\ NH_2}}{\overset{\substack{H \\ |}}{R-C-COOH}} \quad \xrightarrow{\quad PQQ \ / \ 1/2 \ O_2 \quad} \quad R-C\overset{\displaystyle O}{\underset{\displaystyle H}{\diagup}} \quad + \ CO_2 \ + \ NH_3 \qquad [1]
$$

Since this reaction comprises a catalytic cycle with PQQ as a catalyst (see below), it will be referred to as the cyclic pathway.

On the other hand, Bruice and coworkers reported the conversion (of a PQQ-analogue) to redox inactive oxazoles in this reaction [2]:

$$
\underset{\substack{| \\ NH_2}}{\overset{\substack{H \\ |}}{H-C-COOH}} \ + \ PQQ \quad \xrightarrow{\quad 1/2 \ O_2 \quad} \quad PQQ\text{-oxazole} \ + \ CO_2 \qquad [2]
$$

This reaction will be referred to as the linear pathway.

In view of these controversial findings, and the expectation that the outcome of the results might have significance for the aims indicated above, it was attempted to analyze the products formed from PQQ in the linear pathway, and to search for (optimal) conditions for either of the pathways. In order to shed light on the presence of PQQ adducts in nature, we developed methods to convert the presumed condensation products into PQQ (a detailed report will be presented elsewhere).

RESULTS AND DISCUSSION

Model reaction of PQQ with glycine

Immediately after addition of glycine to the mixture, an intensely yellow

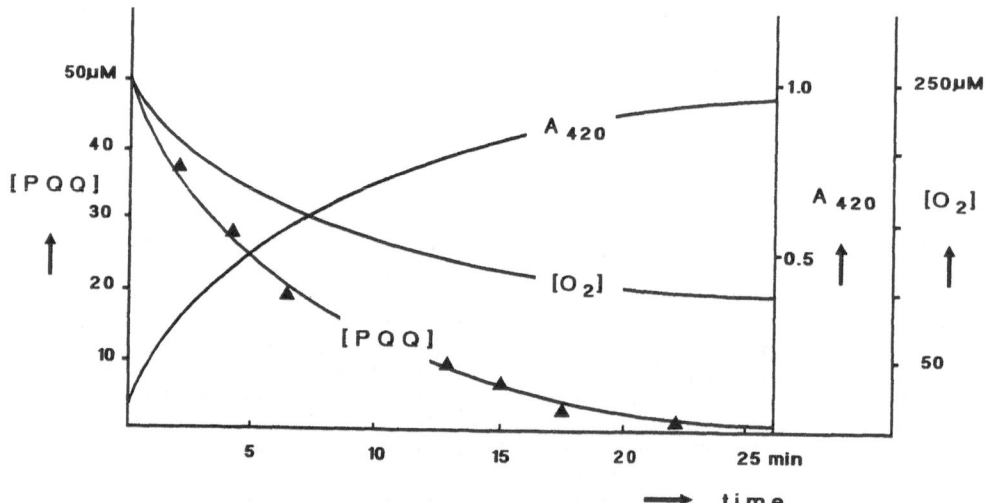

Fig. 1: Aerobic oxidation of glycine with PQQ. A_{420} is indicative for the amount of oxazole 1 formed.

coloured compound (λ_{max} = 420 nm) was formed (Fig. 1), the oxazole as product of the linear pathway (see below). Also oxygen consumption (as measured with a Clark electrode) and a decrease in PQQ level (as measured with an apo-quinoprotein, Groen et al., 1986) were observed. Since glycine was present in large excess, the reaction is pseudo-first-order in PQQ (Fig. 1). Only 25 μmoles of oxygen are necessary to convert 50 μmoles of PQQ into oxazole [2], so that it can also be deduced from Fig. 1 that the majority of oxygen is consumed via the cyclic pathway. Thus, in the case of glycine, a cyclic as well as a linear pathway for amino acid conversion exists.

Standard conditions for the reaction of PQQ with glycine were: 0.03 M Tris/HCl, pH 8.3; 0.3 M NH_4Cl; 0.05 M Glycine; 50 μM PQQ, and 20 μM Ca^{2+}.

The products of the linear pathway

Identification of the products of the linear pathway, formed with different amino acids, occurred with HPLC coupled to a photodiode array spectrophotometer (which revealed the retention times and absorption spectra of the eluted compounds). ^{1}H-NMR spectroscopy was used to obtain structural information. Three different types of oxazoles can be distinguished (Table 1, Fig. 2). ^{1}H-NMR analysis revealed that glycine, L-serine, and L-threonine

Table 1: Different types of oxazoles. Samples were taken from the standard mixture after incubation for 2 h and subjected to HPLC. Products were detected by monitoring the eluate by UV detection at 250 nm. Spectra were taken with the photodiode array detector at the apex of the peaks. Yields were calculated using a molar extinction coefficient at 420 nm of the glycine oxazole ($\epsilon = 15,000 \ M^{-1}.cm^{-1}$) as a standard. Structures of the oxazole types are presented in Fig. 2.

Amino acid	Retention time t_R (min)	Yield (%)	Oxazole type
Glycine	17.7	100	1
L-serine	17.6	75	1
L-threonine	17.7	86	1
L-tryptophan	17.5	70	1
L-tyrosine	18.3	9	1
L-alanine	17.5	5	1
	15.4	8	2
L-valine	18.0	6	2
L-leucine	17.7	10	2
L-isoleucine	18.7	10	2
L-phenylalanine	19.8	24	2
L-methionine	18.0	40	2
L-glutamate	15.8	13	2
L-glutamine	15.5	32	2
L-aspartate	15.9	22	2
L-asparagine	15.6	30	2
L-histidine	14.9	9	2
Glycine methyl ester	16.9	54	3
Glycine ethyl ester	17.7	56	3

220

all yielded an oxazole with the same structure, in which R =H (Fig. 2, oxazole 1). This product was also formed from PQQ and L-tryptophan or L-tyrosine. With most of the other amino acids, oxazoles were formed with the residue (R) being the R of the amino acid used (R-CH(NH$_2$)-COOH, Fig. 2, oxazole 2).

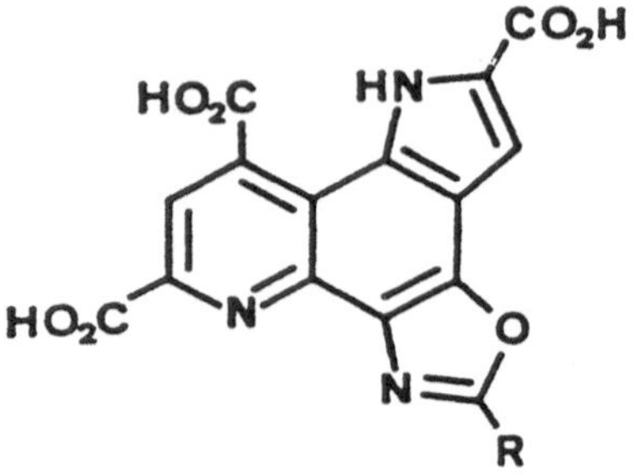

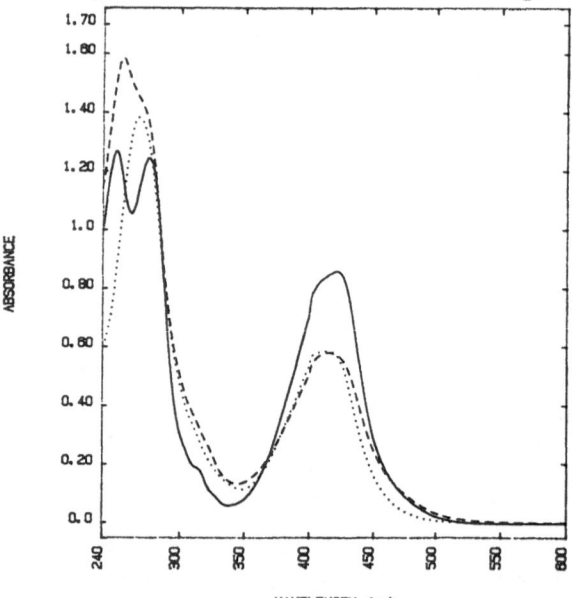

Fig. 2. Structure of PQQ-oxazoles: oxazole 1, R=H; oxazole 2, R = the R-group of the amino acid; oxazole 3, R = COOC$_n$H$_{2n+1}$.

^{1}H-NMR analysis of the product of glycine methyl ester revealed that the ester group is retained in the resulting oxazole. This compound is therefore indicated as oxazole 3 in which R = RCOOC$_n$H$_{2n+1}$ (Fig. 2). The absorption spectra of different types of oxazoles are shown in Fig. 3.

Fig. 3: Absorption spectra of different types of oxazoles. Spectra were taken after virtually complete conversion of PQQ into oxazole in the standard mixture, containing either glycine (——), L-methionine (– – –), or glycine methyl ester (·····).

Reaction rates of the cyclic and linear pathways with different amino acids and related compounds

Initial reaction rates of oxazole formation (linear pathway) and oxygen consumption (sum of both pathways), as well as the proportional contribution of the linear pathway are given in Table 2. In contrast with glycine, L-serine and L-threonine are predominantly converted via the linear pathway, and initial reaction rates are much lower. With the other α-amino acids, very low rates were found, and no reliable estimate for the contribution of either of the pathways was possible. The basic amino acids behaved as the other α-amino acids in the presence of Ca^{2+} ions, but in the presence of Cu^{2+} ions, high rates of oxygen consumption were found, while no oxazoles were formed. Under the latter condition, probably not the α-, but the ω-amino group is oxidized, just as in the case of primary amines. The presence of the α-amino group however seems to be essential, since ω—amino carboxylic acids were not oxidized at all by PQQ (Table 2). Glycine esters showed low oxygen consumption rates, but a high rate of oxazole formation.

Table 2: Oxygen consumption and oxazole formation rates for several amino acids and related compounds.

Amino acid	Stimulation by		O_2 consumption rate v_{init} (nmol.1^{-1}.s^{-1})	Rate of oxazole formation v_{init} (nmol.1^{-1}.s^{-1})	Linear pathway (%)
	NH_4^+	cations			
Glycine	+	Ca^{2+}	245	130	15
L-serine	+	Ca^{2+}	16	26	67
L-threonine	+	Ca^{2+}	15	28	75
L-alanine	+	Ca^{2+}	6	1.1	-
L-leucine	+	Ca^{2+}	8	2.5	-
L-lysine	-	Ca^{2+}	12	-	-
		Cu^{2+}	152	0	0
L-ornithine	-	Ca^{2+}	23	-	-
		Cu^{2+}	245	0	0
δ-aminovalerate			0	0	
ϵ-aminocaproate			0	0	
Glycine methyl ester	+	-	21	95	75

The stimulatory effects of NH_4^+ and divalent cations are also shown in Table 2. NH_4Cl (0.3 M) stimulated the oxygen consumption and oxazole formation about 30-fold. Only with basic amino acids in the presence of Cu^{2+} ions, NH_4^+ salts did not stimulate, which is an additional indication for a different mechanism in the case of this type of amino acid.

As we found that the oxidation of amino acids was inhibited completely on addition of metal chelating agents (e.g. EDTA), various divalent cations were tested. In the case of α-amino acids, Cu^{2+} was able to replace Ca^{2+} effectively. No metal ion requirement was found with the glycine esters. The primary action of the metal ions therefore seems to be related to a depolarizing effect on the charged carboxyl group. The results with the basic amino acids in the presence of Cu^{2+} ions strongly favour the idea that metal ion chelates also play an important role in this reaction. It should be mentioned that metal ions are supposed to catalyze the in vitro reaction of amino acids with pyridoxal phosphate according to similar mechanisms (Greenstein and Winitz, 1961).

Reaction under anaerobic conditions

As both the linear as well as the cyclic pathway proceed with oxygen consumption, the operation of the pathways under anaerobic conditions was questionable. However, when PQQ was incubated with glycine under anaerobic (standard) conditions, $PQQH_2$ and oxazole 1 were found in high yield in a 1:1 ratio, indicating that the linear pathway is operating. An explanation for this finding is given in the next paragraph.

Reaction scheme for the aerobic oxidation of glycine with PQQ

From the aforementioned results, it follows that the oxidation of amino acids by PQQ comprises both a cyclic pathway, apparent in the catalytic degradation of amino acids, and a linear pathway, resulting in the formation of stable PQQ-amino acid condensation products, mainly the oxazole. The proposed intermediates of the two pathways are shown in Fig. 4.

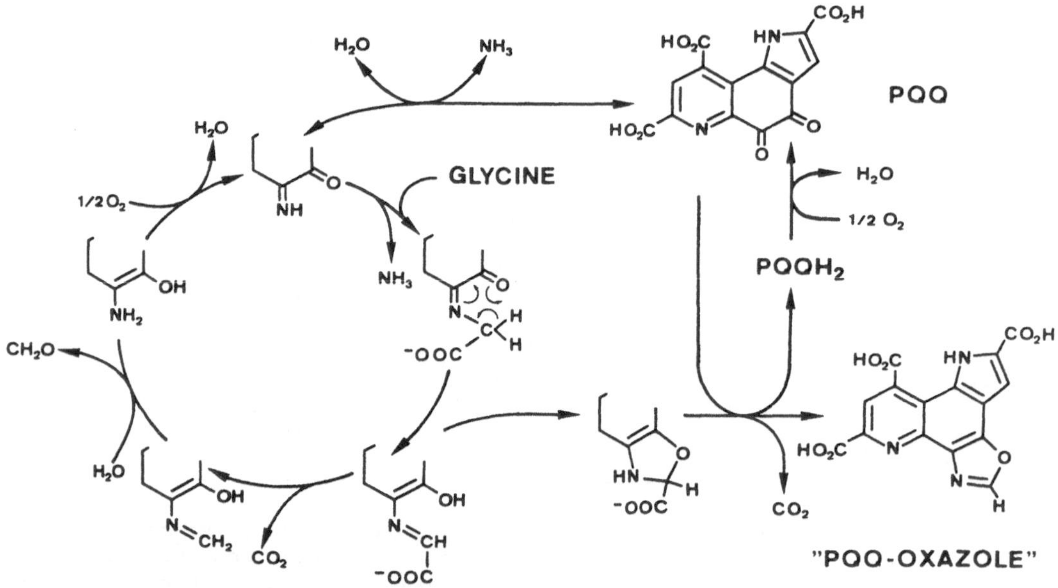

Fig. 4: Reaction scheme for PQQ and glycine in the cyclic and linear pathway.

The final step in the linear pathway is the conversion of PQQ-oxazoline into PQQ-oxazole. It has been shown that oxazole formation from quinones and quinone imines proceeds via an oxazoline (Cornforth, 1957). Another molecule of the quinone is assumed to function as oxidant in the final dehydrogenation step (Cornforth, 1957). The formation of high amounts of PQQH$_2$ and oxazole 1 in the anaerobic reaction of glycine with PQQ (in a 1:1 ratio) can aptly be explained by this mechanism: one molecule of PQQ is converted into oxazole 1, and a second one is reduced to PQQH$_2$ in the dehydrogenation step.

The scheme in Fig. 4 also takes into account the high stimulation observed by NH$_4^+$ salts. Since the 5-imino form of PQQ is formed on addition of ammonia to PQQ (Dekker et al., 1982), this stimulation can be explained on the basis of the strong electrophilic character of the C$_5$-imino group. Especially the addition of the amino group of α-amino acids will take place more readily on C$_5$-imino PQQ than on the parent C$_5$-carbonyl PQQ. This is in accordance with the observation that the oxidation of basic amino acids by PQQ is not dependent on ammonia.

The quinone-imine formed after addition of the amino acid to C$_5$-imino PQQ is proposed to tautomerize to the Schiff base (Fig. 4) (evidence for this type of rearrangement has been obtained recently by Klein et al. (1988) for the reaction of o-quinones with sec-alkyl primary amines). From this point on the two pathways diverge. In the cyclic pathway, decarboxylation is followed by carbon-nitrogen double bond hydrolysis of the aldimine formed. In the case of glycine, PQQ-aminophenol and formaldehyde are formed. In the linear pathway, ring closure of the Schiff base leads to PQQ-oxazoline which is dehydrogenated by PQQ to oxazole, as already discussed. Finally, under aerobic conditions, the reduced forms of PQQ from the linear and cyclic pathways (PQQH$_2$ and PQQ-aminophenol, respectively) are regenerated by oxygen.

Not only the course of the reaction but also the type of product is governed by the amino acid used. Thus, with the glycine esters the ester group is retained in the oxazole (type 3). This indicates that decarboxylation is not a prerequisite for operation of the linear pathway. It also suggests that the Schiff base might be the branching point, the properties of the constituing amino acid determining whether CO$_2$ is split off (leading to the cyclic pathway) and/or cyclisation occurs (the linear pathway). If that view is correct, another possibility for decarboxylation should exist in the linear pathway leading to oxazole type 1. For the moment, the most likely step where this occurs is in the conversion of the oxazoline into the oxazole.

Acid hydrolysis of PQQ-oxazoles

When oxazoles 1 or 2 were hydrolyzed in 2 M HCl at 100°C for 2 h, PQQ was routinely recovered from the mixture in a 70-100% yield, even when the concentration of oxazole in the hydrolysis mixture was as low as 0.5 μM. When PQQ (5-50 μM) was converted completely to oxazoles in nutrient broth (10 g/l) under standard conditions, however, no PQQ was detected after subsequent hydrolysis. With a synthetic mixture of amino acids, some PQQ could be detected (10-50% recovery depending on the PQQ concentration). Since oxazoles were obviously formed and degraded, but recovery of PQQ was variable, the stability of PQQ itself under hydrolysis conditions was checked (Table 3). These experiments showed that severe losses occurred in the presence of amino acids in 2 M HCl. Testing the amino acids separately,

224

L-tryptophan was found to be the only causative factor. No PQQ could be detected after incubation of a 1 μM PQQ solution with 0.1 % L-tryptophan for 2 h at $100°C$ in either 2 M or 6 M HCl. With the other amino acids no losses of PQQ were observed.

Table 3: Stability of PQQ under strongly acidic conditions in the presence of amino acids.

Amino acid	[PQQ] (μM)	remaining activity (%)
---	1-50	100
Glycine (0.2%)	1-50	100
Amino acid mixture	5.0	79
"	2.0	70
"	1.0	45
"	0.5	29
Tryptophan (0.1%)	1.0	0

On reaction in 2 M HCl, only one product was formed from PQQ and L-tryptophan, as revealed by HPLC analysis (in 6 M HCl, several products were observed). ^{1}H-NMR spectroscopy of the product revealed that the aromatic protons of PQQ, as well as the protons of the indole ring of L-tryptophan are retained. Therefore, a tentative structure is proposed as depicted in Fig. 5.

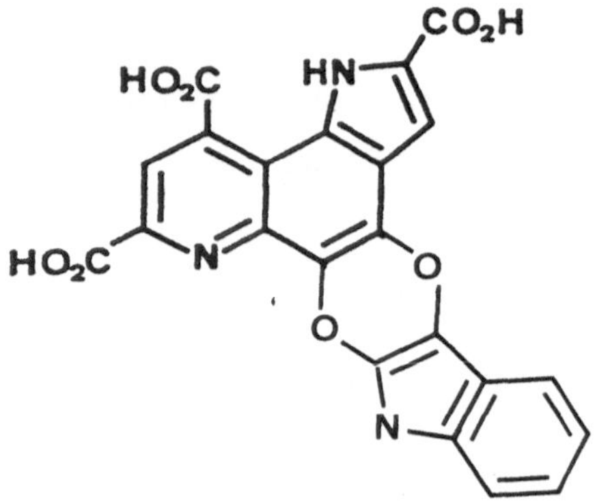

Fig. 5 : Proposed structure of the PQQ-tryptophan product formed in 2 M HCl.

CONCLUSIONS

1. Depending on the amino acid used:
   - conversion occurs via a linear and/or cyclic pathway.
   - different PQQ-oxazoles are formed.
   - ammonium salts and divalent cations have a significant effect on the reaction rate.

2. Under anaerobic conditions, PQQ functions as oxidant in the final dehydrogenation step of PQQ-oxazoline to PQQ-oxazole.

3. It is not possible to quantify or to detect PQQ in a protein hydrolysate, or other complex samples containing amino acids, because:
   - at physiological pH (proteolysis), biologically inactive oxazoles are formed.
   - in 6 M HCl (acid hydrolysis), PQQ reacts with L-tryptophan to biologically inactive, and as yet unidentified compounds.
4. Further studies on the catalytic role of metal ions (and NH_4^+ salts) could be worthwhile for understanding the mechanism of metallo-quinoprotein enzymes.

ACKNOWLEDGMENTS

We thank Ms. N.J. Viveen for technical assistance.

REFERENCES

Cornforth JW, 1957. In: Heterocyclic Compounds, (Elderfield, R.C., ed.), vol. 5, pp 418-450, Wiley, New York.

Dekker RH, Duine JA, Frank Jzn.J, Verwiel PEW and Westerling J, 1982. covalent addition of H_2O, enzyme substrates and activators to pyrroloquinoline quinone, the coenzyme of quinoproteins. Eur.J.Biochem. 125: 69-73.

Duine JA, Frank J and Verwiel PEJ, 1980. Eur.J.Biochem. 108, 187-192.

Duine JA, Frank Jzn.J and Jongejan JA, 1986. Enzymology of quinoproteins. Adv. Enzymol. 59: 169-212.

Duine JA, Frank Jzn.J and Jongejan JA, 1986. PQQ and quinoprotein enzymes in microbial oxidations. FEMS Microbiol.Rev. 32: 165-178.

Flückiger R, Woodtli T and Gallop PM, 1988. The interaction of amino groups with pyrroloquinoline quinone as detected by the reduction of nitroblue tetrazolium. Biochem. and Biophys.Res.Commun. 153: 353-358.

Greenstein JP and Winitz M, 1961. Chemistry of the amino acids, Vol I, pp 601-604, Wiley, New York.

Groen BW, Kleef MAG van, and Duine JA, 1986. Quinohaemoprotein alcohol dehydrogenase from Pseudomonas testosteroni. Biochem.J. 234: 611-615.

Itoh S, Kato N, Oshiro Y and Agawa T, 1984. Oxidative decarboxylation of α-amino acids with coenzyme PQQ. Tetrahedron Lett. 25: 4753-4756.

Itoh S, Katamura Y, Oshiro Y and Agawa T, 1986. Kinetics and mechanism of the oxidative deamination of amines by coenzyme PQQ. Bull.Chem.Soc.Jpn. 59: 1907-1910.

Itoh S, Kato N, Mure M and Oshiro Y, 1987. Kinetic studies on the oxidation of thiols by coenzyme PQQ. Bull.Chem.Soc.Jpn. 60: 420-422.

Klein RFX, Bargas LM, Horak V and Navarro M, 1988. Spontaneous rearrangement in Corey's reaction. Tetrahedron Lett. 29: 851-852.

Lobenstein-Verbeek CL, Jongejan JA, Frank J. and Duine JA, 1984. Bovine serum amine oxidase: a mammalian enzyme having covalently-bound PQQ as prosthetic group. FEBS Lett. 170: 305-309.

Meer RA van der, Jongejan JA and Duine JA, 1987. Phenylhydrazine as probe for cofactor identification in amine oxidoreductases. FEBS Lett. 221: 299-304.

Oshiro Y, Itoh S, Kurokawa K, Kato J, Hirao T and Agawa T, 1983. Micelle enhanced oxidation of amines by coenzyme PQQ. Tetrahedron Lett. 24: 3465-3468.

Sleath PR, Noar JB, Eberlein GA and Bruice TC, 1985. Synthesis of 7,9-didecarboxymethoxatin and comparison of its chemical properties with those of methoxatin and analogous o-quinones. Model studies directed toward the action of PQQ requiring bacterial oxidoreductases and mammalian plasma amine oxidase. J.Am.Chem.Soc. 107: 3328-3338.

INHIBITION OF COLLAGEN HYDROXYLASES BY PQQ REVEALS ITS DOMAIN STRUCTURE

V. Günzler[#], H.M. Hanauske-Abel[$], J.A. Duine[&], K.I. Kivirikko[#], E.J. Corey[*]

[#]Department of Medical Biochemistry, University of Oulu, Finland; [$]The Children's Hospital, Harvard Medical School, and [*]Department of Chemistry, Harvard University, Boston, USA, and [&]Department of Microbiology and Enzymology, Technical University Delft, The Netherlands

INTRODUCTION

Prolyl 4-hydroxylase (Procollagen-L-proline, 2-oxoglutarate: oxygen oxidoreductase (4-hydroxylating), EC 1.14.11.2.) catalyzes the hydroxylation of peptide-bound proline residues, exclusively at the N-terminus of glycine (1-3). The first step of the enzymatic reaction, the oxidative decarboxylation of 2-oxoglutarate by one atom of a dioxygen molecule, has been suggested to proceed as a ligand reaction in the coordination sphere of enzyme-bound Fe^{2+} (4), generating succinate, carbon dioxide, and a highly reactive iron-oxygen atom complex. This ferryl ion subsequently hydroxylates an appropriate proline residue, probably by an abstraction-recombination mechanism (4). The oxidative decarboxylation of 2-oxoglutarate can proceed without subsequent hydroxylation in so-called uncoupled reaction cycles

227

J. A. Jongejan and J. A. Duine (eds.), PQQ and Quinoproteins, 227–232.
© 1989 by Kluwer Academic Publishers.

(5-7), in which ascorbate is utilized as a specific alternative oxygen acceptor (8,9). In the absence of this reducing agent, prolyl 4-hydroxylase is rapidly inactivated by self-oxidation (10,11).

The thermal stability of the triple helix of collagenous proteins is crucially dependent upon the intramolecular hydrogen bonds involving the 4-hydroxyproline residues synthesized by prolyl 4-hydroxylase, making this enyzme a possible target for therapeutic antifibrotic agents (4). Competitive antagonists of 2-oxoglutarate (12) and ascorbate (13) have been shown to inhibit prolyl 4-hydroxylase in vitro. It was recently pointed out that the most potent of these inhibitors represent fragments of the PQQ molecule, and it was predicted that PQQ should inhibit prolyl 4-hydroxylase activity (14; Fig. 1). We now report the effects of PQQ and a structural analogue on the activity of prolyl 4-hydroxylase and the closely related enzyme lysyl hydroxylase in vitro.

METHODS

Chick embryo prolyl 4-hydroxylase and lysyl hydroxylase were purified and their activity was tested according to published protocols (15,16).

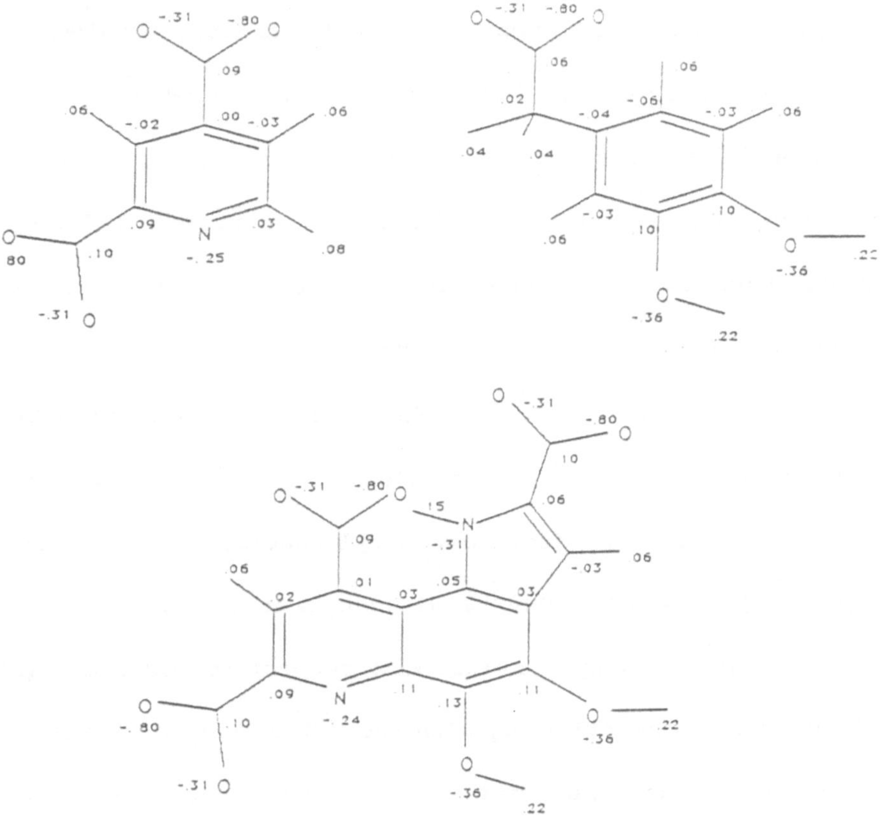

Fig. 1. Structural similarity between PQQH$_2$, pyridine-2,4-dicarboxylate, and 3,4-dihydroxyphenyl acetate (modified from ref. 14). Charge distributions were calculated using the method of Gasteiger and Marsili (18).

RESULTS AND DISCUSSION

PQQ was found to inhibit prolyl 4-hydroxylase (app. K_i 25 μM) and lysyl hydroxylase (app. K_i 50 μM). Kinetic studies with prolyl hydroxylase revealed that at low metal concentrations, PQQ was competitive with respect to divalent iron (K_i 15 μM), but showed nonlinear inhibition at higher concentrations. Identical findings have been reported for pyridine-2,6-dicarboxylate, a terdentate ligand able to form high spin 2:1 complexes with Fe^{2+} in solution (17). At 50 μM Fe^{2+}, PQQ showed competitive inhibition with respect to ascorbate (K_i 11 μM), a finding attributed to the 3,4-dihydroxyphenyl acetate - like domain of PQQ (Fig. 1). 3,4-Dihydroxyphenyl acetate is a known competitive inhibitor of prolyl 4-hydroxylase with respect to ascorbate (13). PQQ behaved non-competitively with respect to 2-oxoglutarate and the enzyme's peptide substrate. 5-Methoxy, 4-deoxy - PQQ which lacks the iron chelating- and ascorbate-like domains, but shows the 2-oxoglutarate like domain, was not inhibitory.

These studies indicate that PQQ can act as a terdentate chelator in a biological sysytem. Moreover, PQQ fragments may function as modulators of connective tissue formation in vivo. The study also points to the significance of PQQ fragments and

PQQ analogues as tools for the elucidation of structure-activity relationships and biological functions.

REFERENCES

(1) Prockop, D.J., Berg, R.A., Kivirikko, K.I., and Uitto, J. (1976) in Biochemistry of Collagen (Ramachandran, G.N. and Reddi, A.H. eds) pp. 163-273, Plenum Press, New York

(2) Kivirikko, K.I. and Myllylä, R. (1980) in The Enzymology of Post-translational Modification of Proteins (Freedman, R.B. and Hawkins, H.C. eds) pp. 53-104

(3) Kivirikko, K.I. and Myllylä, R. (1986) Ann. N. Y. Acad. Sci. 460, 187-201

(4) Hanauske-Abel, H.M., and Günzler, V. (1982) J. Theor. Biol. 94, 421-455

(5) Tuderman, L., Myllylä, R. and Kivirikko, K.I. (1977) Eur. J. Biochem. 80, 341-348

(6) Rao, N.V. and Adams, E. (1978) J. Biol. Chem. 253, 6327-6330

(7) Counts, D.F., Cardinale, G.J., and Udenfriend, S. (1980) Proc. Natl. Acad. Sci. USA 75, 2145-2149

(8) Myllylä, R., Majamaa, K., Günzler, V., Hanauske-Abel, H. and Kivirikko, K.I. (1984) J. Biol. Chem. 259, 5403-5405

(9) DeJong, L., and Kemp, A. (1984) Biochim. Biophys. Acta 787, 105-111

(10) Myllylä, R., Kuutti-Savolainen, E.-R., and Kivirikko, K.I. (1978) Biochem. Biophys. Res. Commun. 83, 441-448

(11) DeJong, L., Albracht, S.P.L., and Kemp, A. (1982) Biochim. Biophys. Acta 704, 326-332

(12) Majamaa, K., Hanauske-Abel, H.M., Günzler, V., and Kivirikko, K.I. (1984) Eur. J. Biochem. 138, 239-245

(13) Majamaa, K., Günzler, V., Hanauske-Abel, H.M., Myllylä, R. and Kivirikko, K.I. (1986) J. Biol. Chem. 261, 7819-7823

(14) Hanauske-Abel, H.M., Tschank, G., Günzler, V., Baader, E. and Gallop, P. (1987) FEBS Lett. 214, 236-243

(15) Berg, R.A., and Prockop, D.J. (1973) J. Biol. Chem. 248, 1175-1182

(16) Turpeenniemi-Hujanen, T.M., Puistola, U., and Kivirikko, K.I. (1980) Biochem. J. 189, 247-253

(17) Günzler, V., Majamaa, K., Hanauske-Abel, H.M., and Kivirikko, K.I. (1986) Biochim. Biophys. Acta 873, 38-44

(18) Gasteiger, J., and Marsili, M. (1980) Tetrahedron 36, 3219-3228

REACTION OF 2,7,9-TRICARBOXY-PQQ WITH NUCLEOPHILES

William S. McIntire, Molecular Biology Division, Veterans Administration Medical Center San Francisco, CA 94121, and Department of Biochemistry and Biophysics, University of California, San Francisco, CA 44143, U.S.A.

Key words: PQQ, 2,7,9-tricarboxy-PQQ, PQQ reaction with nucleophiles, PQQ redox potentials, pH titrations of PQQ, 5-imino-PQQ, PQQ-sulfite adduct.

Abstract: The K_d values for the SO_3^{-2}-complex with 2,7,9-tricarboxy-PQQ were determined from pH 5.5 to 10.5. It was found that SO_3^{-2} and not HSO_3^- is the nucleophile in this reaction. The midpoint potentials (E_o') were also measured in this pH range for this quinone. A plot of E_o' vs. pH produced a $pK_a = 8.44$ for the dihydroquinone. As expected, a plot of E_o' vs. the K_d for the PQQ-sulfite complex was linear. When NH_4^+ was included, the sulfite adduct rapidly formed, however, a slow appearance of an unidentified species was evident. E_o' for tricarboxy-PQQ + NH_4^+ were measured. pH titrations were carried out for oxidized tricarboxy-PQQ ± NH_4^+ and for the dihydroquinone form of this cofactor. The resulting pK_a values are compared to corresponding literature values for PQQ analogs.

INTRODUCTION

As an undeniably important enzyme cofactor, the chemical, spectra, and redox properties of protein-free 2,7,9-tricarboxy-PQQ are of great interest. In order to gain information concerning these properties, studies of the reaction of this quinone with the nucleophile SO_3^{-2} were initiated, followed by the determination of its E_o' values as a function of pH. Many quinoproteins oxidize amines, e.g., methylamine dehydrogenase (MADH) and the copper amine oxidases, or can be "activated" by NH_4^+, e.g., methanol dehydrogenase and MADH; thus, these studies with tricarbox-PQQ were repeated in the presence of NH_4^+. To complement this work, spectrophotometric pH titrations for derivatives of this quinone are being carried out, and the preliminary results of these experiments are presented.

MATERIALS AND METHODS

2,7,9-Tricarboxy-PQQ was from Fluka Chemical Corp. All other chemicals were reagent grade. All experiments were done at 25 °C. Fresh solutions of Na_2SO_3 were made before each experiment. The anaerobic pH titration of reduced PQQ was done under Ar in a cuvette fitted with a gas tight syringe containing 4 M KOH as titrant, and a glass Ag/AgCl microprobe combination electrode, which was submerged in the solution. Glucose, and catalytic amounts of catalase and glucose oxidase were in the solution to scavenge oxygen. UV-visible spectra were obtained with a Hewlett-Packard 8451A diode array spectrophotometer. Time courses were monitored with a Kontron 810 spectrophotometer. Values for E_o' were obtained with 2 mM anaerobic solutions of the quinone in 0.25 M NaOAc, pH 5.5, 0.25 M NaP_i, pH 5.5 to 8.5, and 0.25 M $NaPP_i$, pH 8.5 to 10.5 buffers, which included 1 M KCl or 1 M NH_4Cl, using a Bioanalytical Systems CV-27 cyclic voltammograph. The working, auxiliary, and reference electrodes were glassy carbon, platinum, and Ag/AgCl, respectively. The scan rates were 5 or 50 mV/sec. At the 5 mV/sec rate the $E_{Pa} - E_{Pc} \leqslant 60$ mV, and it is assumed that the reduction is a 2 e^- process (1-3). No changes occurred through multiple cycles at all the pH values, insuring that chemical processes, other than reversible

233

electron transfer, were not taking place. All potentials are relative to the normal hydrogen electrode. K_d and pK_a values were determined from the computer fit of the data by non-linear regression analysis.

RESULTS

Reaction of 2,7,9-tricarboxy-PQQ with sulfite. Forty μM solutions of the quinone were titrated with 10 or 100 mM Na_2SO_3 from pH 4.5 to 10.5 (0.3 M Na acetate, pH 4.5 and 5.5; 0.3 M NaP_i, pH 5.5 to 8.5; 0.2 M $NaPP_i$, pH 8.5 to 10.5). Ten to 15 min were required to reach endpoints after each addition of sulfite. Spectra were recorded for each step and the absorbances at 284 nm were noted. The K_d values for the sulfite adduct were determined from the equation $\Delta A = \Delta A_{max}[S]/(K_d + [S])$, where $[S] = [HSO_3^{-1}] + [SO_3^{-2}]$. A plot of the log K_d values vs. pH produced a curved line. When the K_d values are corrected so that sulfite is the only reactant, a linear plot of log K_d vs. pH results. This proves that SO_3^{-2} is the major nucleophile in the reaction, as it is with flavin analogs (4).

When the titration was attempted at pH 7.5 in the presence of 4 M NH_4Cl, the fast formation of the sulfite adduct was followed by a slower spectral change not observed in its absences. If 4 M KCl is substituted for NH_4Cl, only the "stable" sulfite adduct formed, indicating that the slow reaction is not ionic strength related. The time course for the slow reaction at pH 9.5 yielded a rate constant of 0.05 min^{-1} with 20, 40, or 80 mM sulfite. The slowly formed species is as yet unidentified. Whether NH_4^+ was include or not, the spectra always reverted back to that of un-adulterated oxidized PQQ on overnight incubation, due to decomposition of the sulfite. The spectrum of tricarboxy-PQQ is changed by addition of 4 M NH_4^+ at pH 7.0. This change may be partly due to an ionic strength effect, since 4M KCl also alters the spectrum somewhat.

Redox chemistry of 2,7,9-tricarboxy-PQQ. The plot of the E_o' vs. pH gave slopes of 60 mV/pH below and 30 mV/pH above pH 8.44. This indicates that the pK_a for the dihydroquinone is 8.44, a value lower than the ionization of the phenolic OH group of the 2,7-dicarboxy-quinol (pK_a = 9.31; ref. 5). As expected, the plot of E_o' vs. the log K_d for the sulfite complex is linear. Below pH 7.0, the plot of E_o' vs. pH for the tricarboxy-PQQ in the presence of NH_4^+ is essentially the same as in its absent. Above pH 7.0, the plot is significantly different (slope = 37 mV/pH), and difficult to interpret due to formation of 5-imino-PQQ on reaction with NH_3, and the pH transitions for the oxidized and reduced forms of the quinone. At pH values > 8, the E_o' are ~ 30 mV higher when NH_4^+ is involved.

The only unusual observation in these studies occurred for the cyclic voltammetry measurements with PQQ at pH 9.5 and 10.5. At the higher scan rate, only a single cathode and single anode current wave were produced, whereas at 5 mV/sec a second small cathode wave was seen at potentials 58 mV (pH 9.5) and 80 mV (pH 10.55) higher then the larger cathode wave. This pattern was unchanged through several cycles, and when again scanned at 50 mV/sec, the "normal" voltammogram was seen. At present this phenomenon is unexplained. This type voltammogram was not seen when NH_4^+ was included.

pH titrations. The pH titration of a 45 μM solution of oxidized 2,7,9-tricarboxy-PQQ was started at pH 3.5 using 2 N KOH as the titrant. Plots of A_{272} or A_{372} vs. pH showed a single transition with a pK_a = 10.73 ± 0.01. This value is identical to that found for the ionization of hydrated form of 2,7-dicarboxy-PQQ (5). A solution of 32.5 μM PQQ and 2.5 M NH_4Cl was titrated from pH 4.30 with 2 M KOH. Three transitions were seen at all wavelengths. The low pH transition is due to formation of the 5-imino-quinone on reaction with NH_3, and the other two due to the formation of unidentified species. The mechanism and equation for the computer analysis are given here:

$$P + NH_3 \xrightleftharpoons[K_d]{-H_2O} P-NH \underset{H^+}{\overset{K_1}{\rightleftharpoons}} P-NH^{-1} \underset{H^+}{\overset{K_2}{\rightleftharpoons}} P-NH^{-2}$$

$$K_a = 10^{-9.24}, H^+ \Big\downarrow\Big\uparrow$$

$$NH_4^+$$

$$A = \frac{A_o + K_a[NH_4^+]/\{K_d([H^+]+K_a)\}\{A_1 + A_2K_1/[H^+] + A_2K_1K_2/[H^+]^2\}}{1 + K_a[NH_4^+]/\{K_d([H^+]+K_a)\}\{1 + K_1/[H^+] + K_1K_2/[H^+]^2\}} .$$

From data at 330 nm, the values for K_d, pK_1, and pK_2 are 41.1 ± 4 mM, 9.08 ± 0.04, and 11.86 ± 0.02, respectively.

An anaerobic 31.3 µM solution of 2,7,9-tricarboxy-PQQ at pH 3.97 was treated with a small amount of $NaBH_4$ to produce the dihydroquinone (6), and titrated with KOH. Two pH transition were apparent at all wavelengths. The higher pK_a (8.46 ± 0.01, 300 nm; 8.28 ± 0.03, 414 nm) is equal to the value determined by cyclic voltammetry for the dihydroquinone. The lower pK_a (4.34 ± 0.02, 300 nm; 4.70 ± 0.01, 414 nm) is attributed to the ionization of a carboxyl group. The corresponding pK_a values for 2,7-dicarboxy-PQQ are 4.56 and 9.31 (5).

CONCLUSIONS

Although the redox potentials of 2,7,9-tricarboxy-PQQ have been reported at pH 7.0 (1) and lower (1-3), and over the pH range from 0-14 for 2-carboxy-PQQ (2), the values for E_o' in the "physiologically relevant" pH range are not available. This study provides these values from pH 5.5 to 10.5, as measured by the cyclic voltammetry technique. SO_3^{-2} reacts with 2,7,9-tricarboxy-PQQ to presumably form an adduct at the 5-position. As with riboflavin analogs, the E_o is a linear function of the log K_d for this adduct. This relationship should provide a convenient method for the determination of E_o' for PQQ analogs as well. It is evident form the effects of pH on reactivity with SO_3^{-2}, E_o', and the spectra that PQQ reacts with NH_3, not NH_4^+, to form what is assumed to be 5-imino-2,7,9-tricarboxy-PQQ.

Acknowledgements: This work was supported by the Veterans Administration, and Program Project Grant HL 16251 from The National Institutes of Health.

REFERENCES

1. Duine, J.A., Frank, J., and Verwiel, P.E.J., 1981. Characterization of the second prosethetic group in methanol dehydrogenase from *Hyphomicrobium* X. *Eur. J. Biochem.* **118**, 395-399.
2. Sleath, P.R., Noar, J.B., Eberlein, G.A., and Bruice, T.C., 1985. Synthesis of 7,9-didecarboxymethoxtin (4,5-Dihydro-4,5-dioxo-1H-pyrrolo-[2,3f]quinoline-2-carboxylic acid) and comparison of its chemical properties with those of methoxatin and analogous o-quinones. Model studies directed toward the action of PQQ requiring bacterial oxidoreductases and plasma amine oxidase. *J. Amer. Chem. Soc.* **107**, 3328-3338.
3. Eckert, T.S., Bruice, T.C., Gainor, J.A., and Weinreb, S.M., 1982. Some electrochemical and chemical properties of methoxatin and analogous quinones. *Proc. Natl. Acad. Sci. USA* **79**, 2533-2536.
4. Muller, F. and Massey, V., 1969. Flavin-sulfite Complexes and their structure. *J. Biol. Chem.* **244**, 4007-4016.
5. Rodriguez, E.J., Bruice, T.C., and Edmondson, D.E., 1987. Studies on the radical species of 9-decarboxymethoxatin. *J. Amer. Chem. Soc.* **109**, 532-537.
6. Itoh, S., Ohshiro, Y., and Agawa, T., 1986. Reaction of reduced PQQ (PQQH_2) and molecular oxygen. *Bull. Chem. Soc. Japan* **59**, 1911-1914.

REACTION OF PQQ AND ITS ANALOGUES WITH NADH; APPLICATION TO THE HLADH-CATALYZED OXIDATION OF ALCOHOLS

Shinobu ITOH, Masashi KINUGAWA, and Yoshiki OHSHIRO*
Department of Applied Chemistry, Faculty of Engineering,
Osaka University, Yamada-oka 2-1, Suita, Osaka 565, Japan

Abstract - Redox reaction between PQQ and NADH occurs readily to give reduced PQQ (quinol) and NAD^+, respectively. When the reaction is carried out under *aerobic conditions*, NADH is converted into NAD^+ almost quantitatively even in the presence of a catalytic amount of PQQ (1 mol %). The catalytic reaction has therefore been applied to the HLADH-catalyzed oxidation of alcohols. Cyclohexanol (2.0 mM), for example, is oxidized to cyclohexanone in good yield (75 %) when reacted with HLADH (5.0 x 10^{-7}M), NAD^+(0.1 mM), PQQ (0.1 mM), and catalase at pH 8.3 for 24 h. As a result, we find that PQQ acts as an efficient catalyst to regenerate NAD^+ with the aid of molecular oxygen. Catalytic efficiency of PQQ analogues is also examined in this system.

The interaction between PQQ and nicotinamide coenzyme (NADH) has the potential to be important and widespread in various enzymatic systems (Duine *et al.*, 1984). However, the role of PQQ in such systems has not been clearly demonstrated yet.

On the other hand, enzyme-catalyzed organic synthesis has been given much attention these days because of its inherent high selectivity. Particularly, NAD(H)-requiring oxido-reductases are potentially useful catalysts in organic asymmetric synthesis (Jones, 1985).

We have already demonstrated that PQQ acts as an efficient electron transfer catalyst from several organic substrates to molecular oxygen constructing quinoprotein model reactions (for example; Ohshiro *et al.*, 1983). In the course of our investigations of PQQ, we found that PQQ also reacts with a NADH model compound (BNAH) to give $PQQH_2$ (quinol) and BNA^+, respectively (Itoh *et al.*, 1986). In this paper, we would like to study the reaction between PQQ and NADH in some more detail and to apply this reaction to HLADH-catalyzed oxidation of alcohols (regeneration of NAD^+).

At first, the reaction between PQQ (4.0 x 10^{-5}M) and NADH (4.0 x 10^{-5}M) in 0.1 M phosphate buffer (pH 6.63, $\mu = 0.3$) was examined spectrophotometrically under *anaerobic conditions* at 30°C. Quantitative formation of reduced PQQ was clearly indicated by increase of absorption at 303 nm and also by

236

J. A. Jongejan and J. A. Duine (eds.), PQQ and Quinoproteins, 236–238.

regeneration of PQQ by aeration of the final reaction mixture. The formation of NAD^+ was confirmed by HPLC analysis. Furthermore, the reaction followed second-order kinetics (k_2 = 26.9 $M^{-1}s^{-1}$). These results indicate the stoichiometry of the reaction was 1:1 as expected (eq. 1).

$$PQQ \quad + \quad NADH \quad \xrightarrow{\quad H^+ \quad} \quad PQQH_2 \quad + \quad NAD^+ \qquad (1)$$

The reaction mechanism of NADH or its model compounds and several benzoquinone derivatives has been investigated in detail, but there is still controversy between direct hydride transfer mechanism and stepwise electron-proton-electron transfer mechanism (Carlson *et al.*, 1985; Fukuzumi *et al.*, 1984).

When NADH (8.0 x 10^{-3}M) was treated with a catalytic amount of PQQ (8.0 x 10^{-5}M) under *aerobic conditions* at pH 6.85 (0.1 M phosphate), NADH was oxidized to NAD^+ easily (initial rate; 1.5 x $10^{-6}Ms^{-1}$). Hydration of NADH was relatively slow under the same conditions (1.9 x $10^{-7}Ms^{-1}$). Thus, it is clear that PQQ acts as an efficient catalyst in the NADH-autoxidation (Figure).

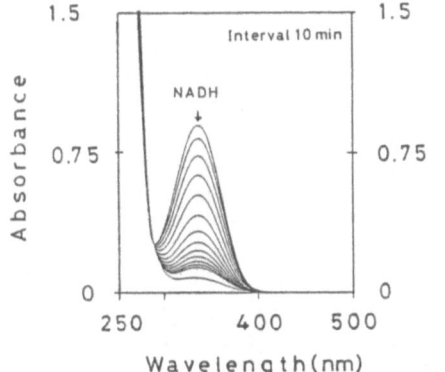 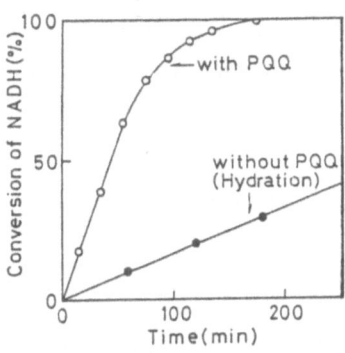

[PQQ] = 8.0 x 10^{-5}M, [NADH] = 8.0 x 10^{-3}M, 0.1M Phosphate buffer, aerobic conditions, under dark, 30°C, pH 6.85

This catalytic system was applied to regeneration of NAD^+ in HLADH-catalyzed oxidation of alcohols. In Table 1, the results including the catalytic efficiency of PQQ model compounds and other catalysts are shown.

So far, FMN and some dyes such as PMS, DCIP, and methylene blue have been examined as a catalyst to regenerate NAD^+ in HLADH-catalyzed alcohol oxidation (Lee *et al.*, 1985). As shown in Table 1, PQQ and its analogues such as 9-decarboxy PQQ and phenanthroline-5,6-quinone derivatives are found to be much more efficient catalysts in this system to compare with others.

The present system could provide some information for the NAD(H)-dependent quinoprotein alcohol dehydrogenase (Duine *et al.*, 1984).

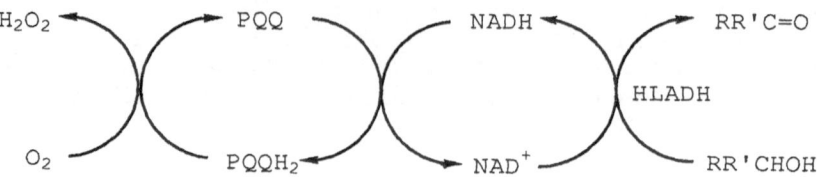

Table 1. HLADH-catalyzed oxidation of cyclohexanol

Catalyst	Cat.: NADH	Yield (%)	TN[d] of NAD$^+$
PQQ	1 : 1	67	13
PQQ	1 : 1[a]	75	15.6
PQQ	1 : 0	6	1
–	0 : 1	8	1.7
PQQ	0.1 : 1[b]	15	3.2
PQQ	1 : 0.1[c]	12	24.6
9-decarboxy PQQ	1 : 1	60	12
1,7-phenanthroline-5,6-quinone	1 : 1	69	15
1,10-Phenanthroline-5,6-quinone	1 : 1	67	13
phenanthrenequinone	1 : 1	29	6
2-carbomethoxy-indole-4,5-quinone	1 : 1	20	4
2-carbomethoxy-5-methyl-quinoline-7,8-quinone	1 : 1	25	5
FMN	1 : 1	20	4
PMS	1 : 1	72	15
DCIP	1 : 1	19	4
Methylene Blue	1 : 1	36	7
Methyl viologen	1 : 1	6	1

[Catalyst] = [NADH] = 1.0 x 10^{-4}M, [HLADH] = 5 x 10^{-7}M,
[cyclohexanol] = 2.0 x 10^{-3}M, 0.1 M phosphate buffer
pH 8.1 ~ 8.4, 30°C, 24 h, under aerobic conditions.
[a] Catalase (5 mg) was added. [b] [PQQ] = 1.0 x 10^{-5}M,
[c] [NADH] = 1.0 x 10^{-5}M, [d] The turnover number

References

Carlson, B. W. & Miller, L. L. (1985) *J. Am. Chem. Soc.*, *107*, 479-485.
Duine, J. A., Frank, J., & Berkhore, M. P. J. (1984) *FEBS Lett.*, *168*, 217.
Fukuzumi, S., Nishizawa, N., & Tanaka, T. (1984) *J. Org. Chem.*, *49*, 3571-3578.
Itoh, S., Ohshiro, Y., & Agawa, T. (1986) *Bull. Chem. Soc., Jpn*, *59*, 1911-1914.
Jones, J. B. (1985) in *"Asymmetric Synthesis"*, Morrison, J. D. Ed, Academic Press, New York.
Lee, L. G. & Whitesides, G. M. (1985) *J. Am. Chem. Soc*, *107*, 6999-7008, and references cited therein.
Ohshiro, Y., Itoh, S., Kurokawa, K., Kato, J., Hirao, T., & Agawa, T. (1983) *Tetrahedron Lett.*, *24*, 3465-3468.

CHEMICAL BEHAVIOR OF COENZYME PQQ TOWARD AMINE HOMOLOGS

Yoshiki Ohshiro, Minae Mure, Kazumi Nii, Shinobu Itoh
Department of Applied Chemistry, Osaka University
Yamadaoka 2-1, Suita, Osaka 565, Japan

key words; Pyrroloquinoline Quinone, ethylenediamine, aminoguanidine

Nonenzymatic oxidation of several amine homologs was carried out under mild conditions. Aliphatic amines were efficiently oxidized by PQQ in CTAB micellar system. In the case of ω, ω'-diamines, length of the carbon chain was significant; the reaction rate became larger with the increase in the carbon number. On the contrary ethylenediamine showed competitive inhibition forming the cyclic adduct, a fused diazine. Carbonyl reagents such as hydrazines and aminoguanidine were also oxidized by PQQ, but the latter compound deactivated PQQ by forming the 1,2,4-triazine derivative under catalytic conditions or at lower pH region.

1. Introduction

Cu^{2+}-containing amine oxidases (E.C.1.4.3.6) have been found in mammals, plants, fungi, yeasts and bacteria, and these enzymes convert amines into aldehydes, NH_3, and H_2O_2. According to the roles of PQQ which is probably bound covalently to the enzyme, PQQ has recentry been elucidated as an organic cofactor[1]. In this study, we wish to clarify the oxidizing ability of PQQ toward several amine homologs. Furthermore the reaction of hydrazine which is known as an irreversible inhibitor of the enzymes containing PQQ is also investigated.

2. Results and Discussion

2-1. The Reaction of PQQ with Monoamines

As a model system of amine oxidase, we have already demonstrated the micellar enhanced oxidative deamination of benzyl- and phenethylamines by using CTAB micellar system (Eq. 1)[2].

$$RR'CH_2NH_2 + O_2 + H_2O \xrightarrow[\text{CTAB}]{\text{PQQ}} RR'C=O + NH_3 + H_2O_2 \quad (1)$$

When we applied this system to the oxidation of several aliphatic monoamines, the oxidative deamination reaction proceeded efficiently in the CTAB micellar system. From the results of the kinetic study, incorporation of the substrate into micelles was found to be the most important effect for the increase in the reaction rate[3].

2-2. The Reaction of PQQ with Diamines

When 1,4-diaminobutane (4.0×10^{-4} mol) was treated with a catalytic amount of PQQ (1 mol%) in the presence of CTAB (4.0×10^{-5} mol) under anaerobic conditions (at pH 10.0) at 30 °C for 24h, 4-aminobutylaldehyde was formed in 1028% yield (based on PQQ)[4].

The progress of the reaction was followed by monitoring the apperarance of reduced PQQ under anaerobic pseudo-first-order conditions. The plots of the pseudo-first-order rate constant (k_{obsd}) versus the concentration of several diamine showed Michaelis-Menten type saturation phenomena. Reactivity of diamines increased markedly with the increase in the carbon number (Fig.1).

239

J. A. Jongejan and J. A. Duine (eds.), PQQ and Quinoproteins, 239–241.
© *1989 by Kluwer Academic Publishers.*

These results show that hydrophobic binding, namely incorporation of the substrate into micelle, is largely responsible for the increase in the rate of the oxidation reaction. It is also clear that the extent of incorporation of diamines into micelles is less than that of corresponding monoamines.

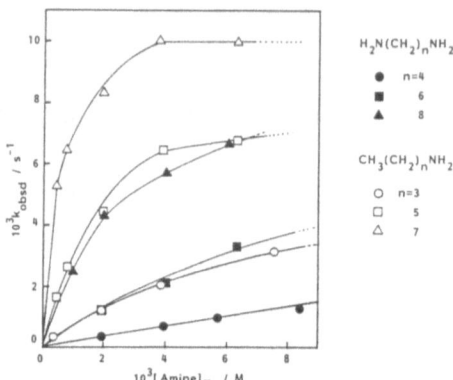

Fig. 1
$[PQQ] = 4.0 \times 10^{-5}M$, $[CTAB] = 2.0 \times 10^{-3}M$
$[Buffer] = 0.05 M(\mu = 0.2$ with KCl$)$
pH 10.01 ~ 10.36, 35°C, anaerobic conditions

In the case of ethylenediamine, however, two competing reaction took place. Figure 2 shows a spectral change along the progress of the reaction of PQQ (4.0 x 10^{-5}M) and ethylenediamine (4.0 x 10^{-3}M) in 0.05M phosphate buffer(pH 6.9, μ = 0.2 with KCl) at 35°C under anaerobic conditions. Absorption at 285 nm, 302 nm, and 320 nm increased with proceeding of the reaction. Upon introduction of air into the final reaction mixture, the spectrum of reduced PQQ (λmax 302–320 nm) disappered, and the species absorbing at 285 nm remained unchanged. Thus, at pH 6.9, there are two competing reactions; formation of reduced PQQ and formation of product(s) which has absorption maximum at 285 nm. At lower pH or at higher pH, the latter reaction predominantly occurred. In order to isolate the unknown product(s) the reaction of PQQ (3.3 x 10^{-5} mol) with 10-fold excess of ethylenediamine was performed at pH 4.6, 50 °C for 1h under aerobic conditions. The product was identified as the pirazine derivative of PQQ (a; 82% yield) by ^{1}H-NMR and IR spectrometry.

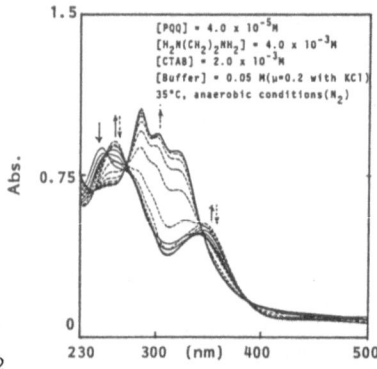

Fig. 2

(a)

Kagan et al[5] have reported that the chain length dependency of the inhibitory strength of alkylenediamines for lysyl oxidase, 2- or 3-carbon diamines indicate optimal interaction with the enzyme, with apparent affinity decreasing markedly as the carbon chain lengthened. From our results, it is proposed that ethylenediamine inhibits enzymes by forming cyclic adducts such as a fused diazine with carbonyl functions of PQQ.

3-3. The Reaction of PQQ with Carbonyl reagents.

The reaction of PQQ with carbonyl reagents such as hydrazines and aminoguanidine was also investigated. Hydrazines (phenylhydrazine and methylhydrazine, etc.) were oxidized in the presence of catalytic amount of PQQ (1 mol%) under aerobic conditions. Under anaerobic conditions, reduced PQQ was obtained quantitatively. The reaction of PQQ with aminoguanidine under the pseudo-first-order conditions of [aminoguanidine]= 4.0 x 10^{-3}M >> [PQQ]= 4.0 x 10^{-5}M (0.1 M Carbonate buffer μ= 0.2 with KCl, 30°C, anaerobic conditions) was monitored by UV spectrometer. The final spectrum was identical to that of reduced PQQ and the spectrum of PQQ was regenerated allowing air to bubble into the final reaction mixture. The reaction was then followed at 320 nm, reasonable pseudo-first-order kinetics was observed. The reaction was first order in [aminoguanidine] and a linear plot was obtained of k_{obsd} versus OH^- over the pH range 9-11. The reaction of PQQ with aminoguanidine thus follows rate Eq. 2. When a pK_{a1} value of 11.04 is used for aminoguanidine, k_1 is calculated as 32 $M^{-1}s^{-1}$.

$$k_{obsd} = k_1 \text{[aminoguanidine]} = k_1 \frac{K_a}{K_a + [H^+]} \text{[aminoguanidine]}_{tot} \qquad (2)$$

Interestingly, we could not observe such redox reaction but an increase of absorbance at 283 nm at lower pH region. When the reaction was investigated aerobically at pH 10.0, the final spectra closely resembled to the spectrum at lower pH region. Thus PQQ is converted to a redox inactive derivative(s) at high pH region in the presence of O_2. In order to investigate this phenomenon and determine the unknown product(s), trimethylester of PQQ was treated with excess amount of aminoguanidine at neutral pH ($CH_3CN/H_2O=1/1$), 30°C for a few hours under aerobic conditions, 1,2,4-triazine derivative of PQQ(b) was yielded, which has characterized by mass-spectrometry (M^+ 410), ^{1}H-NMR, and IR spectrometry. A plausible mechanism for the formation of (b) is provided in Eq. 3.

3. References

1) H. Yamada and K. T. Yasunobu, J. Biol. Chem., 237, 1511 (1962)
2) S. Itoh, Y. Kitamura, Y. Ohshiro, and T. Agawa, Bull. Chem. Soc. Jpn., 59, 1907 (1986)
3) S. Itoh, M. Mure, and Y. Ohshiro, J. Jpn. Oil Chem. Soc. (YUKAGAKU) 36, 90 (1987)
4) 4-aminobutylaldehyde was assayed colorimetrically by modifying the known method; B. Holmstedt, L. Larsson, R. Tham., Biochim et Biophys. Acta., 48, 182 (1961)
5) P. C. Trackman and H. M. Kagan, J. Biol. Chem., 254, 7831 (1979)

THE THREE-DIMENSIONAL STUCTURE OF PQQ AND RELATED COMPOUNDS

H. van Koningsveld[1], J.A. Jongejan[2] and J.A. Duine[2]

[1] Department of Applied Physics, Delft University of Technology,
Lorentzweg 1, 2628 CJ Delft, The Netherlands
[2] Department of Microbiology and Enzymology, Delft University of Technology,
Julianalaan 67, 2628 BC Delft, The Netherlands

SUMMARY

The three-dimensional structures of (non-covalently) hydrated PQQ and a PQQ-potassium complex have been determined by X-ray analysis. Structural features of these compounds are compared with those of PQQ-acetone adduct (Salisbury et al., 1979) and PQQ-dinitrophenylhydrazone triester (van Koningsveld et al., 1985). The observed position of K^+ in the PQQ-potassium complex appears to be excellently suited to accommodate a Cu^{2+} atom. PQQ-dinitrophenyl-hydrazone triester consists of a resonant mixture of 'hydrazone' and 'azo' configurations. The PQQ-acetone adduct shows a major distortion of the central ring. Implications of this finding for steric effects on the electrophilic properties of the $PQQ-C_5$-carbonyl group are discussed.

INTRODUCTION

X-ray crystallography has played an important role in the relatively short history of PQQ and quinoprotein research. Salisbury, Cruse, Forrest and Kennard reported the determination of the structure of PQQ-acetone adduct (Salisbury et al., 1979, Cruse et al., 1980), from which conclusive evidence for the identity of the tricarboxy-pyrroloquinoline structure of PQQ could be derived. Combined with the results on the nature of the ortho-quinone moiety obtained earlier by Duine and coworkers (Duine et al., 1978, 1980a, Westerling et al., 1979), the structure of PQQ as 2.7.9-tricarboxy-1 H-pyrrolo[2,3-f]quinoline-4,5-dione could be ascertained (Duine et al., 1980b). Complete structural proof was subsequently provided by no less than five independent total synthetic routes (Buechi et al., 1985, loc. cit.).

The detection of PQQ, or a closely related compound, as the organic cofactor of plasma amine oxidase (Lobenstein-Verbeek et al., 1984), can be considered as a starting point for the study of quinoproteins present in eukaryotic species. Again, X-ray analysis (of the dinitrophenylhydrazone of PQQ triester, van Koningsveld et al., 1985) formed the basis for the structural characterization of the isolated cofactor hydrazone (Van der Meer et al., 1986).

243

J. A. Jongejan and J. A. Duine (eds.), PQQ and Quinoproteins, 243–251.

X-ray data on methylamine dehydrogenase (MADH) from *Thiobacillus versutus*, a putative quinoprotein (de Beer *et al.*, 1980), have been obtained (Vellieux *et al.*, 1986). Preliminary results at 2.1 Å resolution, reveal the location of the covalently bound cofactor (F.M.D. Vellieux, personal communication). The electron density, however, appears inadequate to accommodate the full three-membered ringsystem of PQQ. As PQQ-phenylhydrazone can be isolated from this enzyme (van der Meer *et al.*, 1987), it is postulated that the cofactor of MADH is a peptide-bound 7-γ-glutamyl-indole derivative (so-called 'pro-PQQ') that is transformed into PQQ-phenylhydrazone upon derivatization and subsequent proteolysis of the peptide. 7-γ-Glutamyl-indole derivatives have been proposed as intermediates of PQQ-biosynthesis (Duine and Jongejan, 1989, Houck *et al.*, 1988, van Kleef and Duine, 1988).

A standing problem of PQQ chemistry, amenable to X-ray crystallography, concerns the structure of PQQ-metal ion complexes (Jongejan *et al.*, 1987, Suzuki *et al.*, 1988). The importance of such complexes is suggested by the combined presence of PQQ and metal ions in several quinoproteins, including copper-containing amine oxidases (van der Meer and Duine, 1986), dopamine β-hydroxylase (van der Meer *et al.*, 1988), galactose oxidase (van der Meer *et al.*, 1989), lipoxygenase (van der Meer and Duine, 1988) and possibly also nitrile hydratase (Nagasawa and Yamada, 1987). To date, no X-ray investigations on PQQ-metal complexes have been reported, probably because of the difficulties encountered in the preparation of suitable crystals of such complexes.

Recently, we succeeded in preparing crystals of PQQ of a quality sufficient for X-ray crystallography. In addition, we obtained a limited amount of crystalline material representing a complex of PQQ and potassium. A detailed description of the structure of both compounds will be given in a forthcoming paper. In the present contribution some outstanding features of the three-dimensional structures of these compounds will be described and compared to those reported earli for PQQ-acetone adduct and PQQ-dinitrophenylhydrazone triester.

MATERIALS AND METHODS

PQQ dimethyl-ethyl-triester was synthesized by the method of Corey and Tramontano (Corey and Tramontano, 1981; Jongejan *et al.*, 1988). Hydrolysis of PQQ triester was conducted by dissolving the ester in excess 0.3 M KOH at 50 °C for 30 min. The clarified solution was acidified (pH 1.5) using 3 M HCl. By dissolving the copious precipitate in hot water (90 °C) and adjusting the pH to 1.5 a deep-red precipitate, containing only minor amounts of contaminating potassium chloride, was obtained. A solution of the purified preparation in water (approx. 1 gram/liter) was kept at 80 °C for several days, during which time deep-red crystals were deposited. Analysis of air-dried material showed the elemental composition to be $C_{14}H_8N_2O_9$. Upon extensive drying (100 °C, P_2O_5, > 6 hrs, in vacuo) 5.12 % of weight was lost, thus confirming the presence of 1 mole of water of hydration (Compound I; $C_{14}H_6N_2O_8 \cdot H_2O$, $PQQ \cdot H_2O$).

By repeated crystallization of approx. 1 gram of this material from large amounts of water at elevated temperature (30 - 40 °C), a few yellow-orange crystals, differing in form and composition from the bulk material, could be obtained. Elemental composition of these crystals was deduced from X-ray analysis (Compound II; $C_{28}H_{29}N_4O_{25}K$, $KH(PQQ)_2.9H_2O$). The presence of potassium in this preparation can be rationalized by the likely entrapment of minor quantities of this ion, contaminating the original preparation.

X-ray data were collected by an Enraf-Nonius CAD4 diffractometer with graphite monochromated Cu K$_\alpha$ radiation. The structures were solved by direct methods (MULTAN; Germain et al., 1971). All calculations were performed on the Delft University Amdahl 470/V7B computer using XRAY72 (Stewart et al., 1972).

RESULTS AND DISCUSSION
Structural properties of PQQ and PQQ-potassium complex
The present report constitutes an unequivocal structure determination of non-covalently hydrated PQQ (compound I). In addition, the determination of the three-dimensional structure of PQQ-potassium complex (compound II) offers the first example of a properly defined PQQ-metal complex. A perspective view of the structures of both compounds is given in Fig. 1 and Fig. 2, respectively. An extensive listing of the final atomic parameters will be published elsewhere (Van Koningsveld and Jongejan, manuscript in preparation).

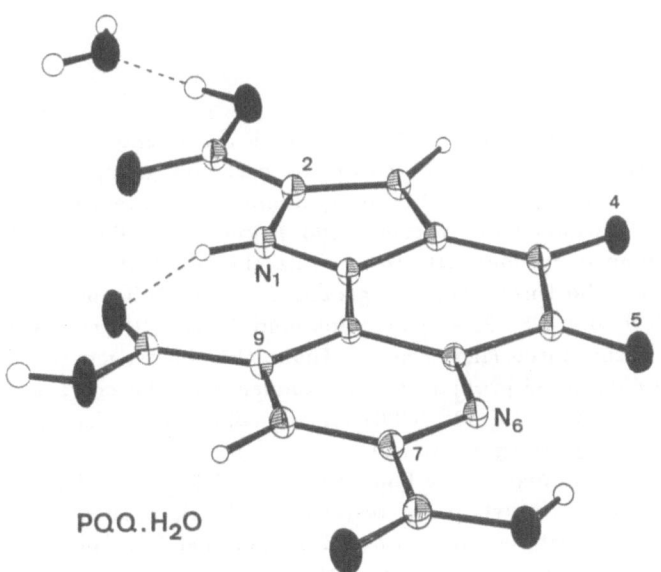

PQQ.H$_2$O

Figure 1. ORTEP-drawing of the structure of compound I

246

Figure 2. ORTEP-drawing of the structure of compound II

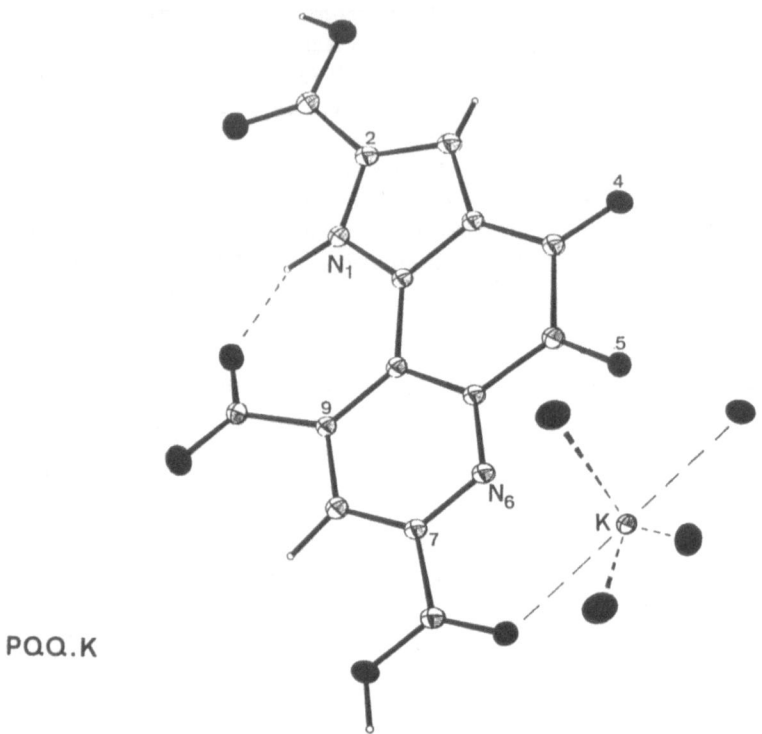

PQQ.K

Distances and angles in both PQQ molecules are found equal within 0.02 Å and 1.5°, respectively, exept for those of the carboxyl groups at C(7) and C(9). The greater part of the pyrroloquinoline quinone ring system of both molecules is planar, with all atoms lying within 0.098 Å for I and 0.045 Å for II of the least-squares plane containing all atoms. This, however, does not apply to the carboxyl O atoms. The three carboxyl groups at C(2), C(7) and C(9) are inclined 6.8 (5.9), 1.2 (3.0) and 7.8 (22.4)° in compound I and (II), respectively, to the plane formed by the three-ring system. Thus, the difference in the rotational orientation of the carboxyl groups of I, as compared to the corresponding groups in II, is 1 (C(2)), 4 (C(7)) and 30° (C(9)) (disregarding the differentiation between carbonyl- and hydroxyl-oxygen atoms).

A comparison of the (exocyclic) bond angles at C(7) in I and II shows a slight displacement of the carboxyl group towards K^+ in II, probably as a result of the strong interaction between the carbonyl oxygen and K^+. Curiously, the charge on K^+ is largely balanced by ionization of the carboxyl group at C(9). The anionic character of this group can be readily deduced from the observed geometry. In the unit cell, comprising two molecules of PQQ⁻ arranged in diad symmetry around the central K^+ atom, the second equivalent of positive charge is accommodated by two (partially protonated) water molecules, each accounting for approximately half a positive charge. From the present data it is not clear whether some anionic character is also resident at the C(7) carboxyl group. If this is the case, the population of hydrogen at the C(7)-carboxy hydroxyl group, which clearly shows up in a difference Fourier map, should be slightly decreased.

As it is mentioned above, the carboxyl group at C(9) in II is completely ionized, while a reasonably strong hydrogen bond is formed involving N(1) and the oxygen atom having the longer C-O bond. The same applies to the non-ionized C(9)-carboxyl group in I, where a hydrogen bond is formed between N(1) and the carbonyl oxygen (having the shorter C-O bond).

PQQ-Potassium complex as a model for PQQ-Copper complexes

A detailed analysis of the coordination around K^+ in II allows the first coordination sphere to be described as a distorted trigonal bipyramid. Both the K^+ atom and an oxygen atom belonging to one of the water molecules at an apical position, are located at the 2-fold symmetry axis. The second coordination sphere, involving the C(5)-carbonyl oxygen and N(6), as well as the two-fold related atoms, is square planar, although a considerable distortion is evident. The observed position of K^+ appears to be excellently suited to accomodate a Cu^{2+} atom, as might be the case for several copper-containing quinoproteins.

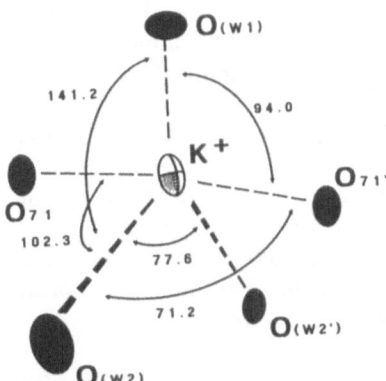

Figure 3. Some characteristics of the K^+-site in compound II

Metal chelation at this site has already been proposed by Noar, Rodriguez and Bruice (Noar *et al.*, 1985), based on the structural similarity of PQQ and pyridine-2-carboxylates, with regard to this particular site; by Jongejan (Jongejan *et al.*, 1987) and Frank (Frank, 1988), based on the structural similarity of PQQ-C$_5$-OH derivatives with 8-hydroxyquinolines and 6-hydroxyflavins; and by Hanauske-Abel and coworkers, who presented convincing evidence for the involvement of this site in the inhibitory action of PQQ with respect to the iron centre in proline hydroxylase (Hanauske-Abel *et al.*, 1987).

However, it may well be that the C(9)-carboxyl group of PQQ forms part of an additional site for copper ligation. The topology of the active site of plasma amine oxidase, as deduced by Williams and Falk (Williams and Falk, 1986), suggests that ligation of the copper ion to PQQ in this enzyme involves the C(9) position. In view of the preferent ionization of the C(9)-carboxyl group, that is .apparent from the present work, it appears likely that the formation of amorphous precipitates upon mixing of equimolar quantities of PQQ and Cu^{2+} could result from the formation of head-to-tail concatamers of PQQ joined by copper atoms ligated to the N(1)-region of one PQQ and the C(9)-region of a second molecule. Addition of sub-stoichiometric amounts of copper affords micro-crystalline precipitates. Crystallization experiments are in progress.

Steric factors involved in the electrophilic character of the PQQ-C$_5$-carbonyl group

The electrophilic character of the C(5)-carbonyl group of PQQ is well documented (Dekker *et al.*, 1982). Reaction of free PQQ with a variety of nucleophiles involves covalent addition at this position. Moreover, all quinoprotein enzymes that have been detected up till now, with the possible exception of lipoxygenase (van der Meer and Duine, 1988), catalyze the conversion of substrates carrying a nucleophilic group. Although convincing evidence has been presented in a few cases only (Dijkstra *et al.*, 1984, Frank *et al.*, 1988), it is commonly accepted that formation of a PQQ-C(5)-substrate adduct comprises the initial stage of quinoprotein catalysis. As argued for the case of quinoprotein alcohol dehydrogenases (Jongejan *et al.*, these proceedings), both inductive, mesomeric, as well as steric effects may contribute to the electrophilic behaviour of the C(5)-carbonyl group.

To assess the importance of geometric factors for the formation of C(5)-adducts, we set out to compare the three-dimensional structures of hydrated and unhydrated PQQ. However, sofar no suitable crystals of the (covalently-) hydrated form of PQQ have been obatained. In addition we realize that it may not be possible to obtain a conclusive answer on this matter from X-ray analysis.

From a comparison of the three-dimensional structures of I and PQQ-acetone adduct, representing C(5)-sp^2 and C(5)-sp^3 hybridized species, respectively, it is clear that a major distortion of the central ring is induced upon adduct formation. It would appear that the slight gain in total bond-energy is largely set off by the loss of aromatic character of the central ring following adduct formation. We feel that the relief of steric strain around N(1) and C(9)-carboxyl in PQQ-acetone adduct as compared to PQQ itself (visualized in Fig. 4), may play a desicive role in adduct formation.

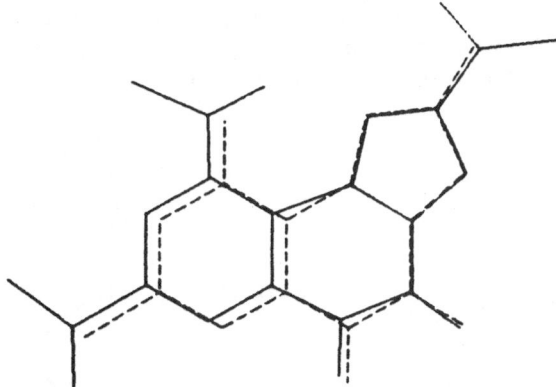

Figure 4. Actual geometry of PQQ (————), relative to the configuration derived
from standard bond lengths and angles (– – – –).

The extent of (covalent) hydration that can be inferred from spectroscopic
data of PQQ derivatives carrying a substituent at N(1), as compared to that of
a C(8)-methylated derivative in which this strain is (partially) removed by an
induced rotation of the C(9)-carboxyl group, appears to support this hypothesis
(Jongejan et al., these proceedings). In this respect, 'tuning' of the electrophilic
character of the PQQ-C(5)-carbonyl in the enzyme-bound cofactor, could well
arise from rotation of the C(9)-carboxylic group.

The structure of PQQ-hydrazine products

Derivatization of free and enzyme-bound PQQ with hydrazines has played an
important role in the detection and characterization of PQQ in quinoproteins
(for a review see R.A. van der Meer et al., these proceedings). Several types
of PQQ-(dinitro-)phenylhydrazine products have been isolated from the reaction
of PQQ-containing enzymes with (dinitro-)phenylhydrazine (Van der Meer et al.,
1987). Curiously, treatment of free PQQ with phenylhydrazine leads to the
predominant formation of PQQH$_2$ (Duine et al., 1981, Itoh et al., 1985). Virtually
no reduction is seen, on the other hand, upon treatment of free PQQ with
dinitrophenylhydrazine; only one type of product, the dinitrophenylhydrazone of
PQQ, can be isolated in good yields. Judged from the spectroscopic properties
of the phenylhydrazine products of PQQ that can be obtained from phenyl-
hydrazine-treated quinoproteins and (as minor products) from free PQQ, it
appears likely that both 'azo' and 'hydrazone' products, as well as a carbinolazine-
adduct are involved. In view of the tautomerism relating 'azo-' and 'hydrazone'
compounds, and the possibility of cis/trans isomerism of the individual tautomers,
we propose that the (cis/trans-)isomers that are initially formed in the reaction
of free PQQ (giving rise to loss of dinitrogen and formation of PQQH$_2$), differ
from those formed upon treatment of PQQ-containing enzymes. The (limited)
stability of the latter compounds could then result from slow isomerization.

Formation of a single (relatively stable) product upon dinitrophenylhydrazine treatment of free PQQ can be rationalized by comparing the three-dimensional structure of PQQ-dinitrophenylhydrazone triester with that of PQQ (compound I).

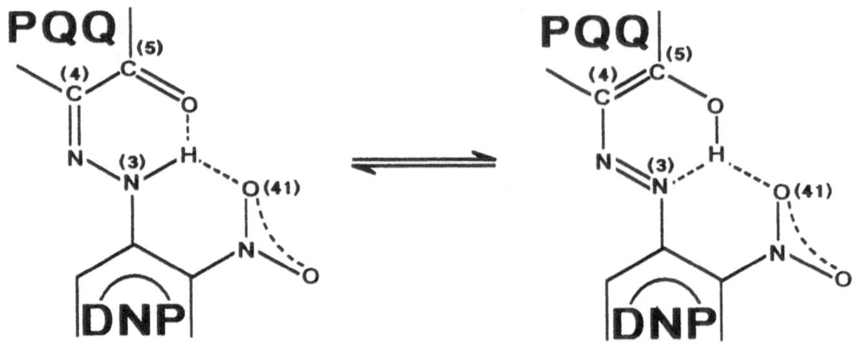

Figure 5. 'Azo'/'hydrazone' tautomers of PQQ-dinitrophenylhydrazone triester.

As is shown in Fig. 5, the presence of a bifurcated hydrogen bond in the former structure adds greatly to the stability of this compound. Comparison of the bond lengths and bond angles with those of PQQ shows that the structure of the dinitrophenyl-'hydrazone' actually represents a resonant mixture of contributions from both 'azo' and 'hydrazone' configurations. In the absence of the bifurcated hydrogen bond, as will be the case for PQQ-phenylhydrazine adducts lacking the nitro group, a similar stabilization cannot be attained.

Stimulating discussions with Prof. Dr. J. Reedijk (Leiden University) are kindly acknowledged.

REFERENCES

de Beer, R., Duine, J.A., Frank Jzn, J. and Large, P.J. (1980), *Biochim. Biophys. Acta,* *622,* 370–374

Buechi, G., Botkin, J.H., Lee, G.C.M. and Yakushijin, K. (1985), *J. Am. Chem. Soc.,* *107,* 5555–5556

Corey, E.J. and Tramontano, A. (1981), *J. Am. Che. Soc., 103,* 5599–5600

Cruse, W.B.T., Kennard, O. and Salisbury, S.A. (1980), *Acta Cryst., B36,* 751–754

Dekker, R.H., Duine, J.A., Frank Jzn, J., Verwiel, P.E.J. and Westerling, J. (1982), *Eur. J. Biochem., 125,* 69–73

Dijkstra, M., Frank Jzn, J., Jongejan, J.A. and Duine, J.A. (1984), *Eur. J. Biochem., 140,* 369–373

Duine, J.A., Frank Jzn, J. and Westerling, J. (1978), *Biochim. Biophys. Acta 524,* 277–287

Duine, J.A. and Frank Jzn, J. (1980a), *Biochem. J.*, *187*, 221-226

Duine, J.A., Frank Jzn, J. and Verwiel, P.E.J. (1980b), *Eur. J. Biochem.*, *108*, 187-192

Duine, J.A., Frank Jzn, J. and Verwiel, P.E.J. (1981), *Eur. J. Biochem.*, *118*, 395-399

Duine, J.A. and Jongejan, J.A. (1989), in: *Vitamins and Hormones*, *46*, in press

Frank, Jzn, J., Dijkstra, M., Duine, J.A. and Balny, C. (1988), *Eur. J. Biochem.*, *174*, 331-338

Frank, J., Jzn, Thesis (1988), Delft, Krips Repro Meppel, pp. 90-91

Germain, G., Main, P. and Woolfson, M.M. (1971), *Acta Cryst.*, *A27*, 368-376

Hanauske-Abel, H.M., Tschank, G., Guenzler, V., Baader, E. and Gallop, P. (1987), *FEBS Lett.*, *214*, 236-243

Houck, D.R., Hanners, J.L. and Unkefer, C.J. (1988), *J. Am. Chem. Soc.*, *110*, 6920-6921

Itoh, S., Mure, M., Ohsiro, Y. and Agawa, T. (1985), *Tetrahedron Lett.*, *26*, 4225-4228

Jongejan, J.A., van der Meer, R.A., van Zuylen, G.A. and Duine, J.A. (1987), *Recl. Trav. Chim. Pays-Bas*, *106*, 365

Jongejan, J.A., Bezemer, R.P. and Duine, J.A. (1988), *Tetrahedron Lett.*, *42*, 3259-3268

van Kleef, M.A.G. and Duine, J.A. (1988), *FEBS Lett.*, *237*, 91-97

van Koningsveld, H., Jansen, J.C., Jongejan, J.A., Frank Jzn, J. and Duine, J.A. (1985), *Acta Cryst.*, *C41*, 89-92

Lobenstein-Verbeek, C.L., Jongejan, J.A., Frank Jzn, J. and Duine, J.A. (1984), *FEBS Lett.*, *170*, 305-309

van der Meer, R.A., Jongejan, J.A., Frank Jzn, J. and Duine, J.A. (1986), *FEBS Lett.*, *206*, 111-114

van der Meer, R.A. and Duine J.A. (1986), *Biochem. J.*, *239*, 789-791

van der Meer, R.A., Jongejan, J.A. and Duine, J.A. (1987), *FEBS Lett.*, *221*, 299-304

van der Meer, R.A., Jongejan, J.A. and Duine, J.A. (1988), *FEBS Lett.*, *231*, 303-307

van der Meer, R.A. and Duine, J.A. (1988), *FEBS Lett.*, *235*, 194-200

van der Meer, R.A., Jongejan, J.A. and Duine, J.A. (1989), *J. Biol. Chem.*, accepted for publication

Nagasawa, T. and Yamada, H. (1987), *Biochem. Biophys. Res. Commun.*, *147*, 701-709

Noar, J.B., Rodriguez, E.J. and Bruice, T.C. (1985), *J. Am. Chem. Soc.*, *107*, 7198-7199

Salisbury, S.A., Forrest, H.F., Cruse, W.B.T. and Kennard, O. (1979), *Nature (London)*, *280*, 843-844

Stewart, J.M., Kruger, G.J., Ammon, H.L., Dickinson, C.W. and Hall, S.R. (1972); *The XRAY72 system. Tech. Rep. TR-192.* Computer Science Center, Univ. of Maryland, College Park, Maryland

Suzuki, S., Sauray, T., Itoh, S. and Ohshiro, Y. (1988), *Inorg. Chem.*, *27*, 591-592

Vellieux, F.M.D., Frank Jzn, J., Swarte, M.B.A., Groendijk, H., Duine, J.A., Drenth, J. and Hol, W.G.J. (1986), *Eur. J. Biochem.*, *154*, 383-386

Westerling, J., Frank Jzn, J. and Duine, J.A. (1979), *Biochem. Biophys. Res. Commun.*, *87*, 719-724

Williams, T.J. and Falk, M.C. (1986), *J. Biol. Chem.*, *261*, 15949-15954

COPPER(II) COMPLEXES CONTAINING REDUCED PQQ AS A MODEL
FOR THE ACTIVE SITE OF COPPER-REQUIRING AMINE OXIDASE

Shinnichiro Suzuki, Takeshi Sakurai, Shinobu Itoh,[+] and Yoshiki Ohshiro[+]

Institute of Chemistry, College of General Education, Osaka University,
Toyonaka, Osaka 560, JAPAN
[+]Department of Applied Chemistry, Faculty of Engineering, Osaka University,
Suita, Osaka 565, JAPAN

Key words: coenzyme, PQQ, reduced PQQ, copper complex

The ternary Cu(II) complex containing reduced coenzyme PQQ ($PQQH_2$) and
2,2'-bipyridine (bpy), 1,10-phenanthroline (phen), or 2,2':6',2"-ter-
pyridine (terp) was characterized by electronic absorption and electron
paramagnetic resonance (EPR) spectroscopy. The binding sites of $PQQH_2$ to
Cu(II) were proposed to be N(6) and carboxylate(7) groups. The oxidation
of the $PQQH_2$ ligand was accelerated by complex formation, compared with
that of free $PQQH_2$.

INTRODUCTION

Cu-requiring amine oxidase contains nonblue copper(II) and an organic
cofactor which was recently reported to be PQQ or its derivatives (Duine et
al., 1987). The nonblue site in the enzyme indicated a tetragonal geometry
with three imidazole-like nitrogen ligands and one oxygen ligand in the
equatorial plane. However, the structural and the functional relationships
between Cu(II) and the organic cofactor remain unclear. We prepared
several ternary Cu(II) complexes containing PQQ (Suzuki et al., 1988a) or
$PQQH_2$ (Suzuki et al., 1988b) as a model for the active site of amine
oxidase. The present paper details the property of the ternary Cu(II) complex containing $PQQH_2$ and bpy, phen, or terp from the standpoint of coordination chemistry of coenzyme PQQ.

MATERIALS AND METHODS

$PQQH_2$ was prepared by the reduction of PQQ commercially available with
sodium dithionate in an aqueous solution (Itoh et al., 1987). The crude
product was recrystallized from DMSO-acetonitrile solvent. The ternary
Cu(II) complexes of $PQQH_2$ and bpy, phen, or terp were obtained by addition
of a small amount of DMSO containing $PQQH_2$ to an aqueous solution of
Cu(II)(bpy), Cu(II)(phen), or Cu(II)(terp) under Ar atmosphere at pH 5.5
($PQQH_2$/Cu = 1). The oxidation of $PQQH_2$ with dioxygen was carried out at pH
5.6 by adding 0.01 ml of DMSO containing $PQQH_2$ (0.060 μmol) to an aqueous
solution (3ml) of Cu(bpy), Cu(phen), or Cu(terp) (0.060 μmol). The reaction rate was monitored by the disappearance of the 310-nm band of $PQQH_2$ at
25°C.

J. A. Jongejan and J. A. Duine (eds.), PQQ and Quinoproteins, 253–255.
© 1989 by Kluwer Academic Publishers.

RESULTS

Figure 1 shows the electronic absorption spectra of Cu(PQQH$_2$)(phen), Cu(PQQ)(phen), and free PQQH$_2$. The absorption band of free PQQH$_2$ at around 310 nm does not shift in the presence of Cu(II)(phen), whereas the 401-nm band was shifted to shorter wave lengths (Figure 1, solid line). The similar blue shifts were also found in Cu(PQQH$_2$)(bpy) and Cu(PQQH$_2$)(terp) (Suzuki et al., 1988b). These findings suggest the coordination of PQQH$_2$ to Cu(II) ion. The Cu(PQQH$_2$)(phen) complex affords the absorption spectrum of the ternary complex containing PQQ (oxidized form) when exposed to air (Figure 1, chain line). Two absorption bands near 350 and 500 nm are characteristic of the ternary Cu(II) complexes of PQQ (Suzuki et al., 1988a). The spectral change of the oxidation of Cu(PQQH$_2$)(terp) shows in Figure 2. The appearance of the isosbestic points at 333 and 401 nm implies that there are only two species (Cu(PQQH$_2$)(terp) and Cu(PQQ)(terp)) and the species containing Cu(I) ion is not formed. The oxidation rates of PQQH$_2$ in the ternary complexes are ≥ 50 (Cu(PQQH$_2$)(bpy) and Cu(PQQH$_2$)-(phen)) and 3 (Cu(PQQH$_2$)(terp)) times as fast as that of free PQQH$_2$.

The EPR parameters of the ternary complex of PQQH$_2$ quite resemble those of the corresponding complex containing PQQ (Table 1), indicating the similarity of the coordination geometry and the donor set of the complex of PQQH$_2$ to those of the corresponding complex containing PQQ.

DISCUSSION

On the basis of the spectroscopic data of the ternary Cu(II) complexes containing PQQH$_2$, the structures of the complexes are proposed (Figure 3). PQQH$_2$ could be coordinated to copper by N(6) and carboxylate(7). The redox reaction between PQQH$_2$ and Cu(II) ion in an aqueous solution takes place under anaerobic conditions (Jongejan et al., 1987), but Cu(II) ions in the ternary complexes of PQQH$_2$ are not reduced by PQQH$_2$ because of higher redox potential of PQQ than those of Cu(bpy), Cu(phen), and Cu(terp) (Suzuki et al., 1988b). These electrochemical features of the binary Cu(II) complexes coincide with the fact that Cu(II) ion in amine oxidase is

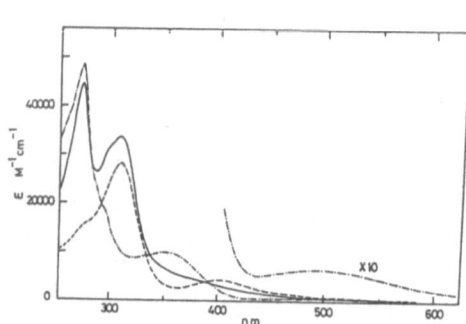

FIGURE 1. Absorption spectra of Cu-(PQQH$_2$)(phen) (——) and PQQH$_2$(----) under Ar atmosphere and Cu(PQQ)(phen) (—·—·—) at pH 5.5.

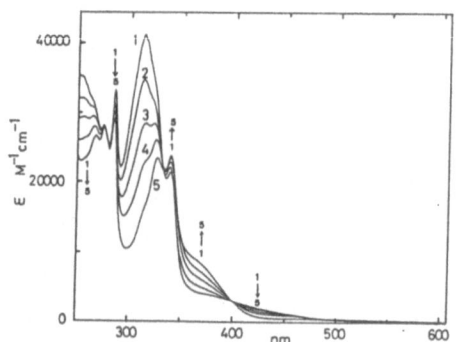

FIGURE 2. Absorption spectra in the course of the oxidation of Cu(PQQH$_2$)(terp) (spectrum 1) at pH 5.5. spectrum 5:Cu(PQQ)(terp)

TABLE 1. EPR parameters of the ternary
Cu(II) complexes of $PQQH_2$ or PQQ.

	$g_{//}$	$g_{\perp}$	$A_{//}(G)$
Cu($PQQH_2$)(bpy)	2.28	2.07	164
Cu(PQQ)(bpy)	2.28	2.07	165
Cu($PQQH_2$)(phen)	2.29	2.07	161
Cu(PQQ)(phen)	2.29	2.07	162
Cu($PQQH_2$)(terp)	2.25	2.06	168
Cu(PQQ)(terp)	2.25	2.06	173

pH 5.5, 77K

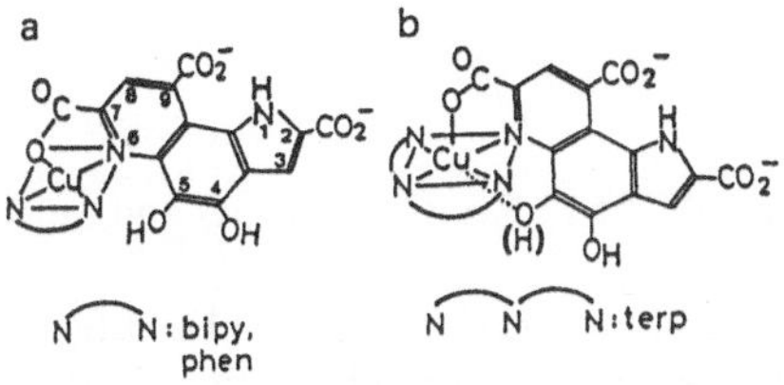

FIGURE 3. Proposed structures
of ternary Cu(II) complexes
containing $PQQH_2$.

not reduced even under the anaerobic reduction of the organic cofactor with
a substrate (Suzuki et al., 1983). The structures in Figure 3 are
analogous to those of the corresponding complexes of PQQ (Suzuki et al.,
1988a). The EPR parameters in Table 1 suggest more distorted tetragonal
geometry of Cu(II) in the complexes of phen than that in the complexes of
bpy (Peisach & Blumberg, 1974).

The oxidations of $PQQH_2$ were promoted in three ternary complexes,
which might be due to the electron-withdrawing effect of Cu(II). High
oxidation rates of Cu($PQQH_2$)(bpy) and Cu($PQQH_2$)(phen) are probably at-
tributable to the structural unstabilization of the ternary complexes by
the repulsion between $PQQH_2$ and bpy or phen coordinated to equatorial plane
of Cu(II) (Figure 3, a). On the other hand, such a repulsion does not occur in Cu($PQQH_2$)(terp) where two aromatic planes of PQQ and terp are per-
pendicular to each other.

REFERENCES

Duine JA, Frank jzn J and Jongejan JA, 1987. Enzymology of quinoproteins.
    Adv. Enzymol. Relat. Areas Mol. Biol. 59: 169-212.
Itoh S, Ohshiro Y and Agawa T, 1986. Reaction of reduced PQQ ($PQQH_2$) and
    molecular oxygen. Bull. Chem. Soc. Jpn. 59: 1911-1914.
Jongejan JA, van der Meer RA, van Zuylen GA and Duine JA, 1987.
    Spectrophotometric studies on pyrroloquinolinequinone-copper(II)
    complexes as possible models for copper-quinoprotein amine oxidases.
    Recl. Trav. Chim. Pays-Bas 106: 365.
Peisach J and Blumberg WE, 1974. Structural implications derived from the
    analysis of electron paramagnetic resonance spectra of natural and
    artificial copper proteins. Arch. Biochem. Biophys. 165: 691-708.
Suzuki S, Sakurai T, Nakahara A, Manabe T and Okuyama T, 1983. Effect of
    metal substitution on the chromophore of bovine serum amine oxidase.
    Biochemistry 22: 1630-1635.
Suzuki S, Sakurai T, Itoh S and Ohshiro Y, 1988a. Preparation and Charac-
    terization of ternary copper(II) complexes containing coenzyme PQQ and
    bipyridine or terpyridine. Inorg. Chem. 27: 591-592.
Suzuki S, Sakurai T, Itoh S and Ohshiro Y, 1988b. Characterization of
    ternary copper(II) complexes containing reduced PQQ ($PQQH_2$) and
    bipyridine or terpyridine. Chem. Lett.: 777 780.

^{13}C and ^{1}H NMR Studies of PQQ and Selected Derivatives

David R. Houck and Clifford J. Unkefer
Los Alamos National Laboratory INC-4
University of California
Los Alamos, NM, USA 87545

Introduction

The ortho-quinone structure of PQQ is famous for its reactivity with nucleophilic species of carbon, nitrogen, and oxygen(Duine et. al. 1987). In fact, the crystal structure of PQQ was solved in the form of the C-5 acetone adduct(Salisbury et. al 1979). The propensity of the ortho-quinone to accept nucleophiles is the chemical basis of the function of PQQ at enzyme active sites. The present study focuses on the NMR of PQQ and various derivatives formed with oxygen and nitrogen nucleophiles. Our goals are to assign the ^{1}H, ^{13}C, and ^{15}N NMR spectra and to rigorously confirm the structures of the adducts. Once the NMR data of the relevant adducts are well defined, we will use ^{13}C-and ^{15}N-labeled substrates to probe the active sites of PQQ-containing enzymes.

Materials and NMR Spectroscopy

Biosynthetic PQQ was isolated(modification of Ameyama et.al. 1984) from the culture broth of *Methylobacterium* AM1 which had been grown on methanol or ethanol in mineral-salts medium(Beardsmore et.al. 1982). Commercial material was purchased from Fluka Chemical Co. For NMR, 25 mg of PQQ was dissolved in 2.3 ml d_6-DMSO and placed in a 10-mm NMR tube. Derivatives were prepared by adding a reagent directly to the NMR sample. Spectra were obtained at 50.3 MHz on a Bruker AM200 under the following conditions: 45° pulse, 8200 Hz sweep width, 16K data points, with the decoupler gated off for 2 to 10 sec.

Results and Discussion

Assignment of the ^{13}C NMR spectrum of PQQ A prerequisite to studies on derivatives was the unambiguous assignment of each of the fourteen ^{13}C NMR signals of PQQ (Table 1). This was achieved by analysis of the ^{1}H-^{13}C coupling patterns ($^{1}J_{CH}$ and $^{3}J_{CH}$) and carbon-carbon correlations. These data agree with the partial assignments made by Duine and coworkers(1981). Carbon-carbon couplings were observed using a sample of [U-^{13}C]PQQ (90+ %^{13}C) isolated from cultures grown on [^{13}C]methanol (99.7%). The complete assignment was achieved by selecting for one-bond ^{13}C coupling interactions ($^{1}J_{C-C}$=55 Hz) in ^{13}C COSY experiments .

Reaction of PQQ with oxygen nucleophiles. A solution of PQQ in DMSO was treated with 50% [^{18}O]H$_2$O (4% v/v). The rate of exchange of ^{18}O into the ortho-quinone carbonyls was followed by ^{13}C NMR (Fig 1). Substitution of ^{18}O for ^{16}O at either of the ortho-quinone carbonyls yields resolvable resonances (0.04 -0.05 ppm upfield shift); therefore, the ratio of signals from ^{18}O and ^{16}O species can be used to monitor the

256

J. A. Jongejan and J. A. Duine (eds.), PQQ and Quinoproteins, 256–258.
© 1989 by Kluwer Academic Publishers.

exchange. C-5 equilibrated with H_2O within 20 h, whereas C-4 did not reach equilibrium even after 36 h.

The reaction of methanol with PQQ was also monitored by ^{13}C NMR. A solution of PQQ was treated with methanol (20% v/v in d_6-DMSO). The spectrum contained 28 resonances, indicating the sample was a mixture of PQQ and its methanol adduct. The methanol derivative of PQQ had two new resonances (93.7 ppm and 186.4 ppm) that could be assigned to C-4 and C-5. The chemical shift at 93.7 is typical of a hemiketal carbon; the resonance at 186.4 would arise from the neighboring carbonyl. To determine which of the possible hemiketals was present, this experiment was repeated using [5,9a-[13]C_2]PQQ. The resulting ^{13}C spectrum contained four resonances: two from PQQ (179.2 and 126.1 ppm) and two from the methanol adduct. The hemiketal carbon was labeled with [13]C (93.7 ppm), but the resonance of the downfield carbonyl(186.4 ppm) was not observed. This provides direct physical evidence that the methanol adduct is the C-5 hemiketal.

Reaction with nitrogen nucleophiles. PQQ was treated with a 10% molar excess of d_5-[1-^{15}N]phenylhydrazine·HCl in d_6-DMSO. The ^{1}H NMR of the reaction mixture contained two resonances: the signal of H-8 was shifted 0.05 ppm upfield, and the signal of H-3 was shifted 0.3 ppm downfield from the corresponding signals of PQQ. Only 12 signals were observed in the ^{13}C NMR spectrum, even when using long repetition times(up to 15 sec). No lines in the spectrum showed coupling to ^{15}N and the signals for C-4 and C-5 were not observed. The compound was isolated by reverse-phase flash chromatography (0.005 M HCl followed by 75% aqueous methanol). The ^{1}H and ^{13}C NMR spectra of the resulting compound were identical to those of the original reaction mixture. Therefore, reaction with phenylhydrazine produces a single derivative of PQQ. The absence of ^{13}C NMR signals from C-4 and C-5 of this adduct is very likely due to azo-hydrazone tautomerism. The azo-hydrazone tautomeric equilibrium has been determined for several systems by ^{15}N NMR(Lycka et.al. 1981 and references therein), and we plan to use such methods to characterize the equilibrium for the PQQ-phenylhydrazone adduct.

Table I. ^{13}C NMR Chemical Shift Assignments of PQQ

Carbon	δ,ppm	$^1J_{CC}$, Hz($^2J_{CC}$, Hz)	$^1J_{CH}$, Hz
1a	136.7	61,62	
2	127.6	65,87	
2'	161.3	87	
3	113.8	61,63	178
3a	123.4	60,63	
4	173.4	58,61	
5	179.2	60(6.5)	
5a	148.1	59	
7	146.5	80,59	
7'	165.4	80	
8	130.3	57,58	170
9	142.2	58,59,65	
9'	167.2	65	
9a	126.1	56(6.5)	

258

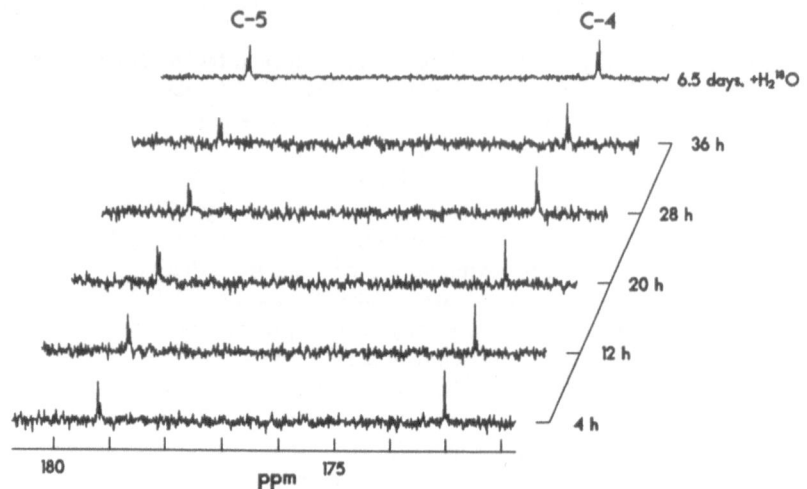

Figure 1. ^{13}C NMR spectra of PQQ detailing the exchange of the o-quinone oxygens with $H_2{}^{18}O$.

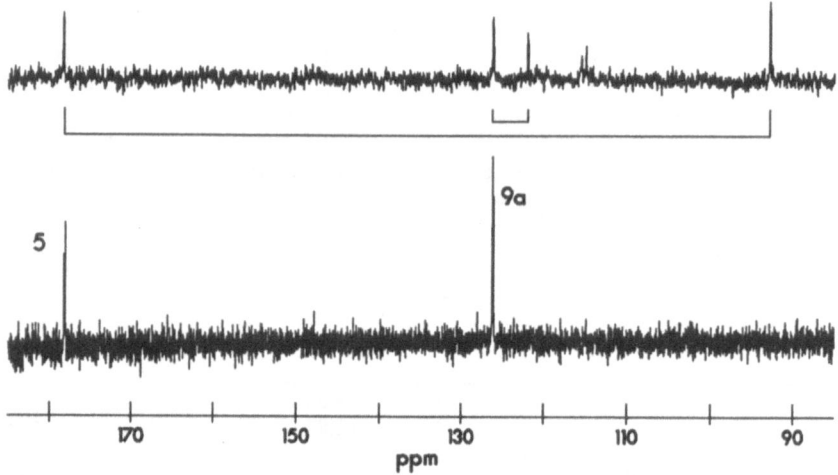

Figure 2. ^{13}C NMR spectra of [5,9a-C2]PQQ and its C-5 methanol hemiketal.

References

Ameyama, M., Shinagawa, E., Matsushita, K. and Adachi, O., 1984. *Agric. Biol. Chem.* **48**, 2909-2911.

Beardsmore, A.J., Aperghis, P.N.G. and Quayle, J.R., 1982. *J. Gen. Microbiol.* **128**, 1423-1439.

Duine, J.A., Frank, J. Jzn, and VerWiel, P.E.J., 1981. *Eur. J. Biochem.* **118**, 395-399.

Duine, J.A., Frank, J. Jzn, and Jongejan, J.A., 1987. *Advances in Enzymology and Related Areas of Molecular Biology* (Meister A Ed.) 169-211.

Lycka, A., Snobl, S., Machacek, V. and Vecera, M., 1981. *Org. Mag. Resonance* **16**, 17-19.

Salisbury, S.A., Forrest, H.S., Cruse, W.B.T. and Kennard, O., 1979. *Nature (London)* 843-44.

Mechanism of the Methylamine Dehydrogenase Reductive Half Reaction

Robert B. McWhirter and Michael H. Klapper, The Division of Biological Chemistry, Department of Chemistry, The Ohio State University, Columbus, OH 43210, USA

SUMMARY

Bacterial W3A1 methylamine dehydrogenase (MADH) catalyzes the reductive deamination of primary amines. With no electron acceptor present, the amine reduces the enzyme bound methoxatin (PQQ) to a "hydroquinone" species, and product aldehyde, but no ammonia, is released. In stopped flow experiments, we observe 2 kinetically significant intermediates before and 1 after the functionally irreversible PQQ reduction, a reduction that displays an exceptionally large H/D kie. We observe no other substantial H/D kie in the reductive half-reaction. The reaction that follows PQQ reduction, the release of aldehyde from the enzyme, is either fully or partially rate limiting in the catalytic turnover of methylamine and propylamine with phenazine methosulfate as the 1-electron acceptor. Finally, we have been able to reconstruct visible absorption spectra of two proposed catalytic intermediates. Our data suggest that the MADH reductive half reaction may have a mechanism similar to that proposed by Eckert and Bruice (J. Am. Chem. Soc. 105 4431, 1983) for amine reduction of model $\underline{o}$-quinones.

INTRODUCTION

The importance of the methoxatin (PQQ) containing enzymes as a new class of protein catalysts becomes clearer as more of these quinoproteins are found in an ever widening range of organisms. However, there is still no detailed catalytic mechanism based on a comprehensive set of experimental results for any one quinoenzyme. Our goal is to obtain such mechanistic detail with methylamine dehydrogenase (MADH), the PQQ enzyme that catalyzes the oxidative deamination of primary amines.

$$RCH_2NH_3^+ + 2A + H_2O \longrightarrow RCHO + 2A^- + NH_4^+ + 2H^+ \tag{1}$$

When we began, our understanding of the MADH mechanism was sketchy. Steady-state kinetic studies indicated a ping-pong scheme between primary amine and electron acceptor (Eady and Large, 1971; McIntire, 1987). Kenny and McIntire (1983) reported that methylamine reduces W3A1 MADH to its semi-quinone form. While this observation suggested the possible catalytic importance of 1-electron transfer from amine to enzyme bound PQQ, we know now that primary amines effect the 2-electron reduction of the enzyme bound PQQ to its "hydroquinone" form (Husain, et al., 1987; McWhirter and Klapper, 1987). The semiquinone is formed from amine reduction when a mixture of oxidized and fully reduced protein slowly disproportionates. The stoichiometric evidence that has been accumulated supports the following ping-pong

J. A. Jongejan and J. A. Duine (eds.), PQQ and Quinoproteins, 259–268.
© 1989 by Kluwer Academic Publishers.

catalytic scheme:

$$RCH_2NH_3^+ + E\text{-}Q(ox) \longrightarrow RCHO + (NH_2)\cdot E\text{-}Q(red) \qquad (2a$$

$$(NH_2)\cdot E\text{-}Q(red) + 2A \longrightarrow NH_4^+ + 2A^- + E\text{-}Q(ox) \qquad (2b$$

We here show that the reductive half-reaction [2a] is itself composed of a number of kinetically distinguishable steps.

MATERIALS AND METHODS

Preparations of MADH from W3A1 by the method of Chandrasekar and Klapper (1986) are mixtures of oxidized, semiquinone and fully reduced protein species. We, therefore, routinely oxidize MADH with a slight excess of ferricyanide and separate protein from ferri- and ferrocyanide by passage through a Dowex 1-X8 column (Klapper and Klotz, 1968). Solutions of the fully oxidized MADH are stable at -20° C for months. MADH concentrations were determined from the solution absorbance at 444 nm using an extinction coefficient of 8340±230, a value obtained by titration of MADH with phenylhydrazine. α,α-Dideutero-$\underline{n}$-propylamine was prepared by reduction of freshly distilled propionitrile with LiAlD$_4$ and then converted to the HCl salt (Soffer and Katz, 1956). Perdeuteromethylamine and LiAlD$_4$ were purchased from Aldrich (Milwaukee, WI). All other chemicals were purchased from commercial sources and used with no further purification.

All UV-visible spectra were obtained with a Uvikon 820 (Kontron, Chicago, IL). Kinetic experiments were performed with a Dionex (Sunnyvale, CA) or a Kinetic Instruments (Ann Arbor) stopped flow spectrophotometer, both equipped with OLIS (Jefferson, GA) data acquisition systems. We are grateful to Professors D. Busch and R. Hille for permitting us the use of these instruments. We converted all data acquired on these machines into ASCII files that were then edited with PC-Write (Quicksoft, Seattle, WA) for subsequent nonlinear fitting with the software package MINSQ (MicroMath, Salt Lake City, Utah).

RESULTS AND DISCUSSION

MADH has three spectrally distinct, relatively stable oxidation states: the oxidized E-Q, the semiquinone E-QH$\cdot$, and the fully reduced E-QH$_2$ species (Figure 1). Addition of a primary amine to the enzyme in the absence of an electron acceptor reduces E-Q to the 2-electron product E-QH$_2$ with no thermodynamically significant intermediate found (Husain, et al., 1987; McWhirter and Klapper, 1987). The stoichiometry indicated in equation [2a] was established for the primary amine consumption in both these laboratories; we have also determined the stoichiometry of aldehyde production and the absence of ammonium release (McWhirter and Klapper, submitted for publication).

With the stochiometry of the reductive half-reaction [2a] established we turned to the kinetics of this process. Figure 2 contains the results of two stopped-flow experiments with the reductant propylamine saturating. In the left panel is the reaction time profile obtained at 500 nm where we expect an absorbance decrease on MADH reduction (see Figure 1). The

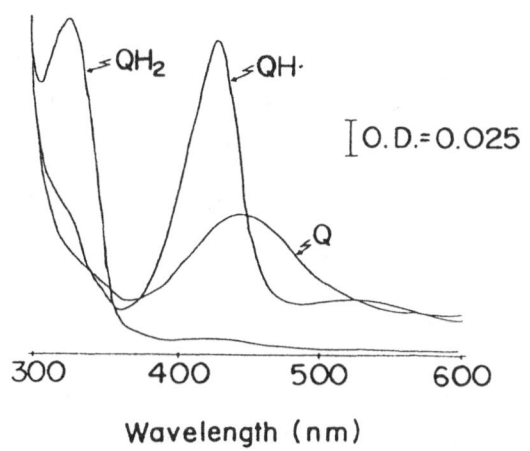

Figure 1. Redox states of MADH

transition with the 23 s^{-1} rate constant meets this expectation, but there is also an earlier absorbance increase. From kinetic evidence not illustrated here, we know that there are a minimum of two processes in this faster phase. In the right panel is the time profile of the same reaction but monitored at 350 nm where we expect an absorbance increase (Figure 1). Once again our expectation is met by a process with a rate constant of 23 s^{-1}, but at this wavelength we also find a slower absorbance decrease. Based on the spectral changes at these two and other wavelengths, we propose that the "23 s^{-1}" reaction is the conversion of E-Q to E-QH$_2$. Kinetic isotope effects (kie) to be described shortly support this proposal. We shall also prove that the slow "4 s^{-1}" process is the release of propanal from the enzyme.

The apparent 1st-order rate constant of the reaction that we propose to be PQQ reduction has a propylamine concentration dependence (Figure 3) that can be fit to the equation for the simple model:

$$S + E\text{-}Q \underset{l_i}{\overset{k_1}{\rightleftharpoons}} X \xrightarrow{k_{red}} \text{"E-QH}_2\text{"} \tag{3}$$

with the second reaction rate determining. When the substrate is dideuterated α,α-D2-propylamine we find a similar concentration dependence. The calculated dissociation constants are closely similar for protio- and dideuteropropylamine, but the rate-determining step is very sensitive to the substitution of deuterium on the propylamine α-carbon with an apparent kie of 16.5±0.4. It should be recognized that this number must reflect both a primary and a secondary effect since two deuteriums are on the alpha-carbon. Nonetheless, the primary kie must be very large; and, hence, this step must involve proton abstraction from the propylamine alpha-carbon, consistent with an intramolecular redox reaction. (We shall not discuss the significance of this very large kie, other than to note that Palcic and Klinman (1983) have observed a very large kie with the PQQ enzyme plasma amine oxidase.)

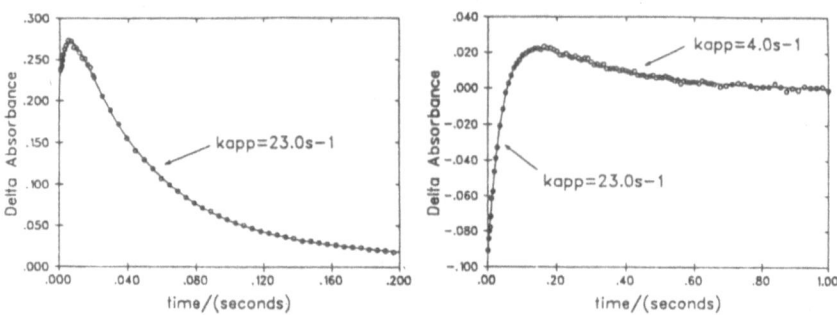

Figure 2. Propylamine reduction of MADH: kinetics

Reaction conditions were those of Table I with a substrate concentration of 9.3 mM. The circles are data points, the lines are calculated from the constants obtained in a nonlinear least squares fit to an equation of 2 exponentials.

We can show that the slow "4 s^{-1} process during the reductive half-reaction is propanal release from the enzyme by coupling with NADH oxidation in the horse liver alcohol dehydrogenase (LADH) catalyzed reaction:

$$NADH + CH_3CH_2CHO + H^+ \xrightarrow{\text{LADH}} NAD^+ + CH_3CH_2CH_2OH \tag{4}$$

With LADH and NADH present in large excess, aldehyde is converted to alcohol as fast as it is released from the enzyme; the rate of NADH oxidation, monitored as an absorbance decrease at the MADH isosbestic point of 356 nm, measures the rate of propanal release. Under these conditions and with propylamine saturating, MADH reduction, measured from the absorbance decrease at 440 nm, has the associated 1st-order rate constant of 23 s^{-1}, as expected; propanal release, measured from the absorbance decrease at 356 nm, has the associated rate constant of 4 s^{-1} (Figure 4). Hence, the "4 s^{-1}" process involves aldehyde release from the quinoprotein.

In the data presented so far, we have dealt only with two fragments of the overall reduction. We turn now to the complete scheme for amine reduction of MADH, and we propose the following set of reactions, a proposal that is based on many kinetic experiments with both methyl- and n-propylamine as reductants in the absence of an added electron acceptor.

$$E_Q + RCH_2NH_3^+ \underset{k_d}{\overset{K_d}{\rightleftharpoons}} X1 \underset{l_2}{\overset{k_2}{\rightleftharpoons}} X2 \underset{l_{red}}{\overset{k_{red}}{\rightleftharpoons}} X3$$

$$\xrightarrow{k_4} X_4 \xrightarrow{k_5} E_{QH_2} + RCHO \tag{5}$$

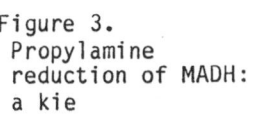
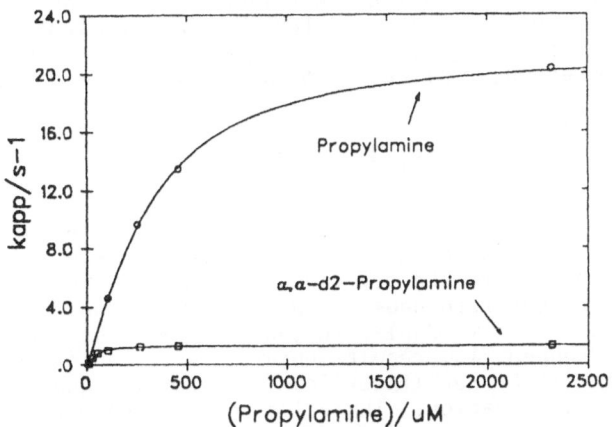

Figure 3.
Propylamine
reduction of MADH:
a kie

Reaction conditions are those of previous experiments with PQQ reduction
measured at 500 nm. The measured pseudo 1st-order reduction rate constants
are plotted against the concentrations of either protio or dideutero propylamine. The k_1, l_1 and k_{red}, of the simple scheme [3] were obtained by
nonlinear least squares fitting.

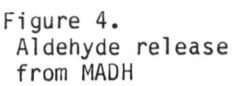
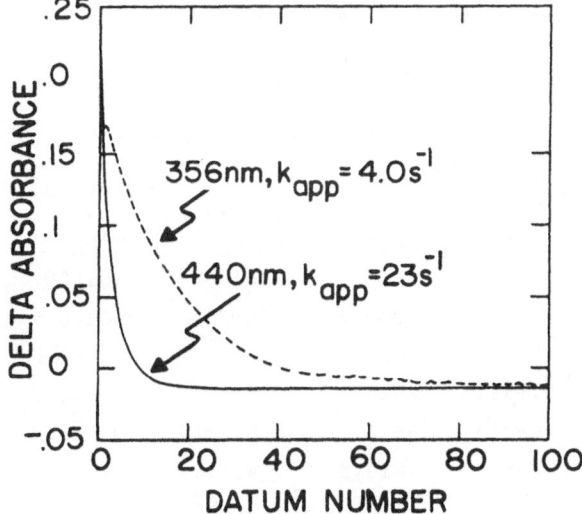

Figure 4.
Aldehyde release
from MADH

Reaction conditions were those of previous experiments except for the
addition of LADH (50 µM) and NADH (0.1 mM)
    ⎯⎯ absorbance change at 440 nm; enzyme reduction
    - - - absorbance change at 356 nm; NADH oxidation

Both the transient and steady-state equations based on scheme [5] can be derived under the assumptions that i) there is a rapid equilibrium forma- tion, with the dissociation constant K_d, of the Michaelis-Menten complex X_1, and ii) amine concentration is much greater than enzyme concentration. The transient equation, applicable to the stopped-flow data, is:

$$A(t) = a_0 + \sum_{i=1}^{3} a_i \exp[-\alpha_i t] \qquad (6$$

in which $A(t)$ is the time dependent absorbance. The a_i reflect the magni- tudes of the absorbance changes associated with the various observed transi- tions, and the α_i are the negative roots of polynomials, the power of which depend on the substrate under study. These α_i contain the equilibrium constant K_d, the individual rate constants and the substrate concentra- tion. The stopped-flow results obtained with various concentations of the amine can be fit by non-linear least squares to equation [6] and its associ- ated polynomial equation. The values of the scheme [5] constants obtained by these fits are presented in Table I for both methyl- and n-propylamine.

We shall consider a few of the conclusions we have drawn from the numbers in Table I. First, the calculated propylamine H/D kie is the same, within experimental error, as that obtained when the simpler scheme [3] is applied to the data of Figure 3. (The larger estimated standard deviation in the second, more complete analysis arises in part from the many more constants required by the more complex model.) Therefore, it is improbable that the very large kie would become much smaller were the model for the analysis made more complex. A very large kie is also associated with the methylamine reduction of MADH, as seen in Table I.

TABLE I

Stopped Flow Rate Constants

	methyl- amine	perdeutero methyl- amine	kie	propyl- amine	dideutero propyl- amine	kie
$K_d{}^{a}\cdot$	800±200	1000±300	0.8±0.3	3300±960	2500±600	1.3±0.5
$k_2{}^{b}\cdot$	1066±88	850±133	1.3±0.2	296±23	240±40	1.2±0.2
l_2	38±34	37±11	1.0±0.9	17±9	3.8±1.0	4.5±2.6
k_{red}	193±15	9.9±1.7	20±4	23±3	1.3±0.2	18±4
l_{red}	4±160	--c.	--	0.1±10	0.1±0.2	1±100
k_4	22±3	22±5	1.0±0.3	4.1±0.2	--c.	--
k_5	118±10	--c.	--c.	--c.	--c.	

a. The unit is μmolar.
b. The units for all the rate constants are s^{-1}.
c. Could not be detected.

TABLE II

Steady-State Kinetic Constants

| | Experimental | | Calculated | |
	methyl- amine	propyl- amine	methyl- amine	propyl- amine
$k_{cat}(s^{-1})$	15.4(0.2)	3.1(0.1)	16.4	3.4
K_m (μM)	26 (2)	57 (10)	15	67
k_{cat}/K_m (μM^{-1}s^{-1})	0.60(0.03)	0.054(0.010)	1.1	0.051

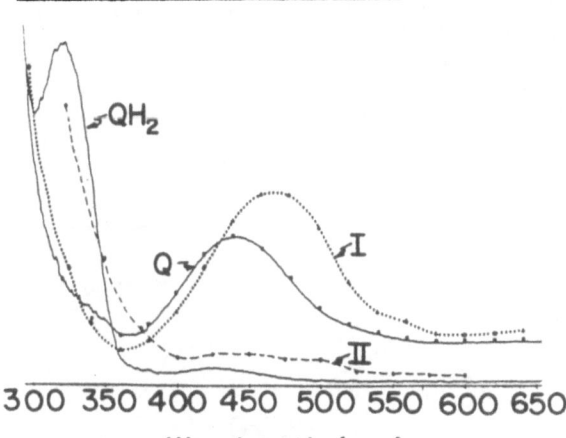

Figure 5. Estimated Spectra of Kinetically Significant Reaction Intermediates

Q - Spectrum of oxidized MADH - the solid line is the spectrum of the oxidized protein; the points were calculated from the stopped-flow data.

QH$_2$- Spectrum of amine reduced MADH.

I - Spectrum of an α,α-dideuteropropylamine accumulate before PQQ reduction - the arbitrarily drawn curve connects points calculated from the kinetic data and normalized to the QH$_2$ spectrum.

II - Spectrum of a methylamine intermediate that accumulates after PQQ reduction and before formaldehyde release - the arbitrarily drawn curve connects points computed from the kinetic data and normalized to the QH$_2$ spectrum

A second conclusion is suggested by the comparison of experimental and calculated steady-state constants. From the steady-state solution to the reaction scheme [5] we have obtained the expressions for k_{cat} and K_m as functions of K_d and the various rate constants.

$$K_m/k_{cat} = K_d/k_2 + K_dK_2/k_3 + K_dK_2K_3/k_4 \qquad\qquad K_i = k_i/l_i$$

$$1/k_{cat} = 1/k_2 + (K_2 + 1)/k_3 + (K_2K_3 + K_3 + 1)/k_4 + 1/k_5$$

(7

We can, therefore, calculate k_{cat} and K_m from the constants of Table I. A comparison of these calculated numbers with the experimentally measured steady-state numbers show that the two sets are very close if not identical (Table II). This camparison, thus, shows that reoxidation [2b] must be faster than reduction [2a] and supports strongly the validity of scheme [5] as a model of the reductive half-reaction.

Our third conclusion, based on the rate constants of Table II, is the expectation that because PQQ reduction by enzyme bound dideuteropropylamine is slower than formation of the species X2, X2 should accumulate transiently. From the reaction time profiles at different wavelengths, we can construct the spectrum of an intermediate (Figure 5, spectrum I), presumably X2, that does accumulate before PQQ reduction. For comparison purposes, we include the spectrum of the MADH/ammonia adduct in Figure 6. The qualitative resemblance of the two spectra suggests that X2 may be an amine adduct.

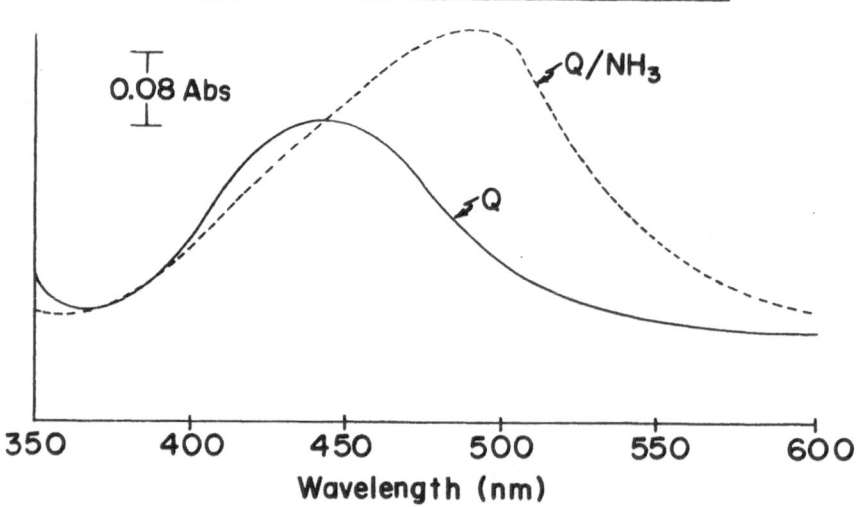

Figure 6. Spectrum of MADH - ammonium adduct

Q   - Spectrum of oxidized MADH
Q/NH_3 - Spectrum of MADH plus 0.21 M ammonium sulfate
Conditions are those of Figure 1 with ammonium sulfate added

Finally, we see from the constants of Table II that formaldehyde release from MADH is approximately 5-fold slower than methylamine reduction of the coenzyme (k_4 vs. k_{red}). We would expect, therefore, and do find an additional spectral intermediate (Figure 5) presumably X3, the species that contains both covalently bound formaldehyde and the bound methylamine nitrogen.

DISCUSSION

The stoichometric and kinetic data we present here is insufficient for the unambiguous assignment of structures to the proposed enzyme inter-mediates. The reduction half-reaction occurs with the release of the alde-hyde product but not of the amine nitrogen. A large kinetic isotope effect associated with spectral changes that signal a 2-electron reduction of the enzyme bound PQQ suggests a proton abstraction during an intramolecular electron redox process. There is no evidence for semiquinone formation at any point during the reaction [2a]. But while a role for the semiquinone appears ruled out in this step, the stability of the MADH semiquinone together with MADH utilization of 1-electron acceptors for reoxidation suggest that the semiquinone may be an intermediate in the oxidative half-reaction [2b].

It is too early to assign a mechanism to the reduction reaction [2a]. Our data is, however, consistent with one proposed for o-phenanthroline-quinone (a methoxatin model) catalyzed amine oxidations (Figure 7; Eckert and Bruice, 1983). Were this the enzymatic mechanism, then species X2 of equation [5] could be a quinoneamine, species X3 an aldimine and the fully reduced MADH species the aminoquinol.

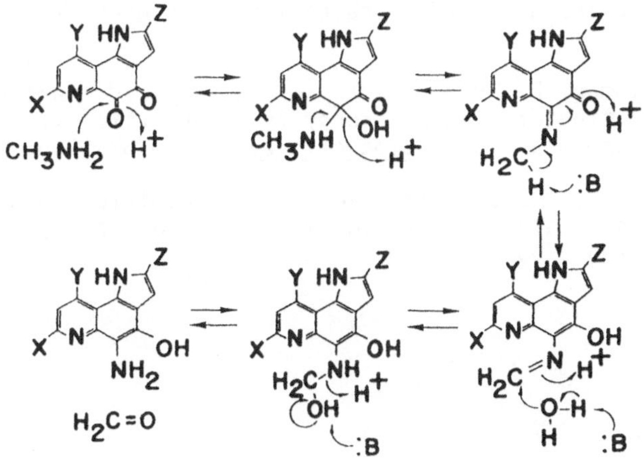

Figure 7. Consistent Mechanism for MADH Reductive Half-Reaction

REFERENCES

Chandrasekar R. and Klapper MH, 1986. Methylamine dehydrogenase and cytochrome C_{552} from the bacterium W3A1. J. Biol. Chem. 261: 3616-3619.

Eady RR and Large PJ, 1971. Microbial oxidation of amines. Spectral and kinetic properties of the primary amine dehydrogenase of Pseudomonas AM1. Biochem. J. 123: 757-771.

Eckert TS and Bruice TC, 1983. Chemical properties of phenanthroline-quinones and the mechanism of amine oxidation by o-quinones of medium redox potential. J. Am. Che. Soc. 105: 4431-4441.

Husain M, Davidson VL, Gray KA and Knaff DB, 1987. Redox properties of the quinoprotein methylamine dehydrogenase from Paracoccus denitrificans. Biochemistry 26: 4139-4143.

Kenney WC and McIntire W, 1983. Characterization of methylamine dehydro-genase from bacterium W3A1. Interaction with reductants and amino contain-ing compounds. Biochemistry 22: 3858-3868.

Klapper MH and Klotz IM, 1968. Cooperative interactions and determination of protein association-dissociation equilibria. Hemerythrin. Biochemistry 7: 223-231.

McIntire WS, 1987. Steady-state kinetic analysis for the reaction of ammonium and alkylammonium ions with methylamine dehydrogenase from bacterium W3A1. J. Biol. Chem. 262: 11012-11019.

McWhirter RB and Klapper MH, 1987. Methoxatin containing methylamine dehydrogenase: enzyme reduction by primary amines. in: Edmondson DE and McCormick DB. Flavins and Flavoproteins 1987. de Gruyter, Berlin. pp 709-712.

Palcic MM and Klinman JP, 1983. Isotopic probes yield microscopic constants: separation of binding energy from catalytic efficiency in the bovine amine oxidase reaction. Biochemistry 22: 5957-5966.

Soffer LM and Katz M, 1956. Direct and reverse addition reactions of nitriles with lithium aluminum hydride in ether and tetrahydrofuran. J. Am. Chem. Soc. 78: 1705-1709.

ELECTRON TRANSPORT FROM METHYLAMINE TO OXYGEN IN THE GRAM-NEGATIVE BACTERIUM *THIOBACILLUS VERSUTUS*

J.E. van Wielink, J. Frank Jzn and J.A. Duine

Department of Microbiology & Enzymology, Technical University Delft
Julianalaan 67, 2628 BC Delft, The Netherlands

INTRODUCTION

Some Gram-negative bacteria are able to grow on methylamine as a carbon and energy source by virtue of the periplasm-located quinoprotein methylamine dehydrogenase (MADH) (Burton *et al.*, 1983; Kasprzak and Steenkamp, 1983; Husain and Davidson, 1985). MADH catalyses the oxidation of methylamine to formaldehyde, ammonia, protons and electrons. Formaldehyde can be assimilated into cell material, while by electron transport from methylamine (and also from formaldehyde) to oxygen the energy is provided for the synthesis of ATP from ADP and inorganic phosphate (Anthony, 1982, 1988). Controversion exists, however, on which periplasmatic components are involved as intermediates in the electron transfer from MADH to the cytochrome aa_3-type oxidase, the final component in the electron transport chain. As has been proposed for *Paracoccus denitrificans* and *Methylobacterium extorquens* AM1, formerly known as *Pseudomonas* AM1 or *Methylobacterium* AM1, the electrons flow from MADH to amicyanin (a Type I blue copper protein) and subsequently to a cytochrome *c* (Tobari and Harada, 1981; Husain and Davidson, 1985; Gray *et al.*, 1986, 1988). On the other hand, Fukumori and Yamanaka (1987) have found that a cytochrome *c* with a high isoelectric point (c_H) was an efficient electron acceptor for the MADH from *M. extorquens* AM1. Although addition of both amicyanin and a cytochrome *c* of low isoelectric point (c_L) gave an increase of the reaction rate, the former observations cast some doubt on the view that amicyanin is the primary electron acceptor for MADH. In this respect, an even better illustrative example is bacterium W3A1, an organism in which amicyanin has not been found and where cytochrome c_{552} has been proposed to be the primary electron acceptor for MADH (Chandrasekar and Klapper, 1986). Apart from the uncertainty on the nature of the primary electron acceptor for MADH, the question can be posed whether these *in vitro* observations have any significance for the *in vivo* situation, in other words, are the observed rates high enough to explain the respiration rates of whole cells? Methylamine grown *Thiobacillus versutus* contains MADH (Vellieux *et al.*, 1986), amicyanin (Van Houwelingen *et al.*, 1985) and two *c*-type cytochromes (cytochrome c_{550} and cytochrome c_{552}, see Van Wielink *et al.*, 1988). To obtain insight into the sequence of electron transfer from MADH, the rate constants were determined for electron transfer between the components. To check the relevance of this for the *in vivo* situation, a model was developed using the measured rate constants and the calculated overall rate was compared with the respiration rates of whole cells.

Abbreviations: AMI, amicyanine; c_{550}, cytochrome c_{550}; MADH, methylamine dehydrogenase; $MADH_{ox}$.S, the complex of substrate and the oxidized form of MADH; $MADH_{red}$.P, the reduced form of MADH with NH_3 still attached; sem, the semiquinone form; v, reaction rate; V_t, volume reaction vessel; V_c, volume of the cells; V_p, volume of the periplasm; $[\ ]_o$, initial concentration.

J. A. Jongejan and J. A. Duine (eds.), PQQ and Quinoproteins, 269–278.
© 1989 by Kluwer Academic Publishers.

MATERIALS AND METHODS

Cell growth. *Thiobacillus versutus* was grown aerobically in a fed-batch culture with methylamine as carbon and energy source. When the cells were in the late log-phase, they were harvested and subsequenly stored at -20 °C.

Respiration studies with whole cells. The O_2-uptake rates of whole cells were determined polarographically with a Clark-type oxygen electrode (Yellow Springs Instruments, Inc., Yellow Springs, Ohio) at 20 °C in 0.01 M MOPS buffer with KCl, pH 7.0 (ionic strength 0.06).

Purification of the components. After cell disruption, MADH, amicyanin, cytochrome c_{550} and cytochrome c_{552} were purified at medium large scale by means of successively, hydrophobic interaction chromatography, ion-exchange chromatography and gelfiltration. The first two purification steps were performed with radial flow colums. Gelfiltration was performed with a conventional column or with a FPLC-Superose column (Van Wielink *et al.*, 1988).
Oxidized preparations of MADH, amicyanin, cytochrome c_{550} and cytochrome c_{552} were obtained by the addition of Wurster's blue or ferricyanide. Reduced preparations were obtained by the addition of a slight excess of methylamine (MADH), methylamine with a catalytic amount of MADH (amicyanin) or dithionite (cytochromes c). After oxidation or reduction, contaminants were removed from the proteins by gelfiltration.

Fast kinetics. Measurements of electron transfer between methylamine and MADH, and between MADH, amicyanin, cytochrome c_{550} and cytochrome c_{552} were done on a SF53 apparatus from Hi-Tech Scientific (Salisbury) at 20 °C in 0.01 M MOPS buffer, pH 7.0 with KCl added till an ionic strength of 0.06. Rates of absorbance decrease or increase were determined at wavelengths giving minimal interference (See Table 1 for an enumeration of the absorption maxima of the components).
Rate constants were calculated by means of a program written under ASYST (Macmillan Software Company).

For the reaction $A_{red} + B_{ox} \xrightarrow{k} A_{ox} + B_{red}$, in which k is the rate constant:

$$d[A_{red}]/dt = -k\ [A_{red}]\ [B_{ox}]$$

Defining $k' = k\ [B_{ox}]$ as the pseudo first order rate constant:

$$d[A_{red}]/dt = -k'\ [A_{red}]$$

or: $[A_{red}] = [A_{red}]_o \exp(-k't)$, in which $[A_{red}]_o$ is $[A_{red}]$ at t = 0

Simulations. Electron transport from methylamine to oxygen was simulated using STELLA (Software from High Performance Systems, New Hampshire) running on an Apple Macintosh computer (see Hayashi and Sakamoto, 1986).

RESULTS AND DISCUSSION

Experimental O_2-uptake rates. Suspensions of *Thiobacillus versutus* cells (14 mg wet weight/1.8 ml buffer) gave an O_2-uptake rate of 20 ± 2 nmol O_2/min in the presence of 10 mM deuterated methylamine (CD_3NH_2). Increasing the concentration of CD_3NH_2 to 20 mM gave no effect, but extra addition of CH_3NH_2 (10 mM) consistently increased the rate to 25 ± 2 nmol O_2/min. The latter value was also observed with only 10 mM CH_3NH_2.

TABLE 1. Some characteristics of MADH, amicyanin, cytochrome c_{550} and cytochrome c_{552} of *Thiobacillus versutus*. Data from Van Houwelingen *et al.* (1985), Vellieux *et al.* (1986) and Van Wielink *et al.* (1988).

	M_r	$A_{280}^{1\ mg/ml}$	λ_{max} (nm)		E_o'(mV)
			oxidized	reduced	
MADH	123,500	1.14	440	327	n.d.
amicyanin	14,000	1.13	596	----	+260
cytochrome c_{550}	16,800	1.66	----	550	+255
cytochrome c_{552}	25,000	-----	----	552	n.d.

n.d., not determined

Characteristics of MADH, amicyanin, cytochrome c_{550} and cytochrome c_{552}. The characteristics of the redox proteins, relevant for the present work, are given in Table 1. MADH and amicyanin were homogeneous, as judged from their $A_{440\ nm}/A_{280\ nm}$ ratio (Vellieux *et al.*, 1986) and $A_{596\ nm}/A_{280\ nm}$ ratio (Van Houwelingen *et al.*, 1985), respectively. The cytochrome *c* preparations were also homogeneous, as revealed by electrophoresis and protein staining (Van Wielink *et al.*, 1988). The concentrations of the proteins were determined on the basis of their absorbance at 280 nm and "$A_{280}^{1\ mg/ml}$"-value.

Rate constants. Pseudo first order rate constants for the reaction of MADH$_{ox}$ with varying methylamine concentrations (CH$_3$NH$_2$ or CD$_3$NH$_2$) were determined from the decrease in absorbance at 440 nm (Fig. 1). At high substrate concentrations, saturation behaviour was observed. The maximum k_{obs} for CH$_3$NH$_2$ (approximately 150 s^{-1}) was much lower in the case of CD$_3$NH$_2$ (approx. 5 s^{-1}). For CH$_3$NH$_2$ a linear relationship existed in the 0-0.15 mM substrate concentration range (second order rate constant 380×10^3 M^{-1} s^{-1}).

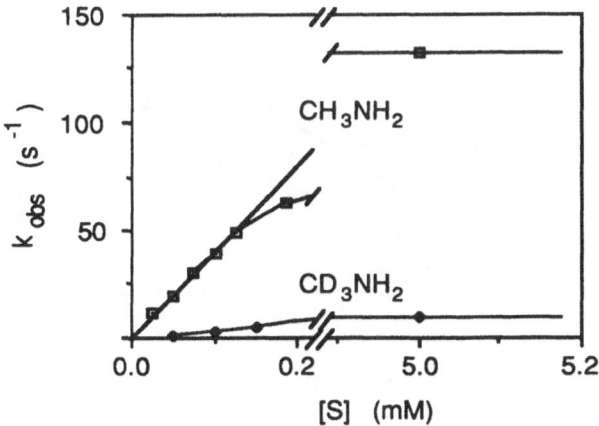

FIGURE 1. Pseudo first order rate constants for the reaction MADH$_{ox}$ (6.25 µM) + S at different substrate concentrations.

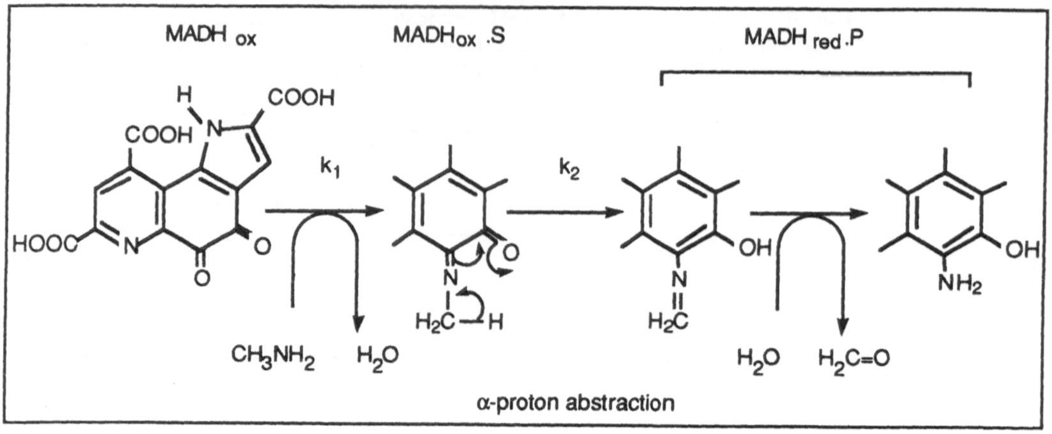

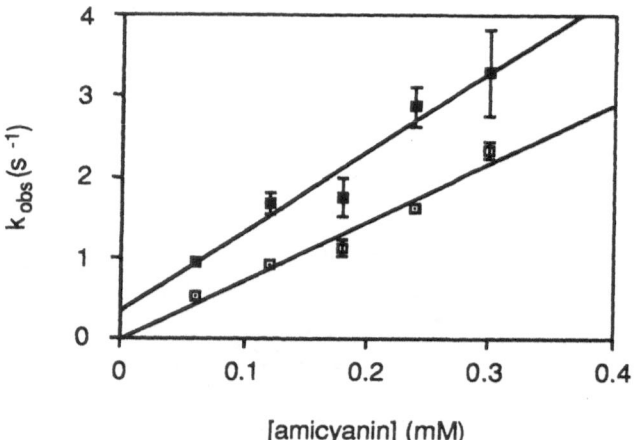

FIGURE 2. The reduction of $MADH_{ox}$ by methylamine.

The phenomena of linearity and saturation can be explained by assuming that first a complex is formed between MA
The phenomena of linearity and saturation can be explained by assuming that first a complex is formed between $MADH_{ox}$ and the substrate with a $k_1=380\times10^3$ M^{-1} s^{-1}, followed by internal conversions from which the rate limiting step (rate constant k_2) determines the maximal k_{obs}. Since CD_3NH_2 gave a much lower maximal k_{obs}, this suggests a rate-limiting α-proton abstraction step, already suggested by Anthony (1982) and observed for model systems (Eckert and Bruice, 1983; Itoh *et al.*, 1986). A scheme for the mechanism is depicted in Fig. 2, where it is also assumed that the end product of the reaction is the 5 amino substituted $MADH_{red}$ ($MADH_{red}.P$), which is converted into $MADH_{ox}$ and NH_3 upon oxidation

FIGURE 3. Pseudo first order rate constants for the reaction of $MADH_{red}P$ (3 μM) with varying amicyanin concentrations. The k_{obs} was calculated on account of the rate of decrease of the A_{596} (black squares) and the increase of the A_{440} (open squares).

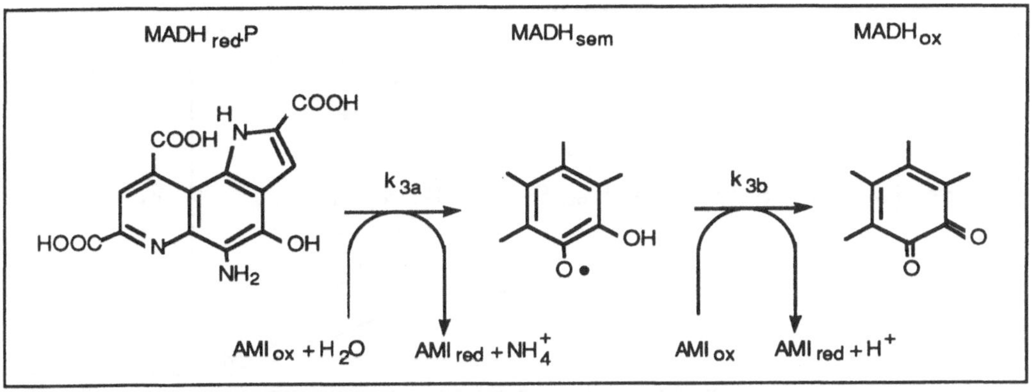

FIGURE 4. The oxidation of $MADH_{red} \cdot P$ by $amicyanin_{ox}$.

$MADH_{red} \cdot P$ was oxidized by $amicyanin_{ox}$. Pseudo first order rate constants for the reaction of $MADH_{red} \cdot P$ with vary
$MADH_{red} \cdot P$ was oxidized by $amicyanin_{ox}$. Pseudo first order rate constants for the reaction of $MADH_{red} \cdot P$ with varying $amicyanin_{ox}$ concentrations were determined from the decrease in absorbance at 596 nm and the increase at 440 nm, attributed to, respectively, the amicyanin reduction and the MADH oxidation (Fig. 3). The linear relationships suggest that $MADH_{red} \cdot P$ is oxidized in two one electron steps, first to the semiquinone and subsequently to the oxidized form (No ternary complex has to be formed, see Fig. 4). The apparent rate constant for $MADH_{ox}$ production (increase of the $A_{440 \, nm}$) is lower than that for $amicyanin_{red}$ production (decrease of the $A_{596 \, nm}$). This difference can be accounted for by assuming that the two steps proceed at similar rates of 20 $mM^{-1} \, s^{-1}$ (k_{3a} and k_{3b} in Fig. 4).

As could be concluded from the increase in absorbance at 550 nm, also the oxidized form of cytochrome c_{550} was able to oxidize $MADH_{red} \cdot P$. The reaction rates ($k_{4a} = k_{4b} = 1.0 \, mM^{-1} \, s^{-1}$) were, however, substantially lower than for amicyanin. In addition, cytochrome c_{550} was able to oxidize $amicyanin_{red}$ ($k_5 = 1.5$ $mM^{-1} \, s^{-1}$). No reaction occurred between cytochrome c_{552ox} and $MADH_{red} \cdot P$ or $amicyanin_{red}$. Thus, if it is assumed that cytochrome c_{550red} transfers its electrons directly to the aa_3-type oxidase, cytochrome c_{552} has no role in electron transfer from methylamine to oxygen. Cytochrome c_{552} may be related to cytochrome c_L, the primary electron acceptor for methanol dehydrogenase in many methylotrophic bacteria and electron donor for cytochrome c_H (cytochrome c_{550} in T. versutus).

<u>Estimated concentrations of MADH, amicyanin and cytochrome c_{550} in the periplasm</u>. Cellular concentrations of MADH, amicyanin and cytochrome c_{550} were determined from the yields after the first chromatography step. To estimate the concentrations of the redox proteins in the periplasm it was assumed that: the cellular volume of 1 g wet weight of cells is 1 ml; the fraction of the volume occupied by the periplasm is between 5 and 20% of the cell volume. A value of 5% can be derived from the 5 nm distance between the cytoplasmic membrane and the outer membrane (Hobot et al., 1984; Oliver, 1987). On account of solute distributions, a value of 20% has been reported (Stock et al., 1977; Oliver, 1987).

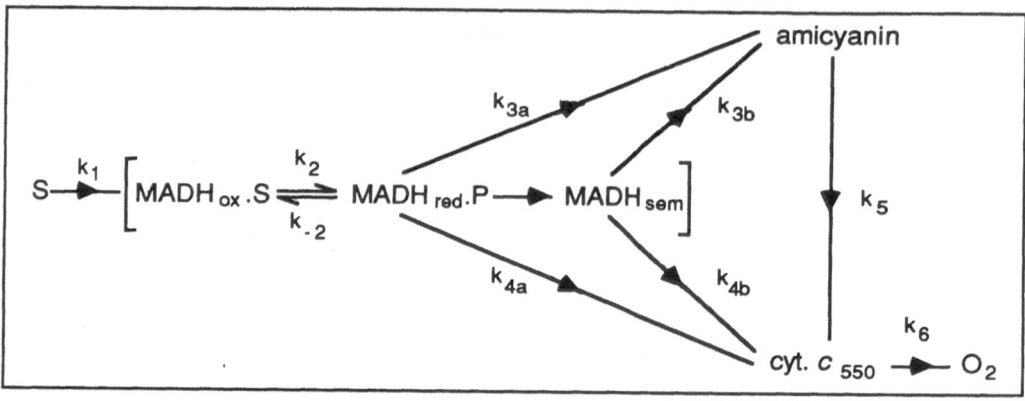

FIGURE 5. Proposed pathways for transfer of reducing equivalents from CH_3NH_2 to O_2.

TABLE 2. Equations, rate constants, initial concentrations, cell volume, volume of the periplasm and volume of the reaction vessel, used to simulate O_2-uptake rates by the cells.

$[S]$	$= [S]_0 + dt\ (-v_1) \times V_t/(V_t-V_c)$	V_t	$=$	1.8×10^{-3}	l
$[MADH_{ox}]$	$= [MADH_{ox}]_0 + dt\ (-v_1+v_{3b}+v_{4b})$	V_c	$=$	$14\ \times 10^{-6}$	l
$[MADH_{ox}.S]$	$= [MADH_{ox}.S]_0 + dt\ (+v_1-v_2+v_{-2})$	$V_p\ (5\%\ V_c)$	$=$	0.7×10^{-6}	l
$[MADH_{red}.P]$	$= [MADH_{red}.P]_0 + dt\ (+v_2-v_{-2}-v_{3a}-v_{4a})$	$[S]_0$	$=$	$10\ \times 10^{-3}$	M
$[AMI_{ox}]$	$= [AMI_{ox}]_0 + dt\ (-v_3+v_5)$	$[MADH_{ox}]_0$	$=$	0.5×10^{-3}	M
$[AMI_{red}]$	$= [AMI_{red}]_0 + dt\ (+v_3-v_5)$	$[MADH_{ox}.S]_0$	$=$	0	M
$[c_{550ox}]$	$= [c_{550ox}]_0 + dt\ (-v_4-v_5+v_6)$	$[MADH_{red}.P]_0$	$=$	0	M
$[c_{550red}]$	$= [c_{550red}]_0 + dt\ (+v_4+v_5-v_6)$	$[AMI_{ox}]_0$	$=$	0.5×10^{-3}	M
mol S-uptake	$= dt\ (v_1) \times V_p$	$[AMI_{red}]_0$	$=$	0	M
mol O_2-uptake	$= dt\ (v_6/4) \times V_p$	$[c_{550ox}]_0$	$=$	1.0×10^{-3}	M
		$[c_{550red}]_0$	$=$	0	M
v_1	$= k_1 [S] [MADH_{ox}]$	k_1	$=$	$380\ \times 10^3$	$M^{-1}\ s^{-1}$
v_2	$= k_2 [MADH_{ox}.S]$	$k_2\ (CH_3NH_2)$	$=$	150	s^{-1}
v_{3a}	$= k_{3a} [MADH_{red}.P] [AMI_{ox}]$	$k_2\ (CD_3NH_2)$	$=$	5	s^{-1}
v_{3b}	$= k_{3b} [MADH_{sem}] [AMI_{ox}]$	k_{-2}	$=$	5	s^{-1}
v_{4a}	$= k_{4a} [MADH_{ox}.P] [c_{550ox}]$	$k_{3a} = k_{3b}$	$=$	$20\ \times 10^3$	$M^{-1}\ s^{-1}$
v_{4b}	$= k_{4b} [MADH_{sem}] [c_{550ox}]$	$k_{4a} = k_{4b}$	$=$	1.0×10^3	$M^{-1}\ s^{-1}$
v_5	$= k_5 [AMI_{red}] [c_{550ox}]$	k_5	$=$	1.5×10^3	$M^{-1}\ s^{-1}$
v_6	$= k_6 [c_{550red}]$	k_6	$=$	100	s^{-1}

Calculated O_2-uptake rates. Simulated O_2-uptake rates for a quantity of 14 mg (wet weight) cells were calculated according to the model presented in Fig. 5. (Equations and parameters are given in Table 2). To explain the O_2-uptake rates with CD_3NH_2 a k_{-2} for the reaction $MADH_{red}.P \rightarrow MADH_{ox}.S$ had to be introduced (See below). The rate of electron transfer to oxygen ($O_2 + 4H^+ + 4c_{550red}$

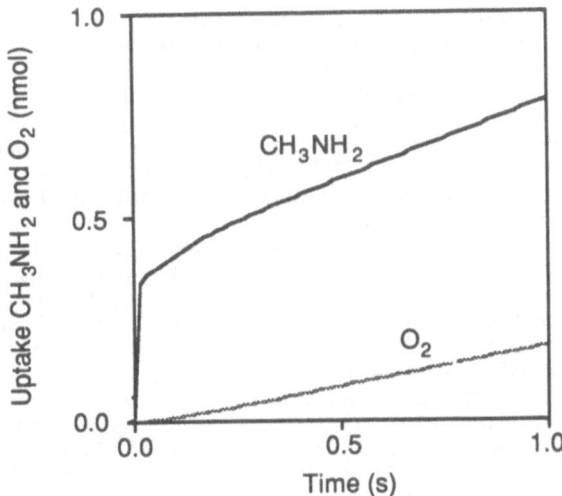

FIGURE 6. Simulated CH_3NH_2-conversion and O_2-uptake rates by cells, according to the model depicted in Fig. 5 (parameter values in Table 2).

TABLE 3. Calculated O_2-uptake rates of cells with CH_3NH_2 (10 mM) or CD_3NH_2 (10 mM) as electron donor. O_2-uptake rates with CD_3NH_2 between brackets.

conc. (mM) in the periplasm			O_2-uptake rate (nmol/min)		
MADH[*)]	AMI	c_{550}	Volume periplasm (% of cell volume)		
			5	10	20
0.5	0.5	1.0	12 (9)	---	---
0.25	0.25	0.5	---	6	---
0.125	0.125	0.25	---	---	3
1.0	1.0	2.0	47 (33)	---	---
0.5	0.5	1.0	---	23	---
0.25	0.25	0.5	---	---	12
0.5	0.0	1.0	5	---	---
0.5	0.0	2.0	10	---	---
0.5	0.0	3.0	15	---	---

[*)], $\alpha\beta$ subunit; ---, not calculated.

$\rightarrow 2H_2O + 4c_{550ox}$), catalyzed by the cytochrome aa_3-type oxidase , is presented by $v_6 = k_6 \times [c_{550red}]$ ($k_6 = 100 \text{ s}^{-1}$, [H$^+$] and [O_2] are assumed to be constant). The result of a simulation is shown in Fig. 6. Starting with an oxidized electron transport chain, there is at first a rapid uptake of reducing equivalents by $MADH_{ox}$. Then after 0.2 s, when all the reactants have attained steady-state concentrations, the electron transport rate becomes constant.

Depending on the assumptions made with respect to the volume of the periplasm and the content of the proteins in the cell, O_2-uptake rates vary between 3 and 47 nmol O_2/min (Table 3.). For a periplasmic space occupying 5% of the cell volume, with 0.5 mM of both MADH and amicyanin and 1.0 mM cytochrome c_{550}, the calculated O_2-uptake rate is 12 nmol/min. Raising the concentrations of MADH, amicyanin and cytochrome c_{550} in the periplasm with a factor two (the cell content of MADH, amicyanin and cytochrome c_{550} may be underestimated), O_2-uptake rates increase with a factor 4. Changing the periplasmic space from 5 to 10% of the cell volume results in a decrease of the O_2-uptake rates with 50%. The O_2-uptake rate is also lowered by 50% if the concentration of amicyanin is made zero. The absence of amicyanin can be compensated, however, by a two times higher concentration of cytochrome c_{550} (2.0 mM).

The O_2-uptake rates are, provided that $k_6 > 5$ s^{-1}, independent of the k_6-value. In addition, O_2-uptake rates are not essentially altered by the introduction of a low k_{-1}-value (this assumption is allowed since the K_m of MADH for methylamine was found to be in the order of 10 µM) or a k_{-5} of the same order as k_5 (taking into account the fact that in *T. versutus* amicyanin and cytochrome c_{550} have almost equal midpoint potentials (Table 1)).

The calculated O_2-uptake rates are in the range of the experimental O_2-uptake rate (25 nmol O_2/min). The experimentally determined 20% lower O_2-uptake rates with deuterated methylamine (20 nmol O_2/min) could be simulated by the model presented in Fig. 3, by virtue of the introduced k_{-2} of 5 s^{-1}. With a k_{-2} of zero the O_2-uptake rates with CD_3NH_2 are only a fraction lower compared to those with CH_3NH_2, in spite of the large deuterium effect on k_2. The reason for this is that, in that case, the steady-state concentration of $MADH_{red}.P$ is very high, both for CH_3NH_2 as for CD_3NH_2. By the introduction of a k_{-2}-value of 5 s^{-1}, the steady-state $[MADH_{red}.P]$ is lower and, hence, more influenced by the nature of the sub-

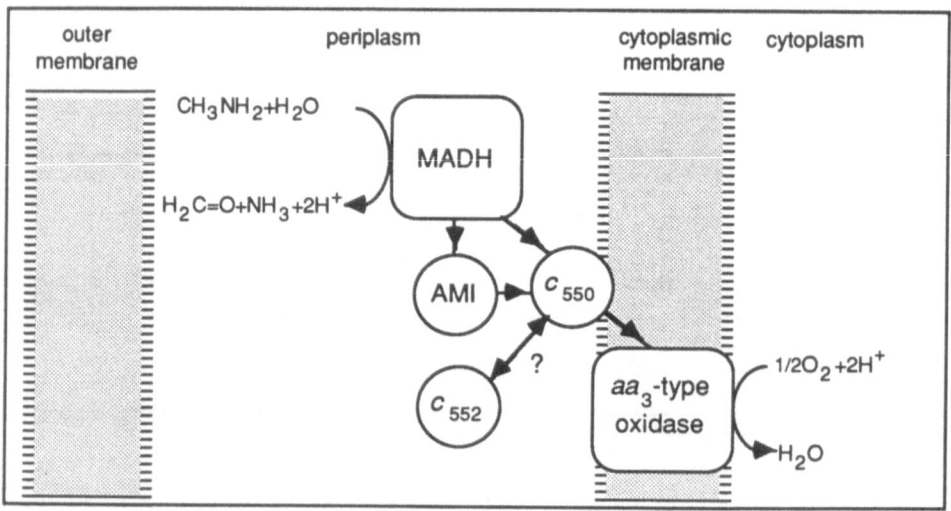

FIGURE 7. Electron transport from methylamine to oxygen in *Thiobacillus versutus*.

strate (CH_3NH_2 or CD_3NH_2). The k_2 is independent of the substrate used since the incoming proton in the reverse reaction derives from H_2O.

It has to be emphasized that the reaction of substrate with $MADH_{ox}$, as depicted in Fig. 5, is an oversimplificatio

It has to be emphasized that the reaction of substrate with $MADH_{ox}$, as depicted in Fig. 5, is an oversimplification. On account of the equilibrium between $MADH_{ox}.S$ and $MADH_{red}.P$, the model predicts that only half of the MADH will be reduced by CD_3NH_2. In reality the equilibrium is far to the right (Reduction almost 100%, results not shown), presumably by the formation of a second "$MADH_{red}.P$" (See Fig. 2).

A shortcoming of the simulation model is the uncertainty with respect to the concentration of the components in the periplasm. As a consequence the possibility of channeling,which means more efficient electron transfer *in vivo*, cannot be ruled out unequivocally. For that reason it will be attempted to determine the concentrations of MADH, amicyanin and cytochrome c_{550} more exactly. Also by direct measurements, with mixtures of MADH, amicyanin, cytochrome c_{550} and cytochrome c_{552}, the possibility of channeling will be investigated.

CONCLUSIONS

Assuming that the pathways for flow of reducing equivalents as proposed in Fig. 5 are correct, the assumptions made in the simulation program appear to be correct since with CH_3NH_2 as electron donor it gives oxygen consumption rates which are in the range of the measured one. Reversibly, assuming that the reaction rate constants *in vitro* are similar to those *in vivo* and the estimated concentrations in the periplasm are realistic, the simulation shows that the pathways presented in Fig. 5 represent the situation as it exists in the cell. The latter statement implies that uncertainty remains with respect to the primary electron acceptor for MADH, both cytochrome c_{550} and amicyanin seem to be suited to play this role, as indicated in the scheme presented in Fig. 7.

Acknowledgements.

Ir. A. Braat, Ir. A. Geerlof, R.P. Stam and F. Zechiël are acknowledged for their technical assistance. Dr. L.F. Oltmann and Professor Dr. A.H. Stouthamer (Department of Microbiology, Vrije Universiteit, Amsterdam, The Netherlands) are acknowledged for providing computer facilities.

REFERENCES.

Anthony C, 1982. The biochemistry of methylotrophs. Academic Press, London.
Anthony C, 1988. Quinoproteins and energy transduction. In: Anthony C. Bacterial energy-transduction. Academic Press, London.
Burton SM, Byrom D, Carver M, Jones GDD and Jones CW, 1983. The oxidation of methylated amines by the methylotrophic bacterium. *Methylophilus methylotrophus*. FEMS Microbiol Lett 17: 185-190.
Chandrasekar R and Klapper MH, 1986. Methylamine dehydrogenase and cytochrome c_{552} from the bacterium W3A1. J Biol Chem 261: 3616-3619.
Eady RR and Large PJ , 1971. Microbial oxidations of amines. Spectral and kinetic properties of the primary amine dehydrogenase of *Pseudomonas* AM1. Biochem J 123: 757-771.

Eckert TS and Bruice TC, 1983. Chemical properties of phenanthrolinequinone: and the mechanism of amine oxidation by *o*-quinones of medium redox potentials. J Am Chem Soc 105: 4431-4441.

Fukumori Y and Yamanaka T, 1987. The methylamine oxidizing system of *Pseudomonas* AM1 reconstituted with purified components. J Biochemistry (Tokyo) 101: 441-445.

Gray KA, Davidson VL and Knaff DB, 1988. Complex formation between methylamine dehydrogenase and amicyanin from *Paracoccus denitrificans*. J Biol Chem 263: 13987-13990.

Gray KA, Knaff DB, Husain M and Davidson VL, 1986. Measurement of the oxidation-reduction potentials of amicyanin and *c*-type cytochromes from *Paracoccus denitrificans*. FEBS Lett 207: 239-242.

Hayashi K and Sakamoto N, 1986. Dynamic analysis of enzyme systems. An introduction. Japan Scientific Societies Press, Tokyo; Springer-Verlag, Berlin, Heidelberg, New York, Tokyo.

Hobot JA, Carlemalm E, Villiger W and Kellenberger E, 1984. Periplasmic gel: new concept resulting from the reinvestigation of bacterial cell envelope ultrastructure by new methods. J Bacteriol 160: 143-152.

Husain M and Davidson VL, 1985. An inducible periplasmic blue copper protein from *Paracoccus denitrificans*. Purification, properties, and physiological role. J Biol Chem 260: 14626-14629.

Itoh S, Kitamura Y, Oshiro Y and Agawa T, 1986. Kinetics and mechanism of the oxidative deamination of amines by coenzyme PQQ. Bull Chem Soc Jpn 59: 1907-1910.

Kasprzak AA and Steenkamp DJ, 1983. Localization of the major dehydrogenases in two methylotrophs by radiochemical labeling. J Bacteriol 156: 348-353.

McIntire WS, 1987. Steady-state kinetic analysis for the reaction of ammonium and alkylammonium ions with methylamine dehydrogenase from bacterium W3A1. J Biol Chem 262: 11012-11019.

McWhirter RP and Klapper MH, 1987. Methoxatin containing methylamine dehydrogenase; enzyme reduction by primary amines. In: Edmondson DE and McCormick DB. Flavins and Flavoproteins. Walter de Gruyter & Co, Berlin, New York.

Oliver DD (1987) Periplasm and protein secretion. In: Neidhardt FC, Ingraham JL, Low KB, Magasanik B, Schaechter M, Umslarger HE. *Escherichia coli* and *Salmonella typhimurium* cellular and molecular biology Vol I. American Society for Microbiology, Washington, DC.

Stock JB, Rauch B and Roseman S, 1977. Periplasmic space in *Salmonella typhimurium* and *Escherichia coli*. J Biol Chem 252: 7850-7861.

Tobari J and Harada Y, 1981. Amicyanin: An electron acceptor of methylamine dehydrogenase. Biochem Biophys Res Commun 101: 502-508.

Van Houwelingen T, Canters GW, Stobbelaar G, Duine JA, Frank JJzn and Tsugita A, 1985. Isolation and characterization of a blue copper protein from *Thiobacillus versutus*. Eur J Biochem 153: 75-80.

Van Wielink JE, Frank JJzn, Braat A, Stam RP, Geerlof A and Duine JA, 1988. The interaction of methylamine dehydrogenase with amicyanin and cytochrome *c* in *Thiobacillus versutus*: EBEC Report 5, 139.

Vellieux FMD, Frank JJzn, Swarte MBA, Groendijk H, Duine JA, Drenth J and Hol WGJ, 1986. Purification, crystallization and preliminary X-ray investigation of quinoprotein methylamine dehydrogenase from *Thiobacillus versutus*. Eur J Biochem 154: 383-386.

COPPER-CONTAINING AMINE OXIDASES.

Alessandro FINAZZI-AGRO'

Dipartimento di Medicina Sperimentale e Scienze Biochimiche, Università degli Studi di Roma "Tor Vergata", Via O. Raimondo, 00173 Roma, Italy.

A. Reaction, distribution and localization.

Amine oxidases catalyze the reaction

$$R\text{-}CH_2NH_3^+ + O_2 + H_2O \longrightarrow R\text{-}CHO + NH_4^+ + H_2O_2$$

where R can be either aromatic or aliphatic. The copper-containing amine oxidases (Cu-AOs) oxidize only primary amines at variance with FAD-containing amine oxidases (eg. mitochondrial MAO). Several trivial names have been used for Cu-AOs, such as benzylamine oxidase, histaminase, diamine oxidase, lysyloxidase, spermine oxidase etc. on the basis of their substrate specificity. They are widespread in nature among eukaryotes though a prokaryotic Cu-AO has been recently described (van Iersel et al., 1986).
This diffusion must correspond to an essential physiological role, which should be related to the degradation of bioactive amines by Cu-Aos. In fact Cu-AOs' level has been found to be modulated in relation to various cellular events like mitosis, transformation and differentiation (Mondovì, 1985).
A striking example from the plant kingdom is the absence of diamine oxidase activity in the seed of leguminosae while a very high activity appears upon germination.
A role of AOs in the regulation of amine concentration in cells and tissues has been proposed to explain the above reported observations. However, as far as mammalians are concerned, no pathology has been clearly associated with Cu-AOs with the significant exception of lysyloxidase, the absence of which causes serious structural problems to collagen and arteries. Cu-AOs are present both inside and outside cells in multicellular organisms (plants and animals). The intracellular enzymes are found free in the cytoplasma or associated to membranes (plasma membrane, endoplasmic reticulum). No nuclear or mitochondrial (or chloroplastic) localization has been reported so far.

B. Structure

The isolated Cu-AOs appear to be rather large proteins (120-250 KDa) composed of two similar or identical subunits. No particular electrophoretic properties have been reported for Cu-AOs except the tendency to give multiple and /or diffuse bands on PAGE. Most of them have isoelectric point around pH 5. The electrophoretic heterogeneity has been ascribed to the glucidic component present both in the extracellular and in the intracellular Cu-AOs. This component may range from 2 to 10% (w/w). The amino acid compositions do not shows particular features.
The subunits can be dissociated by SDS, but isolated subunits have not been yet obtained in the absence of detergent. Some AO seem to undergo association - dissociation phenomena as a function of protein concentration.

J. A. Jongejan and J. A. Duine (eds.), PQQ and Quinoproteins, 279–282.
© 1989 by Kluwer Academic Publishers.

C. Cofactors

Copper has been early recognized (Werle & Hartung, 1956; Hill and Mann, 1962) as an essential costituent of some AOs (Cu-AOs). Two cupric coppers are present per molecule, i.e. one per subunit, though there is no experimental proof that each subunits binds a copper ion. The two coppers have similar (identical) binding sites containing 2-3 N- (possibly histidines) ligands and (a) molecule(s) of water. Subtle differences among the two coppers may sometimes be picked up by magnetic methods (EPR, NMR). A selective removal of one copper only has been reported for bovine serum AO.

The AOs copper belongs to the so-called Type 2 family, i.e. those showing very low visible absorption and "normal" EPR parameters (A parallel>100 x 10^{-4} cm^{-1}).

They do not appear to undergo redox cycles during the catalytic activity. The redox potential must be relatively low as they can be hardly reduced by NADH in the presence of suitable mediators and only slowly by dithionite.

The presence of a second cofactor in Cu-AOs was suggested even earlier than that of copper, on the basis of inibition by carbonyl-directed reagents (Zeller, 1938).

Various experimental evidences brought several authors including myself to suggest the presence of pyridoxal phosphate as the second cofactor. Interestingly enough in at least two different laboratories a pyridoxal-free diet in mice dramatically reduced the level of kidney AO to the same extent as those of well-know pyridoxal-containing enzymes like aspartate and pyruvate transaminases. The discovery of PQQ and subsequently its reported presence in Cu-AOs by Duine, Ameyama and their coworkers, has definitely settled the issue since this molecule appears to fit very well the spectroscopic and mechanistic features of Cu-AOs.

D. Optical spectroscopic features

Cu-AOs are chromoproteins showing a pink-yellow to brick-red color. This color is due to an absorption band at 480-500 nm (ε = 2-3x10^3 $M^{-1}cm^{-1}$) and by a monotonously rising absorption below 400 nm which peaks at about 280 nm i.e. the maximum absorbance of the aromatic amino acids. The presence of a more or less pronounced through at about 400 nm between the two absorption bands gives each AO the characteristic nuance. The absorption at 480-500 nm is sensitive to pH and to the presence of solutes like NH_4^+ or NADH.

The visible CD spectrum shows peaks at wavelenghts different from the absorption indicating the presence of several optical transition under the broad envelope. These transitions are related to PQQ as the removal of copper only slightly affects them. Upon reaction with substrates in anaerobiosis the absorption above 400 nm is either bleached (in mammalian Cu-AOs) or substituted by two sharp peaks at 464 and 430 nm (in plant Cu-AOs). The presence of a strong inhibitor like CN^- makes the animal CU-AOs spectrum in the presence of substrate very like to that of plant AOs. On the contrary upon removal of Cu from plant AOs the 480 nm band is bleached by addition of substrates even in anaerobiosis. These data indicate subtle differences among the AOs that may reflect upon catalytic efficiency.

In the presence of substrate and CN^- several Cu-AOs give rise to a free radical species not related to the substrate. This free radical is most probably the PQQ semiquinone.

Cu-AO-bound PQQ is never fluorescent, at variance with the free cofactor which shows a very broad emission at 480 nm upon excitation at 370 nm (lifetime $\simeq$ 2 ns). Once more this is an indication that the transitions below 400 nm involve states different from those responsible for the absorption at 480-500 nm.

E. Substrate specificity

As reported above each Cu-AO shows a different substrate specificity, which however is far from absolute and can be modified by even slight modification of the incubation medium and / or redox state of the protein. Lysyl oxidase and diamine oxidase have been reported to shift into a monoamine oxidase-like enzyme upon such treatments.

The active site of Cu-AOs must therefore be formed by several recognition sites for the different substrates besides the binding site for the amine to be oxidized. As an example pig kidney AO which oxidizes preferentially histamine, putrescine and cadaverine is thought to have three distinct regions in the active site indicated as A, B and C. A is the binding site for the amino group to be removed, presumably formed by the carbonyl group of the coenzyme. Site C recognizes the second positively charged amino group. Studies with substrates and inhibitors suggest that the distance between A and C could be 6-9 A. In between there is the presumably hydrophobic binding site B with strong affinity for aromatic systems.

F. Kinetics

Cu-AOs catalyze the oxidative deamination of primary amines by the so-called ping-pong bi-ter mechanism. The Michaelis constants for the substrates range between 10 μM and 10 mM. The affinity for oxygen has been determined accurately in few AOs only. The lentil seedling AO has a fairly high affinity for O_2 (upper limit for KmO_2 = 6 μM).

The turnover number of AOs is rather low (< 150s^{-1}). All AOs show a strong inhibition by the amine substrate. The products ammonia and aldehyde are weak inhibitors while hydrogen peroxide irreversibly inactivated the enzymes. Interestingly enough the resting enzyme is not inactivated even at high hydrogen peroxide concentrations.

Pre-steady state kinetics have been studied with several Cu-AOs. In the case of lentil seedling AO which offers the advantage of forming a colored (yellow) intermediate it was found that the enzyme reacts with substrate in a bimolecular reaction ($k \sim$ 10^5 M^{-1} s^{-1}) to give a colorless enzyme - substrate complex. This interconverts to the yellow form with a lag-time which is a function of O_2 present. At high O_2 concentrations no yellow intermediate at all is formed.

The reaction of the reduced enzyme with O is fast (K = 2.5 x 10^7 M^{-1}s^{-1}).

The following kinetic scheme is compatible with the observed data

E+S $\leftrightarrows$ ES(colorless) $\rightarrow$ EP(yellow) $\rightarrow$ ER(yellow) + aldehyde

ER+O2 $\rightarrow$ E+ammonia + hydrogen peroxide

G. Mechanism

Altough the stoichiometry between Cu and PQQ is still under investigation, the available data suggest the presence of only one PQQ every two coppers in several AOs.

This renders unlikely the simple model of equivalent subunits. Furthermore no structural data on the relationship among PQQ and the coppers are available except that each component is not nearest neighbor to any other (Cu to Cu = 6-8 A; Cu to PQQ carbonyl group $\geq$ 10 A).

In any case Cu-AOs appears to contain redox components in excess with respect to a peroxide forming oxidase. EPR evidences have so far excluded redox cycling of coppers, while several evidences show that PQQ in AOs might be reduced to a colorless form which then is reoxidized back by oxygen. In this context it is interesting to note that the removal of copper slows down every catalytic step in lentil AO and dramatically the reoxidation step which in the absence of copper takes hours. Furthermore no yellow intermediate can be seen in the copper-free enzyme.

The role of copper might be the binding of oxygen and/or of its reduced intermediate (superoxide) in analogy with Cu, Zn superoxide dismutase. It should be recalled that the formation of superoxide by pig kidney AO at rather alkaline pH values has been reported. However the spectroscopic features and the redox potentials of coppers in AOs are rather different from those of superoxide dismutase. In fact the copper in superoxide dismutase is reduced during the catalytic cycle. A fast, transient reduction of Cu in AOs cannot however be ruled out.

A non-redox role for copper has been also suggested by Yadav and Knowles (1981) who think that Cu-OH may act as a nucleophile to assist the transfer of hydride from substrate to O_2. There are also claims that only one copper is essential for the catalytic activity. This should involve an asymmetry of subunits only one of which might be supplemented with PQQ and thus functional.

However a definite picture of the catalytic mechanism of AOs must await more thorough understanding of the structural relationship among the three redox components present in the majority of these enzymes.

REFERENCES

Hill BJ and Mann PJG, 1962. The inhibition of pea-seedling diamine oxydase by chelating agents. Biochem. J. 85: 198-207

Mondovi B (ed.), 1985. Structure and functions of amine oxidase. CRC Press, Boca Raton

van Iersel J, van der Meer RA and Duine JA, 1986 Methylamine oxidase from Arthrobacter P1 Eur. J. Biochem. 161: 415-419.

Werle E and Hartung G, 1956. Zur kenntnis der Diamine-oxydase mit besonderer Berucksichtigung des Enzymes and Rotklee. Biochem. Z. 328: 228-237.

Yadav KDS and Knowles Pf, 1981. A catalytic mechanism for benzylamine oxidase form pig plasma. Eur. J. Biochem. 114: 139-144

Zeller EA, 1938. Zur Kenntnis der Diamine-oxydase. Helv. Chim. Acta 21: 1645-1650

ACTIVE SITE STRUCTURES OF COPPER-CONTAINING OXIDASES

P. F. Knowles, I. Singh and K. D. S. Yadav, Department of Biophysics, University of Leeds, UK.

F. E. Mabbs and D. Collison, Chemistry Department, Manchester University, UK.

C. E. Cote, D. M. Dooley and M. A. McGuirl, Chemistry Department, Amherst college, Mass., USA.

KEYWORDS

Pig plasma amine oxidase, galactose oxidase, copper-containing oxidases, active site structures, PQQ.

ABSTRACT

Pig plasma amine oxidase (PPAO; EC 1.4.3.6) has been shown to contain PQQ as well as copper. Experiments are described to quantitate and characterise the functional characteristics of the PQQ centres in PPAO. The results of spectroscopic studies on PPAO are reviewed and a plausible model for the active site copper proposed. There is no evidence from the spectroscopic studies for or against PQQ coordination to the copper. Galactose oxidase (EC 1.1.3.9) has structural and functional similarity to amine oxidases. Experiments to test for the presence of PQQ in galactose oxidase are discussed together with other structural properties of this enzyme.

INTRODUCTION

Copper-containing oxidative enzymes catalyse a variety of oxidations and oxygenation reactions (Table I).

TABLE I

Oxidases	Example
(i) $RH_2 + O_2 \rightarrow R + H_2O_2$	plasma amine oxidase
	galactose oxidase
(ii) $2RH_2 + O_2 \rightarrow 2R + H_2O$	laccase
	cytochrome oxidase
Oxygenases	
Monooxygenases (i) $RH + O_2 + 2e^- + 2H^+ \rightarrow ROH + H_2O$	dopamine-mono-oxygenase
Dioxygenase (ii) $RH + O_2 \rightarrow 'RO_2H'$	quercetinase
Dismutase	
$2O_2^- + 2H^+ \rightarrow O_2 + H_2O_2$	superoxide dismutase

The diversity of the reactions catalysed must reflect differences in structure of the active sites of these enzymes. We are studying two copper

J. A. Jongejan and J. A. Duine (eds.), PQQ and Quinoproteins, 283–288.
© 1989 by Kluwer Academic Publishers.

containing oxidases, pig plasma amine oxidase and galactose oxidase from Dactylium dendroides; some of their structural characteristics will be reviewed.

MATERIALS AND METHODS

Relevant details are given in the figure legends.

RESULTS AND DISCUSSION

Pig plasma amine oxidase (PPAO)

(i) Organic Cofactor

The amine oxidase from pig plasma is one of the better characterised enzymes in this class (Knowles & Yadav 1984). The presence of copper in the enzyme has not been disputed since its first report (Buffoni & Blaschko 1964) though its role in the catalytic mechanism is still unclear. There has, however, been considerable dispute over the nature or even existence of a second cofactor which early reports (Blaschko & Buffoni 1965) indicated to be pyridoxal phosphate. Bioassays for pyridoxal phosphate in PPAO performed in Leeds during 1974 proved negative (Knowles et al. 1987) though quench flow kinetic experiments monitoring the time course of ammonia production (Rius et al. 1984) clearly indicated an aminotransferase mechanism with attendant presence of a carbonyl cofactor. The key reports by Lobenstein-Verbeek et al. 1984 and Ameyama et al. 1984 for pyrroloquinoline quinone (PQQ) being the "carbonyl" cofactor in beef plasma amine oxidase solved this dilemma: we have used resonance raman spectroscopy to substantiate PQQ as the carbonyl cofactor in PPAO (Knowles et al. 1987).

Having established that PQQ is present in PPAO the next question is to quantitate the content. A convenient spectrophotometric assay for free and bound PQQ has been developed based on its reaction with 2-hydrazinopyridine (2-HP). This leads to formation of an azo derivative with a molar absorbance at 415nm of 15,400 litre $mol^{-1}cm^{-1}$. Reaction between PPAO and 2-HP has been found to be biphasic (Figure 1). One PQQ per mole of enzyme reacts rapidly followed by a continued slow increase in the absorbance of the 415nm chromophore up to a plateau value corresponding to two PQQs per mole enzyme.

The activity titrations with 2-HP (Figure 2) confirm the report (Lindstrom & Pettersson 1978) of a single active site in PPAO. What, then, is the role of the second PQQ? It is possible that the enzyme as isolated is not fully active; a procedure developed by S M Jones and J P Klinman (unpublished) has been followed to test whether HPLC of "pure" PPAO on a DEAE column leads to further purification. Preliminary results indicate that a major and minor peak are resolved by HPLC at pH 7.5. The major peak has 20% enhanced activity over that of the enzyme loaded whilst the minor peak has lowered activity. Further refinement of the HPLC procedures may lead to even more highly purified PPAO; it will be interesting to see whether two PQQs react rapidly with 2-HP when HPLC-purified enzyme is analysed.

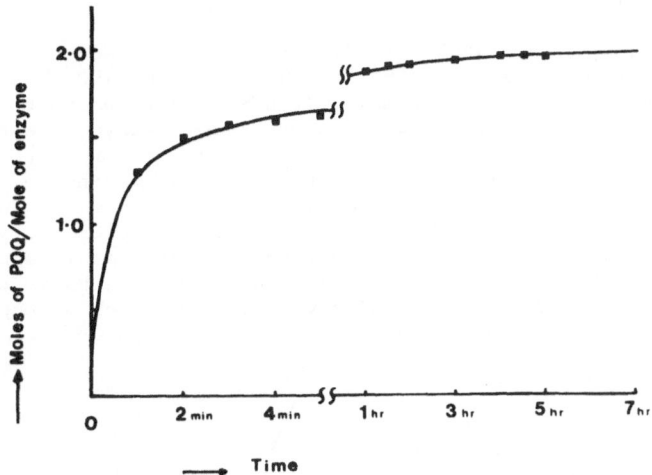

FIGURE 1. Time dependence for reaction of PQQ in pig plasma amine oxidase with 2-hydrazino pyridine at 22°C. Reaction conditions, 20 molar excess of 2-hydrazino pyridine, 40mm phosphate buffer pH 7.0.

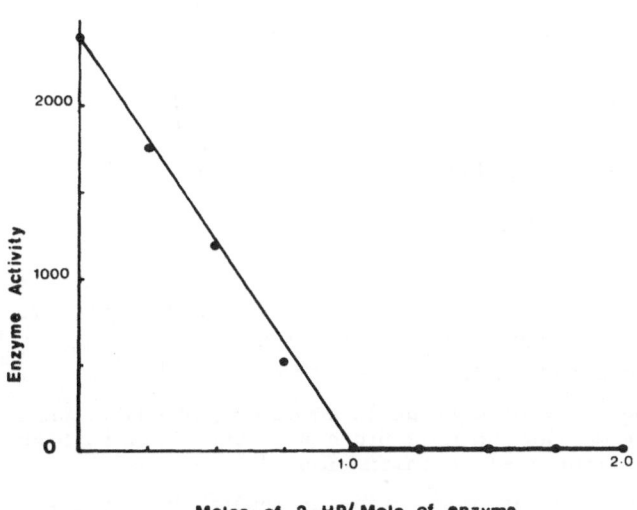

FIGURE 2. Effect of 2-hydrazino pyridine on the activity of pig plasma amine oxidase. Reaction conditions as in Figure 1 with a reaction time of 4 hours for each titration point.

(ii) <u>Active site structure</u>

The procedure developed by Suzuki et al. 1983 has been used to remove copper from PPAO. The copper-free enzyme has minimal activity (50 eu mg^{-1} compared to 1600 eu mg^{-1} for the native enzyme). It has been found that the physical and catalytic properties of the copper-reconstituted enzyme are highly dependent on the method of reconstitution. Reproducible results have been obtained when Cu(NO$_3$)$_2$ is added slowly with efficient mixing to a dilute sample (10mgml^{-1}) of apo-PPAO and the enzyme then incubated at 20°C for 6 days. We have used ESR, NMRD, EXAFS, ENDOR, optical spectroscopy and catalytic activity measurements to compare the properties of copper-depleted and native forms of PPAO. The evidence from all experiments is consistent with a linear recovery of activity and physical properties as cupric ions are added up to a limit of two coppers per mole of enzyme.

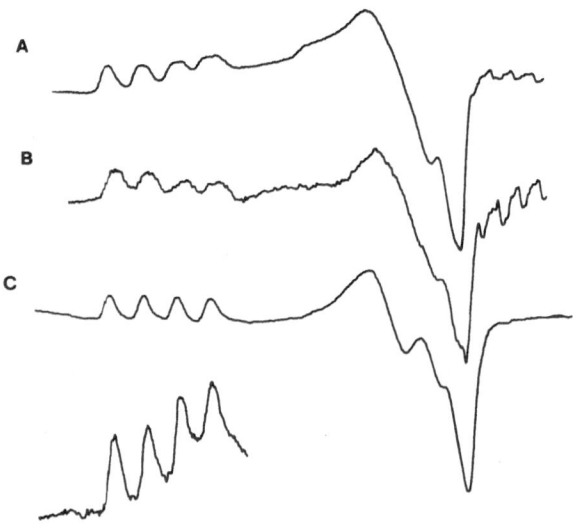

FIGURE 3. 35 GHz ESR on amine oxidases: Varian "Century line" spectrometer, 10dB microwave power, 20 gauss modulation, 150K.

A Native pig plasma amine oxidase

B Pig plasma amine oxidase depleted of copper and reconstituted with ^{63}Cu^{2+}(0.8 gm ions/gm mole protein). Structure to high field is due to a slight Mn^{2+} contamination.

C Native Arthrobacter amine oxidase: insert shows the A$_{11}$ region on expanded scale.

Typical data from 35 GHz ESR studies are shown in Figure 3; comparison of the spectra in Figure 3A and 3B corresponding to native PPAO (two coppers per mole of enzyme) and PPAO reconstituted with a single ^{63}Cu ion per mole of enzyme indicates close similarity. These ESR spectra confirm an earlier report (Barker et al. 1979) that the two copper sites in PPAO are chemically distinct, one site being rhombic, the other axial, and indicate further that copper enters each of these sites during reconstitution with equal facility.

The question whether one or both coppers are required for activity remains unsettled. However, good simulations of the ESR spectrum from benzylamine-reduced enzyme can be achieved by changing the parameters only of the rhombic copper which favours this centre being catalytically active.

The spectroscopic techniques we have applied to study PPAO indicate that the two coppers are in sites of axial and rhombic square pyramidal symmetry with one equatorial and one axial water as coordinated ligands. The other ligands are provided by the protein, prabably histidines though a methionine thio-ether ligand has been suggested from EXAFS studies (Blackburn et al. 1988). Anion inhibitors (N_3^-, CN^-) displace the equatorial but not the axial water and cause the two coppers to become chemically indistinguishable by Q band ESR spectroscopy (Barker et al. 1979). There is no evidence from the above spectroscopic studies for or against an interaction between the PQQ and copper cofactors in the active site of PPAO (Williams & Falk 1986).

A full structural study of PPAO using X-ray crystallographic methods is required to reveal the detailed nature of the active site. However, plasma amine oxidases have 10% of their molecular weight provided by covalently bound carbohydrate (Blaschko & Buffoni 1965) and heterogeneity in the carbo-hydrate accounts for the complex pattern ($\geqslant 7$ bands) found by isoelectric focussing (Falk et al. 1983; Knowles & Yadav unpublished). This hetero-geneity in PPAO would hinder, if not prevent, growth of diffracting crystals. Recent studies at Leeds have shown that PPAO activity is completely lost and the complex isoelectric focussing pattern abolished when the enzyme is incubated with glycopeptidase F which cleaves N-glycans between asparagine and the carbohydrate chain. The protein fragment from this cleavage of PPAO might be crystallisable though its lack of activity makes it less interesting as a subject for X-ray crystallographic study. Probably an amine oxidase from a bacterial source such as Arthrobacter would be better suited for protein crystallography since it may well lack any carbohydrate moiety. The Q band ESR spectrum of Arthrobacter amine oxidase (see Figure 3C) shows that the two copper sites must be closely similar and this is a further promising observation with respect to protein crystallographic study. It also incates that the non-identical nature of the two copper sites found for PPAO is not a common feature of all amine oxidases.

Galactose oxidase

This enzyme is unique amongst metalloenzymes in appearing to catalyse 2-electron redox chemistry at a mononuclear metal site. Recently (Whittacker & Whittacker 1988) evidence has been presented for a non-metal redox grouping in the active site of galactose oxidase though the chemical nature of this species was not defined. It seems possible that PQQ could be the unidenti-fied redox grouping. The spectrophotometric assay using 2-hydrazinopyridine described earlier is well suited to investigate this possibility. Pre-liminary results show that there is no reaction between 2-HP and either native galactose oxidase or enzyme "activated" by treatment with stoichio-metric amounts of potassium ferricyanide. The possibility that lack of reaction is due to any PQQ present being buried within the protein remains to be explored.

Earlier spectroscopic studies on galactose oxidase (Kosman 1984) giving information on the active site structure need to be reappraised in the light of the report (Whittacker & Whittacker 1988) that the native enzyme is a mixture of redox states. However, it seems probable that the copper sites in galactose oxidase, as for PPAO, have axial and equatorial waters co-ordinated (Kosman 1984).

The low molecular weight (68,000) of galactose oxidase makes it an attractive protein for X-ray crystallographic study. At Leeds, a crystal form of native galactose oxidase showing C2 symmetry and with unit cell dimensions 97.8Å x 89.7Å x 86.2Å is being studied in collaboration with Professor A C T North and Dr S E V Phillips.

This work was supported by grants from SERC and NATO. The skilled technical assistance of Mrs V Blakeley is gratefully acknowledged.

REFERENCES

Ameyama M, Hayashi M, Matsushita K, Shinagawa E and Adachi O, 1984. Agric. Biol.Chem. 84, 561-565.

Barker R, Boden N, Cayley G, Charlton SC, Henson R, Holmes MC, Kelly IC and Knowles PF, 1979. Biochem.J. 177, 289-302.

Blackburn NJ, Hasnain SS, Knowles PF and Strange R, 1988. J.Amer.Chem.Soc. accepted.

Blaschko H and Buffoni F, 1965. Proc.Roy.Soc.Ser.B., 163, 45-60.

Buffoni F and Blaschko H, 1964. Proc.Roy.Soc.Ser.B., 161, 153-167.

Falk MC, Staton AJ and Williams TJ, 1983. Biochemistry 22, 3746-3751.

Knowles PF, Pandeya KB, Ruis FX, Spencer CM, Moog RS, McGuirl MA and Dooley DM, 1987. Biochem.J. 241, 603-608.

Knowles PF and Yadav KDS in "Copper Proteins and Copper Enzymes" 1984. Loutie R, editor; CRC press, vol.II, 103-129.

Kosman DJ, 1984. Ibid. p.1-26.

Lindstrom A and Pettersson G, 1978. Eur.J.Biochem. 83, 131-135.

Lobenstein-Verbeek CL, Jongejan JA, Frank J and Duine JA, 1984. FEBS lett. 170, 305-309.

Ruis FX, Knowles PF and Pettersson G, 1984. Biochem.J. 220, 767-772.

Suzuki S, Sakurai T, Nakahara A, Manabe T and Okuyama T, 1983. Biochemistry 22, 1630-1635.

Whittacker MM and Whittacker JW, 1988. J.Biol.Chem. 263, 6074-6080.

Williams TJ and Falk MC, 1986. J.Biochem. 261, 15949-15954.

ON THE ROLE OF COPPER IN PQQ AMINE OXIDASES

Bruno Mondovi and Pierluigi Riccio

Department of Biochemical Sciences and C.N.R. Centre for Molecular Biology, University "La Sapienza", Rome, Italy.

KEY WORDS
Amine oxidase(s); Copper; Spectroscopy; Differential scanning calorimetry (DSC); Pyrroloquinoline quinone (PQQ).

INTRODUCTION

PQQ-dependent amine oxidades (AOs) contain copper as the inorganic cofactor. The presence of this metal ion in this group of AOs has been reported for the first time over 20 years ago, yet its exact role is so far unclear.

Cu-AOs belong to the so called non-blue Cu-proteins, showing an EPR spectrum with large hyperfine splitting constant and an optical absorbance maximum below 500 nm, characteristic of a non-covalent binding of the metal ion with a tetragonal coordination geometry only slightly distorted from square planar (Mondovi et al. 1967; Rotilio 1985; Finazzi-Agro 1985).

The majority of Cu-AOs are dimeric proteins composed of 2 apparently identical subunits, each weighing about 90 Kdaltons and containing a Cu atom in the cupric state. Although apparently symmetrical as it appears from this brief description, the active AO molecule is not, showing most Cu-AOs single carbonyl group reactivity due to the presence of only one PQQ molecule (or a closely similar derivative) per dimer. Very recently, however, two yeast AOs having a Cu:PQQ ratio of 1 have been described (Tur & Lerch 1988). Moreover, the two Cu(II) sites appear to be not identical, as it will be extensively discussed below.

The presence of Cu is essential for catalytic activity. Removal of one or both Cu ions completely abolishes enzyme activity by preventing reoxidation of reduced organic cofactor. In spite of this fact, the role of Cu in AO catalysis can only tentatively be interpreted on the basis of indirect observations, lacking direct evidence of an involvement of the metal cofactor in the redox mechanism.

COPPER LIGANDS

The two Cu environments appear to be asymmetrical: 35 GHz EPR spectroscopy, performed on the pig serum AO, evidenced

J. A. Jongejan and J. A. Duine (eds.), PQQ and Quinoproteins, 289–295.

that the two metal ions are affected differently by the presence of substrate (Grant et al. 1978); the binding of azide or cyanide to Cu (see below) is able to abolish the 35 GHz spectral heterogeneity.

Moreover, it was observed that binding of the carbonyl group reagent phenylhydrazine (PH) to PQQ modifies the EPR spectral parameters relatively to only one Cu ion, in the sense of a smaller $A_{\shortparallel}$ and greater $g_{\shortparallel}$ with respect to the untreated enzyme (Rinaldi et al. 1983).

In addition, the ability to selectively remove a single Cu atom per dimer, obtainable by diethyldithiocarbamate (DDC) treatment under particular conditions (see below), implies a different reactivity or a different accessibility of the two metal ions towards complexing agents, and furtherly supports the non-equivalency of the AO Cu sites.

Substrates do not bind directly to the metal: in fact, EPR spectra recorded either in the presence of of ^{14}N- or ^{15}N-putrescine reveal no significant changes of the hyperfine line patterns, indicating that the substrate's nitrogen is not a Cu ligand (Mondovi et al. 1967).

As regards the nature of the Cu ligands, the g values and binding parameters are consistent with the involvement of nitrogen atoms coordinating the metal ion. (Mondovi et al. 1967). In addition, a recent study performed by pulsed EPR spectrometry (spin echo) revealed the presence of two distinct populations of Cu-coordinated imidazole nitrogens which appear to be in magnetically non-equivalent environments (McCracken et al. 1987). Half Cu-depleted bovine serum AO was employed to clarify whether the two magnetically unequivalent imidazole populations are equally distributed among the two Cu sites: results demonstrated that both Cu centers are equivalent in this respect, being the two populations still distinguishable in the enzyme containing a single Cu ion (Mondovi et al. 1987).

NMR spectroscopy indicated the presence of two water molecules coordinated to Cu, one in axial and one in equatorial positions (Boden et al. 1974). Spin echo measurements suggest that azide or cyanide displace the equatorially-bound water molecule probably without appreciably affecting the axially coordinated one (McCracken et al. 1987; Mondovi et al. 1987). Labile equatorial water coordination position appears therefore to be involved in the catalytic mechanism, as the presence of these anions abolish the enzyme activity.

COPPER-PQQ STRUCTURE RELATIONSHIP

Several efforts have been directed towards the elucidation of structural relationships occurring between metal cofactor and PQQ, but the question on whether PQQ itself might contribute in coordinating Cu or not is still not answered definitively. On one side, X-ray absorption (Scott and Dooley 1985) and electron nuclear double resonance (ENDOR) (Baker et al. 1986) spectroscopic studies would not exclude the possibility of a direct binding of Cu to PQQ; on the other side, a recent fluorescence study would instead seem to rule out the evenience of a PQQ nitrogen coordinating Cu (Lamkin et al. 1988). In any case, all available data preclude the possibility for Cu to lie in the neighborhood of the PQQ carbonyl function.

A clear reduction of Cu has never been observed during catalysis (see below): at any rate, should the electron transfer from substrate to O_2 occur via the enzyme-coordinated Cu, this does not necessarily imply the presence of a stable direct interaction between the metal ion and PQQ. In fact, closing of the cofactors could be obtained by a transient conformational change of the protein structure during the enzyme catalytic cycle. Alternatively, the existance of an intermediate electron carrier between the AO cofactors could be postulated.

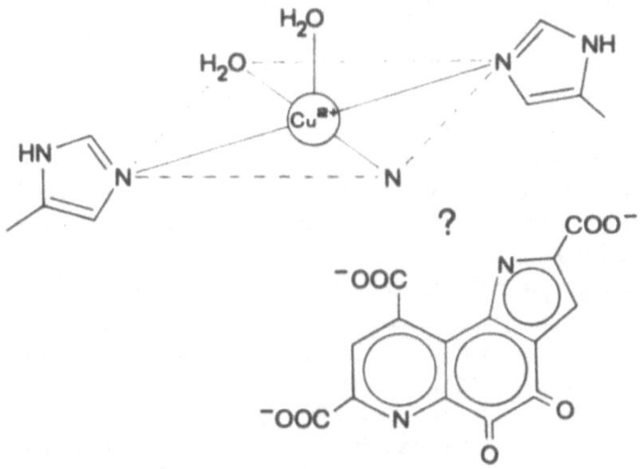

FIGURE 1. Copper-PQQ structure relationship. For details, see text.

ROLE OF THE AO COPPER

The AO Cu could either have structural significance, maintaining the correct tertiary structure of the protein molecule, or it could have functional value, playing a redox role during catalysis.

Although a direct demonstration of the involvement of Cu in the redox reaction has not been obtained to date, yet this possibility cannot be totally ruled out. Actually, practically all known Cu proteins are oxidoreductases, but a clear reduction of the AO Cu(II) has never been observed during the enzyme catalytic cycle. A possible transient redox mechanism was proposed: a scheme admitting partial reduction of Cu was hypothesized (Mondovi et al. 1971) consisting of a mechanism according to which the metal ion is the pathway for the electron transfer. The intermediate Cu(I) species which should appear during catalysis cannot be observed because the reaction $[Cu(II) \; E^{--}] \underset{k_2}{\overset{k_1}{\rightleftharpoons}} [Cu(I) \; E^{-}]$, where E represents the enzyme with bound PQQ, is considerably shifted to the left, i.e. $k_2 >> k_1$ (see below).

It is helpful discussing some matters concerning the preparation of the Cu-free and half Cu-depleted bovine serum AO (Morpurgo et al 1987) because of their effectiveness as models to study the function(s) of Cu. Cyanide and DDC are used as Cu chelating agents, the former for obtaining the metal-free enzyme and the latter (at different concentrations and employing various procedures) to obtain, varying the conditions, half Cu-depleted enzyme or protein-Cu-DDC complexes. In the bovine serum AO, at 1 mM DDC in the presence of dithionite as reducing agent, only about 50 % Cu reacts with the chelating agent, the dithionite-reduced organic cofactor no longer binds PH and nearly full activity can be restored by treatment with excess Cu(II).

To obtain total Cu reaction in the absence of reducing agents, 24 h incubation with 10 mM DDC is required. Freezing-thawing the 10 mM DDC-treated enzyme causes the loss of a single Cu ion removable by centrifugation, while the other ion remains bound to the protein as a Cu-DDC complex. PQQ, being present in the oxidized form, can react with PH. In this case, however, activity is not restored on addition of Cu (Morpurgo et al. 1987).

NMR spectroscopy performed on the pig kidney DAO would seem to rule out the possibility for Cu to bind O_2 (Kluetz & Schmidt 1977). On the other hand, reoxidation of dithionite-reduced AO would suggest a possible role of Cu in O_2 activation (Mondovi et al. 1985): in this case, O_2 binding and positioning for interaction with reduced PQQ could be carried out by Cu. Generation of the oxygen radical superoxide was suggested to occur during the reduction step of molecular oxygen by pig kidney DAO (Rotilio et al. 1970).

The formation of an organic free radical was observed in some plant and animal cyanide-treated AOs upon reaction with substrate (Finazzi-Agro et al. 1984, Dooley et al. 1987). The presence of Cu is essential for the generation of this free radical, which could be interpreted admitting the formation of the semiquinoid form of PQQ. Cyanide should stabilyze Cu in the reduced form Cu(I), which in turn would shift the equilibrium towards the one-electron reduced form of PQQ (Dooley et al. 1987). This model is consistent with the above mentioned transient redox mechanism proposed by Mondovi et al. (1971). It could be postulated that appearance of the organic free radical might be detected in some plant AOs (Hill & Mann 1964; Bellelli et al. 1985) even in the absence of cyanide (Dooley et al. 1987).

Reacting bovine serum AO with the pseudo-substrate benzylhydrazine, the formation of a hydroperoxide was postulated upon reaction with O_2 (Morpurgo et al. 1988). In this case, however, being the catalytic mechanism different from that involving the formation of a Cu(I) intermediate, the structural role of Cu is stressed: the binding of substrate should then be strongly favoured by the presence of Cu in the AO molecule, which would maintain a suitable conformation of the active site (Suzuki et al, 1986).

Further information on the structural role of Cu is provided by differential scanning calorimetry (DSC) studies (Giartosio et al. 1988). DSC patterns obtained with native and Cu-depleted bovine serum AO differ as concerns thermal stability of the proteins. In the absence of Cu, the heat denaturation curve reveals more relevant conformational changes in the protein molecule with respect to the native AO. Four distinct domains can be evidenced in the bovine serum AO, which appear to be organized in two similar sets: the two larger domains should contain the two Cu ions, whereas one of the two smaller ones, which differs in thermal stability from its counterpart, could probably contain covalently bound PQQ. Reaction of the enzyme with small ligands (phenylhydrazine for PQQ and DDC for Cu) does not alter its calorimetric behaviour.

As concerns the possible structural role of Cu, we should remember that total Cu substitution with other metals (e.g. cobalt) is only partially effective in reactivating Cu-depleted AOs (Suzuki et al. 1983). This observation would seem to account against the purely structural role of Cu, but another possibility is to be considered; it is known in fact that the redox potential of protein-bound cofactors is strongly influenced by the surrounding environment; it is therefore likely that Cu may contribute in modifying the redox potential of AO-bound PQQ, thus adjusting it at a level optimal for the kind of reaction catalyzed by this class of oxidades; This would account for a functional significance of Cu even being not this metal ion involved in the electron transfer.

To summarize, both structural and functional roles can be assigned to the AO Cu. Regardless of the mechanism of substrate oxidation, there is no reason to believe that one hypothesis would necessarily exclude the other, as on the basis of the observations reported so far both appear to be equally acceptable. The different reactivity of the two Cu ions, however, would suggest separate roles for each of them, in the sense of a possible functional value in the catalytic reaction of only one Cu ion while the other might play only a conformational role.

ACKNOWLEDGEMENTS

The original studies reported here were in part supported by the C.N.R. Special Project on "Chimica Fine" and by the "Ministero della Pubblica Istruzione".

The authors wish to thank Mr Amleto Ballini for his skilful technical assistance.

REFERENCES

Baker G J, Knowles P F, Pandeya K B, Rayner J B, 1986. Electron nuclear double-resonance (ENDOR) spectroscopy of amine oxidase from pig plasma. Biochem. J. 237: 609-612.

Boden N, Holmes M C, and Knowles P F, 1974. Binding of water to types I and II Cu in proteins. Biochem. Biophys. Res. Commun. 57: 845-848.

Bellelli A, Brunori M, Finazzi-Agro A, Floris G, Giartosio A, Rinaldi A, 1985. Transient kinetics of copper-containing lentil (Lens culinaris) seedling amine oxidase. Biochem. J. 232: 923-926.

Dooley D M, McGuirl M A, Peisach J, and McCracken J, 1987. The generation of an organic free radical in substrate-reduced pig kidney diamine oxidase-cyanide. FEBS Lett. 214: 274-278.

Finazzi-Agro A, Rinaldi A, Floris G, and Rotilio, G, 1984. A free-radical intermediate in the reduction of plant Cu-amine oxidases. FEBS Lett. 176: 378-380.

Finazzi-Agro A, 1985. Optical and spectroscopic properties of copper-containing Amine Oxidases, in: Structure and Functions of Amine Oxidases, Mondovi B, ed. CRC Press, Boca Raton FL, 121-125.

Giartosio A, Agostinelli E, Mondovi B, 1988. Domains in Bovine Serum Amine Oxidase. Biochem. Biophys. Res. Comm. 154: 66-72.

Grant J, Kelly I, Knowles P, Olsson J, Pettersson G, 1978. Changes in the copper centres of benzylamine oxidase from pig plasma during the catalytic cycle. Biochem. Biophys. Res. Comm. 83: 1216-1224.

Hill J M, and Mann P J G, 1964. Further properties of the diamine oxidase of pea seedlings. Biochem. J. 79: 171-180.

Kluetz M D, Schmidt P G, 1977. Proton Relaxation Study of the Hog Kidney Diamine Oxidase Active Center. Biochemistry 16: 5191-5199.

Lamkin M S, Williams T J, Falk M C, 1988. Excitation Energy Transfer Study of the Spatial Relationship between the Carbonyl and Metal Cofactors in Pig Plasma Amine Oxidase. Arch. Biochem. Biophys. 261: 72-79.

McCracken J, Peisach J, Dooley D M, 1987. Cu(II) Coordination Chemistry of Amine Oxidases: Pulsed EPR Studies of Histidine Imidazole, Water, and Exogenous Ligand Coordination. J. Am. Chem. Soc. 109: 4064-4072.

Mondovi B, Rotilio G, Costa M T, Finazzi-Agro A, Chiancone E,

Hansen R E and Beinert H, 1967. Diamine oxidase from pig kidney: improved purification and properties, J. Biol. Chem. 242: 1160-1167.

Mondovi B, Rotilio G, Finazzi-Agro A, Antonini E, 1971. Amine oxidases: a new class of copper oxidases. Franconi C, ed. Gordon and Breach Science Publ. N.Y. London Paris : 232-246.

Mondovi B, Finazzi-Agro A, Rotilio G, and Sabatini S, 1985. Redox titration and reversible removal of copper from Cu-amine oxidases, in: Frontiers in Bioinorganic Chemistry, Xavier A V, ed, VHC Publ. 604- 611.

Mondovi B, Morpurgo L, Agostinelli E, Befani O, McCracken J, Peisach J, 1987. A comparison of the local environment of Cu(II) in native and half-Cu-depleted bovine serum amine oxidase. Eur. J. Biochem. 168: 503-507.

Morpurgo L, Agostinelli E, Befani O, Mondovi B, 1987. Reactions of bovine serum amine oxidase with NN-diethyldi-thiocarbamate. Biochem. J. 248: 865-870.

Morpurgo L, Agostinelli E, Muccigrosso J, Mondovi B, and Avigliano L, 1988. Bovine serum amine oxidase: benzylhydrazine reaction. 1st International Symposium on PQQ and Quinoproteins, Delft 5-7 Sept. 1988.

Rinaldi A, Floris G, Sabatini S, Finazzi-Agro A, Giartosio A, Rotilio G, Mondovi B, 1983. Reaction of beef plasma and lentil seedlings Cu-amine oxidases with phenylhydrazine. Biochem. Biophys. Res. Comm. 115: 841-848.

Rotilio G, Calabrese L, Finazzi-Agro A, and Mondovi B, 1970. Indirect evidence for the production of superoxide anion radicals by pig kidney diamine oxidase. Biochim. Biophys. Acta 198: 618-620.

Rotilio G, 1985. Spectroscopic and chemical properties of the Amine Oxidase Copper, in: Structure and Functions of Amine Oxidases, Mondovi B, ed. CRC Press, Boca Raton FL, 127-134.

Scott R A, Dooley D M, 1985. X-ray Absorption Spectroscopic Studies of the Copper(II) Sites in Bovine Plasma Amine Oxidase. J. Am. Chem. Soc. 107: 4348-4350.

Suzuki S, Sakurai T, Nakahara A, Manabe T, and Okuyama T, 1983. Effect of metal substitution on the chromophore of BSAO. Biochemistry 22: 1630-1635.

Suzuki S, Sakurai T, Nakahara A, 1986. Roles of the Two Copper Ions in Bovine Serum Amine Oxidase. Biochemistry 25: 338-341.

Tur S S, and Lerch K, 1988. 14th International Congress of Biochemistry, Prague, July 10-15, 1988. Poster No. TH:205.

Title: Mechanistic Probes of Copper Amine Oxidases

Authors: Klinman, J. P., Hartmann, C. and Janes, S. M.
Department of Chemistry, University of California, Berkeley, CA
94720

Key Words: Copper amine oxidases, pyrroloquinoline quinone,
enzymatic reaction mechanism

Abstract:
The presence of pyrroloquinoline quinone at the active site of
copper amine oxidases makes numerous mechanistic predictions,
which include the possible formation of covalent adducts between
substrate and cofactor, the transfer of nitrogen from substrate
to cofactor in the reductive half of the overall reaction, and
the oxidation of substrate *via* a proton activation mechanism.
Experimental evidence in support of each of these aspects of
catalysis in bovine plasma amine oxidase is presented. Direct
spectroscopic evidence was also sought for initial formation of a
Schiff base complex, followed by C-H bond cleavage. Rapid scanning stopped flow analysis of the anaerobic reduction of bovine
plasma amine oxidase by benzylamine indicates a rapid formation
of a species at 340 nm (attributed to Schiff base formation),
followed by a slower decrease/increase in peaks at 480/310 nm
(attributed to cofactor reduction). These findings lead to a
working model for enzyme catalysis.

Introduction:
    The mammalian copper amine oxidases have become a subject of
increasing investigation over the past several years. While these
proteins have long been recognized as playing important physio-
logic roles, e.g., the oxidative removal of biogenic amines from
plasma (plasma amine oxidases) and the cross linking of collagen
and elastin (lysyl amine oxidases), mechanistic characterization
has been slow. A major deterrent to such studies has been the
absence of an acceptable working model for the nature of the
active site cofactor. This situation changed in 1984 with the
independent reports of Ameyama (Ameyama et al., 1984) and Duine
(Lobenstein-Verbeek, et al., 1984) that bovine plasma amine
oxidase contains covalently bound pyrroloquinoline quinone (PQQ).
Since 1984 an increasing number of eukaryotic copper amine
oxidases have been implicated as containing PQQ (Hartmann and
Klinman, 1988).
    Examination of the structure of PQQ indicates a unique α-di-
carbonyl functional group which has been incorporated into a
tricyclic aromatic ring structure, **1**.

1

297

J. A. Jongejan and J. A. Duine (eds.), PQQ and Quinoproteins, 297–305.
© 1989 by Kluwer Academic Publishers.

Model studies of anaerobic amine oxidation by PQQ support a direct attack of the substrate nitrogen at the C-5 carbonyl of cofactor, to generate a carbinolamine intermediate. The major pathway for substrate oxidation appears to involve direct oxidation of this carbinolamine, eq (1), with a minor pathway arising *via* Schiff base formation, eq (2).

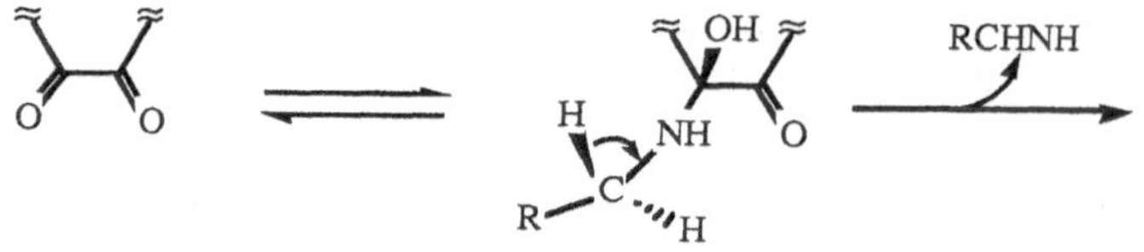

$$(1)$$

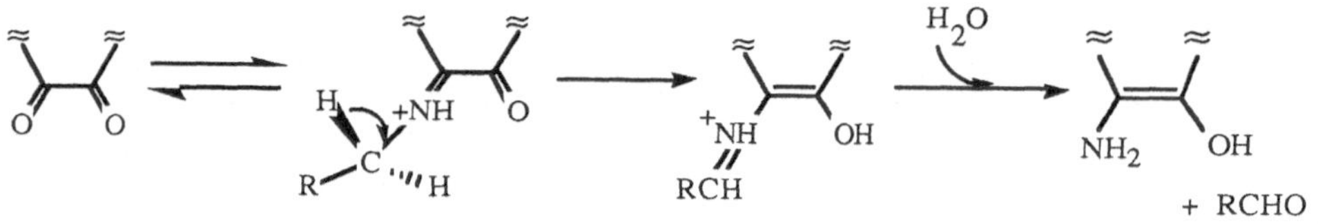

$$(2)$$

An important distinction between eqs (1) and (2) is the transfer of nitrogen from substrate to cofactor in the latter case, leading to an amino-transferase mechanism. To what extent does the mechanism of copper amine oxidases resemble eq (1) or eq (2) above? In an effort to address this question, we have undertaken a detailed mechanistic investigation of the copper amine oxidase from bovine plasma. Four basic questions have been pursued, each of which is outlined below.

Materials and Methods:

All materials were reagent grade unless otherwise indicated. Bovine plasma amine oxidase was prepared according to a modification of the method of Summers et al. (1979). Reductive trapping experiments were conducted as described by Hartmann and Klinman (1987). In anaerobic ammonia and benzaldehyde release studies, enzyme of the highest specific activity was used, Sp. Ac. = 0.4 μmoles min^{-1} mg^{-1}. Anaerobic formation of [^{14}C]-benzaldehyde from [^{14}C]-benzylamine was quantitated following HPLC separation of reactants from products. Ammonia release assays were based on earlier methods used by Levitzki (1970) and Kalb et al. (1978). Measurement of steady state deuterium isotope effects required the synthesis of a range of *para*-substituted benzylamines, according to the methods of Palcic and Klinman (1983). Steady state parameters were obtained by measurement of oxygen uptake using a Yellow Springs Instrument polarographic oxygen electrode or from the differential absorbance of product aldehydes and substrate amines (Palcic and Klinman,1983). Methodologies for anaerobic stopped flow studies at a single wavelength have been described (Palcic and Klinman, 1983). Rapid scanning stopped flow studies were conducted on a Durrum D-110, equipped with a zenon lamp, a Princeton Applied Research OMA-3 multichannel analyzer, 1218 controller and 1214 photodiode array detector. An external time-delay firing circuit was used to ensure that the temporal events of flow stoppage and initiation of data acquisition occurred together in time (2-4 ms after flow had stopped).

Results and Discussion:
Do Covalent Adducts Form Between Cofactor and Substrate?

The involvement of covalent complexes has been inferred from the numerous studies documenting enzyme inhibition by phenyl-hydrazines. We decided to examine whether direct evidence for covalent complexes between substrate and cofactor could be obtained in the course of enzyme turnover. Sodium cyanoboro-hydride was chosen as the trapping reagent, since this relatively mild reductant was expected to irreversibly reduce any Schiff base complex formed between enzyme and substrate, while leaving underivatized cofactor intact. Initial investigations used ^{14}C labelled substrate and unlabelled NaCNBH$_3$. Time dependent incorporation of radiolabelled substrate was found to correlate with enzyme inactivation in a 1:1 manner, such that one mole of ^{14}C per enzyme subunit was incorporated at 100% enzyme inactivation, Figure 1.

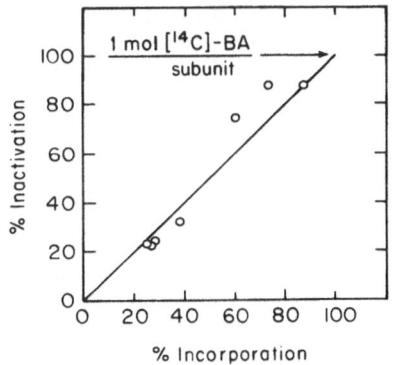

Figure 1: Correlation of Enzyme Inactivation and Radiolabelling by [^{14}C]-Benzylamine in the Presence of NaCNBH$_3$

When this experimental protocol was repeated using unlabelled benzylamine and [^3H]-NaCNBH$_3$, *no incorporation of tritium was observed*. To our knowledge, this ability to reductively trap ^{14}C labelled substrate without a concomitant incorporation of tritium is unprecedented, ruling out a pyridoxal phosphate-like cofactor as well as the non-specific labelling of protein (lysyl) side chains by product aldehyde. Importantly, the structure of PQQ predicts the observed labelling pattern, given the presence of an α-dicarbonyl structure. As shown in eq (3), initial reduction with [^3H]-NaCNBH$_3$ will lead to incorporation of tritium; however, rapid tautomerization of the keto to enol form of cofactor is expected to produce a rapid washout of tritium to solvent.

$$(3)$$

Is Nitrogen Transferred From Substrate to Cofactor in the Reductive Half Reaction?

The ability to reductively trap substrate to cofactor with NaCNBH$_3$ implicates a Schiff base complex during plasma amine oxi-

dase turnover. As shown in eq (2), formation of such an intermediate is expected to lead to an amino-transferase mechanism
involving the transfer of nitrogen from substrate to cofactor in
the reductive half reaction. However, early experiments by Abeles
and co-workers with bovine plasma amine oxidase had indicated a
stoichiometric release of ammonia under anaerobic conditions
(Berg and Abeles, 1980). We therefore reinvestigated this problem using an experimental protocol which permitted the strict
maintenance of anaerobic conditions.

As a control, enzyme was incubated with [^{14}C]-benzylamine to
determine moles of [^{14}C]-benzaldehyde released per enzyme subunit
under anaerobic conditions. Similar to earlier results of Berg
and Abeles (1980), incubation of enzyme led to the release of
approximately one mole of benzaldehyde per subunit (0.89 moles
per subunit). This result supports the catalytic viability of
each subunit, consistent with the reductive trapping experiments
of Hartmann and Klinman (1987). In contrast, previous workers
have found that phenylhydrazine reacts with bovine plasma amine
oxidase with apparent half of the sites reactivity, similar to
amine oxidases from other sources (Suzuki et al., 1986, and Falk,
1983). We therefore examined the reactivity of phenylhydrazine
with our preparations of bovine plasma amine oxidase. Somewhat
unexpectedly, phenylhydrazine incorporation was found to correlate with enzyme specific activity in a linear fashion, leveling
off at a stoichiometry of one mole of phenylhydrazine per subunit
for enzyme of specific activity = 0.45 units mg^{-1}. Thus, three
independent lines of inquiry from this laboratory - reductive
trapping experiments, anaerobic product formation and phenylhydrazine titration - *support the view that bovine plasma amine
oxidase contains two reactive cofactors per dimer*.

The above observations provided the frame of reference for a
quantitative evaluation of ammonia release. Anaerobic reactions
were run by coupling the plasma amine oxidase reaction to glutamate dehydrogenase in the presence of [^{14}C]-α-ketoglutarate, such
that released ammonia would be incorporated into [^{14}C]-glutamate.
Unreacted α-ketoglutarate was subsequently destroyed by hydrogen
peroxide, facilitating the direct analysis of radiolabelled glutamate in product mixtures. Incubation of enzyme of high specific
activity at three levels (10, 20 and 30 nmoles of subunits) led
to no release of ammonia above background. This result is highly
significant, *indicating a quantitative transfer of ammonia from
substrate to cofactor in the reductive half reaction*.

The observed release of ammonia in the experiments of Berg and
Abeles (1980) requires some explanation. Scrutiny of their enzyme
preparation indicates ammonium sulfate precipitation as a final
step. We therefore subjected our enzyme preparation to 75% ammonium sulfate, followed by dialysis against 10 mM phosphate
buffer, pH 7.2, for 24 h to reduce ammonium sulfate contamination
of protein to background. Reexamination of anaerobic ammonia
release indicated a stoichiometry of 0.3 to 0.4 moles of ammonia
per active enzyme subunit. This result suggests that ammonia
incorporation into the carbonyl(s) of cofactor may be the cause
of the anaerobic ammonia release previously reported. Certainly,
Schiff base formation between substrate and an imino-form of
cofactor would be expected to release ammonia in the absence of
oxygen.

Does Substrate Undergo Oxidation by a Proton Activation Mechanism?

Structural features of PQQ indicate a reactive carbonyl in conjugation with a pyridine ring, reminiscent of pyridoxal phosphate. Detailed mechanistic studies of the latter cofactor have provided strong evidence for substrate activation by a proton abstraction mechanism. The mode of C-H bond cleavage in the bovine plasma amine oxidase reaction was investigated through the use of structure reactivity correlations. A series of 9 ring substituted benzylamines and their deuterated analogs were synthesized and studied by steady state techniques, leading to values for V_{max} and isotope effects on V_{max}, $^D(V_{max})$. As described by Miller and Klinman (1985) and Palcic and Klinman (1983), rate constants for individual steps in a complex enzyme mechanism can be calculated from steady state parameters and isotope effects on these parameters, once the value of the intrinsic isotope effect(s) has been obtained. The intrinsic isotope effect is defined as the effect of isotopic substitution on a single chemical step, which in the plasma amine oxidase case is the effect of deuteration on the C-H bond cleavage step, k_3:

$$E_{ox} + RCH_2NH_3^+ \underset{k_2}{\overset{k_1}{\rightleftharpoons}} E_{ox} =\overset{+}{N}H\text{-}CH_2R \xrightarrow{k_3} E_{red} \overset{+}{-}NH=CHR$$

$$(4)$$

Previous stopped flow studies have allowed an estimate of $^Dk_3 = 13.5$ for benzylamine oxidation (Palcic and Klinman, 1983). On the assumption that this value is similar for all benzylamines (c.f. Miller and Klinman, 1985, for a discussion of this point with the enzyme dopamine β-hydroxylase), values of k_3 were calculated for each substrate, Table 1.

Table 1: Values of V_{max} and k_3 for Benzylamine Oxidation with Plasma Amine Oxidase[a]

Ring Substituent	$V_{max}(H)$, s^{-1}	k_3, s^{-1}
H	0.56±0.16	2.5±0.7
Br	2.1±0.3	5.3±0.1
F	2.7±0.3	23±3
CF_3	1.9±0.1	24±2
CH_3	0.44±0.05	1.6±0.1
$N(CH_3)_2$	0.030±0.001	0.46±0.05
$CH(CH_3)_2$	0.28±0.01	0.32±0.04
OH	1.1±0.1	14±0.2
acetyl	1.2±0.1	80±1

[a] Values reflect the average of two to four kinetic analyses.

As seen in Table 1, the rate of C-H bond cleavage increases with electron withdrawing substituent. Analysis of binding constants for the same series of benzylamines has indicated a correlation of substrate affinity with hydrophobicity, consistent with previous evidence for a hydrophobic binding pocket in the active site of plasma amine oxidase (Palcic and Klinman, 1983). Multiple linear regression analysis of the rate constants in Table 1 revealed a similar dependence on hydrophobicity, with the

302

exception that hydrophobicity slows rather than facilitates
catalysis (presumably, through the placement of substrate into a
hydrophobic "hole" which prevents the correct alignment of the C-
H bond of substrate with an active site base). Once correction
was made for this effect, a linear correlation between log k_3 and
electronic substituent constants (σ_p) was observed, yielding a
slope or ρ value of 1.5 ± 0.3. Although carbanion forming reactions frequently show ρ values larger than 1.5 (Kirsch, 1972),
the value we observe for the plasma amine oxidase reaction is
consistent with a (competing) charge delocalization into the tricyclic ring of PQQ as well as the benzene ring of substrate.
*Importantly, the positive sign of ρ indicates significant
negative charge at C-1 of substrate at the transition state,
providing strong support for a base assisted proton abstraction
mechanism in substrate oxidation.*

Does Substrate Oxidation Lead to a Time Dependent Conversion of the Spectrum for PQQ_{ox} to PQQ_{red}?

In a recent study, Itoh et al.(1986) generated the reduced
form of PQQ by $NaBH_4$ reduction. Introduction of oxygen led to
pronounced spectral changes, yielding a large decrease in
absorbance at 310 nm in addition to an increase in absorbance
between 400 and 500 nm. Although previous investigators have
reported changes at 480 nm on anaerobic incubation of copper
amine oxidases with substrate, no evidence has been advanced for
a concomitant change at 310 nm. In an effort to provide direct
evidence for this change and hence the participation of PQQ in
amine oxidation, we undertook rapid scanning stopped flow
spectrophotometric studies. Spectral changes were monitored in
the 300 to 500 nm range, following the anaerobic introduction of
substrate. A full spectrum was collected every 8.5 msec and the
data are plotted as difference spectra in Figure 2.

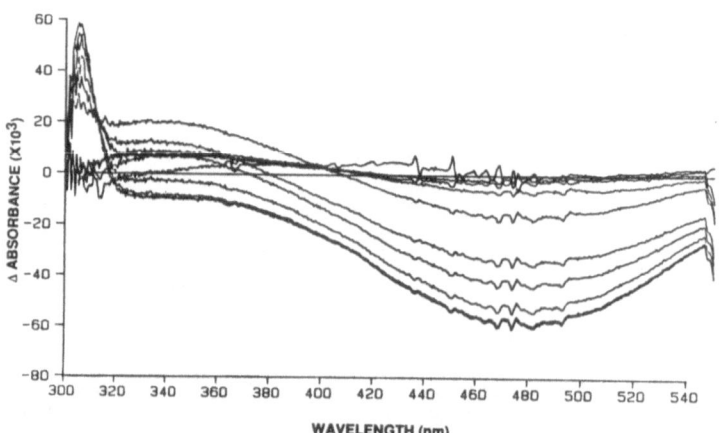

Figure 2: Time Dependent Difference Spectra Obtained From Rapid Scanning
Stopped Flow Studies of the Anaerobic Oxidation of Benzylamine.

Three major spectral changes were observed, with an increase in
absorbance at 340 nm preceeding the changes at 480 and 310 nm;

the observed rise at 340 nm is similar to spectra observed with benzaldehyde and ammonia and, hence, is tentatively attributed to initial formation of a Schiff base complex between cofactor and substrate. A particularly compelling feature of Figure 2 is the large increase in absorbance at 310 nm which accompanies the reduction of cofactor by substrate. This process, which parallels changes at 480 nm, indicates a close correspondence between the behavior of the cofactor at the active site of bovine plasma amine oxidase and PQQ. Quantitative analysis of the transients in Figure 2 was carried out with deuterated benzylamine, in order to uncouple kinetically Schiff base formation from cofactor reduction. This analysis led to rate constants of $25 \pm 4 s^{-1}$ for Schiff base formation (340nm) and $0.17 \pm 0.01 s^{-1}$ for cofactor reduction (310 and 480 nm).

Mechanistic Conclusions:

From the above presented data, central features of bovine plasma amine oxidase catalysis now appear established. First, both reductive trapping and stopped flow studies support the formation of a Schiff base complex between substrate and cofactor prior to oxidation. Subsequent oxidation proceeds via a proton activation mechanism to produce a highly delocalized carbanion. Either this process or, more likely, subsequent rapid proton transfer to the reduced cofactor produces spectral changes at both 480 and 310 nm, as anticipated for the production of reduced PQQ. Finally, hydrolysis of the product imine complex leads to the stoichiometric transfer of ammonia from substrate to cofactor in the absence of oxygen. Each of these features is summarized in the mechanism given in Figure 3.

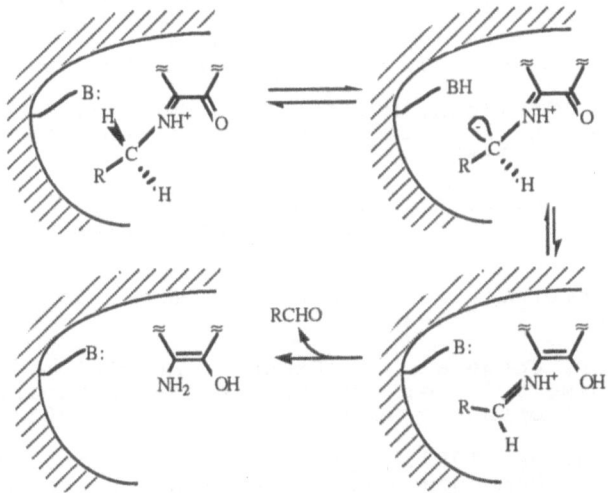

Figure 3: Postulated Reaction Mechanism for Bovine Plasma Amine Oxidase

Although numerous mechanistic aspects of bovine plasma amine oxidase remain to be elucidated - e.g., the isolation and characterization of active site peptide(s) containing covalently linked cofactor, the demonstration and identification of the active site base, the detection of intermediates lying between proton abstraction from substrate and proton transfer to co-

304

factor, and the characterization of the oxidative half reaction involving active site copper – we now have a working hypothesis to guide us in future studies of this fascinating enzyme system.

Acknowledgements:

We acknowledge the National Science Foundation and the National Institutes of Health for support of this project. We are also indebted to Dr. W. McIntire at the V.A. Hospital in San Francisco and Dr. M. Dunn at the University of California, Riverside for assistance in single wave length and rapid scanning stopped flow studies, respectively.

References:

Ameyama, M, Hayashi, M, Matsushita, K, Shingawa, E, and Adachi, O, 1984. Microbial Production of Pyrroloquinoline Quinone. Agric.Biol.Chem. 48: 561-565.

Berg, KA, and Abeles, RH, 1980. Mechanism Action of Plasma Amine Oxidase Products Released under Anaerobic Conditions. Biochemistry 19: 3186-3189.

Falk, MC, 1983. Stoichiometry of Phenylhydrazine Inactivation of Pig Plasma Amine Oxidase. Biochemistry 22: 3740-3745.

Hartmann, C and Klinman, JP, 1987. Reductive Trapping of Substrate to Bovine Plasma Amine Oxidase. J.Biol.Chem. 262:962-965.

Hartmann, C and Klinman, JP, 1988. Pyrroloquinoline Quinone: A New Redox Cofactor in Eukaryotic Enzymes. Biofactors 1: 41-49.

Itoh, S, Ohshiro, Y, and Agawa, T, 1986. Reaction of reduced PQQ (PQQH$_2$) and molecular oxygen. Bull.Chem.Soc.Jpn. 59: 1911-1914.

Kalb, VF, Donohue, TJ, Corrigan, MG, and Bernlohr, RW, 1978. A New and Specific Assay for Ammonia and Glutamine Sensitive to 100 pmol. Anal.Biochem. 90: 47-57.

Kirsch, JF, 1972. In: Advances in linear free energy relationships. Chapman, NB and Shorter, J, eds. Plenum Press, New York.

Levitzki, A, 1970. Determination of Submicro Quantities of Ammonia. Anal.Biochem. 33: 335-340.

Lobenstein-Verbeek, CL, Jongejan, JA, Frank, J and Duine, JA, 1984. Bovine serum amine oxidase: a mammalian enzyme having covalently bound PQQ as prosthetic group. FEBS LETTERS 170:305-309.

Miller, SM and Klinman, JP, 1983. Magnitude of Intrinsic Isotope Effects in the Dopamine β-monooxygenase Reaction. Biochemistry 22: 3091-3096.

Palcic, MM, and Klinman, JP, 1983. Isotopic Probes Yield Microscopic Constants: Separation of Binding Energy from Catalytic Efficiency in the Bovine Plasma Amine Oxidase Reaction. Biochemistry 22: 5957-5966.

Summers, MC, Markovic, R and Klinman, JP, 1979. Stereochemistry and Kinetic Isotope Effects in the Bovine Plasma Amine Oxidase Catalyzed Oxidation of Dopamine.
Biochemistry 18: 1969-1979.

Suzuki, S, Sakurai, T, Nakahara, A, Manabe, T and Okuyama, T, 1986. Roles of the Two Copper Ions in Bovine Serum Amine Oxidase. Biochemistry 25: 338-341.

COPPER-PQQ INTERACTIONS IN AMINE OXIDASES

D. M. Dooley,[1] C. E. Cote,[1] M. A. McGuirl,[1] J. L. Bates,[1] J. B. Perkins,[1] R. S. Moog,[1] I. Singh,[2] P. F. Knowles,[2] and W. C. McIntire[3]. [1]Department of Chemistry, Amherst College, Amherst, MA 01002; [2]Astbury Department of Biophysics, University of Leeds, Leeds LS2 9JT; [3]Molecular Biology Division, Veterans Administration Medical Center, San Francisco, CA 94121

Key Words: Copper/Pyrroloquinoline Quinone/Resonance Raman/EPR

ABSTRACT: Interactions between copper and PQQ in several amine oxidases have been probed by a variety of physical methods. A Cu(I)-semiquinone (PQQ$^{\bullet}$) state can be generated by substrate-reduction in the presence of ligands that stabilize Cu(I). This state has now been characterized in detail for amine oxidases from bovine plasma, porcine kidney, pea seedlings, and _Arthrobacter_ P1. The properties of the semiquinone are independent of the enzyme, substrate, or copper ligand. It is possible that a Cu(I)-PQQ$^{\bullet}$ state may be a catalytic intermediate in the oxidation of the substrate-reduced enzyme by O$_2$. Additional substrate-reduced forms are observed for some amine oxidases; the distribution among these forms is sensitive to copper ligation. Resonance Raman spectroscopy has been used to characterize both native and metal-depleted amine oxidases, which have been derivatized by chromophoric carbonyl reagents. 2-Hydrazinopyridine reacts with all the amine oxidases (and PQQ) to form a hydrazone derivative with $\lambda_{max} \cong 415$ nm; subsequently this converts, in a copper-dependent reaction, to a form with an absorption band at $\cong 520$ nm.

INTRODUCTION

Copper-containing amine oxidases from mammals, plants, and microorganisms require both copper and PQQ for activity. Amine substrates react with PQQ [Farnum and Klinman, 1986; Mondovi, 1985; Knowles, et al. 1987] to generate a two-electron reduced enzyme, which is subsequently reoxidized by O$_2$. Copper may participate in both phases [Barker, et al. 1979; Knowles and Yadav, 1984; Olsson, et al. 1978; Suzuki, et al. 1983; Rinaldi, et al. 1984; Yadav and Knowles 1984]. Clearly, interactions between copper and PQQ may be important to the structure and catalytic mechanism of amine oxidases. Previous work [see Knowles and Yadav 1984; Mondovi 1985, for reviews] has established that the reaction of substrates and carbonyl reagents perturbs the copper site, although the extent of these perturbations among amine oxidases is variable. Substrate-reduction has been shown to affect the copper coordination chemistry in the bovine plasma enzyme [Dooley and Cote 1985]. Evidence for significant copper-PQQ interactions also comes from studies of copper depletion and metal substitution [Suzuki et al. 1983, 1986], sulfide inactivation [Dooley and Cote, 1984], and substrate reduction of enzyme-cyanide complexes [Finazzi-Agro, et al. 1984; Dooley, et al. 1987]. The formation of a Cu(I)-PQQ$^{\bullet}$ (semiquinone) state via internal electron-transfer from reduced PQQ was suggested to account for the properties of the substrate-reduced cyanide complex [Dooley, et al. 1987]. In this paper we describe further studies on the Cu(I)-PQQ$^{\bullet}$ state of several amine oxidases and present some new approaches for investigating copper-PQQ interactions in amine oxidases.

J. A. Jongejan and J. A. Duine (eds.), PQQ and Quinoproteins, 307–316.
© 1989 by Kluwer Academic Publishers.

MATERIALS AND METHODS

Amine oxidases from pea seedlings, porcine kidney, porcine plasma, and bovine plasma were purified by published methods [Kluetz 1980; Dooley and McGuirl, 1986; Rius, et al. 1984;] or by modifications of published procedures [Summers, et al. 1979; Turini, et al. 1982]. Arthrobacter P1 methylamine oxidase was purified by a new procedure, to be published elsewhere. All the enzymes were homogeneous as judged by sodium dodecylsulfate and gradient gel electrophoresis and displayed specific activities and spectroscopic properties consistent with high purity. 2-Hydrazinopyridine (2-HP) was obtained from Aldrich and recrystallized from methanol before use. The 2-HP derivatives of amine oxidases were prepared by adding a 20-fold molar excess (per mole of enzyme) of 2-HP at ambient temperature in 0.1M phosphate buffer, pH 7.0; absorption spectroscopy was used to monitor the reaction. Conversion from the initially formed product ($\lambda_{max} \cong 415$ nm) to the final product ($\lambda_{max} \cong 520$ nm) was achieved by incubating the enzyme-2-HP derivative at ambient temperature or at ~ 40 °C. The phenylhydrazine and 2,4-dinitrophenylhydrazine derivatives were prepared as previously described [Knowles, et al. 1987; Moog et al. 1986]. Cyanide solutions were prepared immediately before use in the appropriate buffer. Anaerobic conditions were achieved by vigorously purging all reagent solutions with Ar; enzyme solutions were flushed with Ar gas for at least one hour in cuvettes fitted with septa (TCS-Medical Products) or stoppered EPR tubes. Solutions were transferred with gas tight syringes. Metal-depleted amine oxidases were produced by anaerobic dialysis against 50 mM NaCN of the dithionite-reduced enzymes, followed by extensive dialysis against buffer. All buffers used with metal-depleted enzymes or samples for EPR spectroscopy were treated with Chelex to remove trace metal ions. Spectra were obtained as described in the following references: resonance Raman [Moog, et al. 1986]; CD [Dooley and McGuirl 1986]; EPR [Dooley, et al. 1987].

RESULTS AND DISCUSSION

We have proposed that a Cu(I)-PQQ$^{\bullet}$ state, with a characteristic EPR signal (Figure 1), can be generated by reducing amine oxidases with substrates in the presence of CN$^-$ [Dooley, et al. 1987]. In fact, the same

Figure 1. 77K X-band EPR spectrum of the cyanide complex of methylamine oxidase following the anaerobic addition of methylamine. Center field marker at 3235 G; sweep width is 200 G.

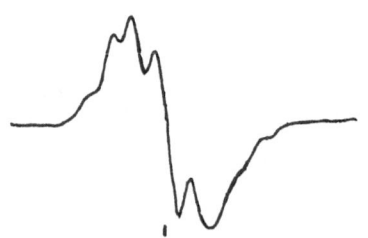

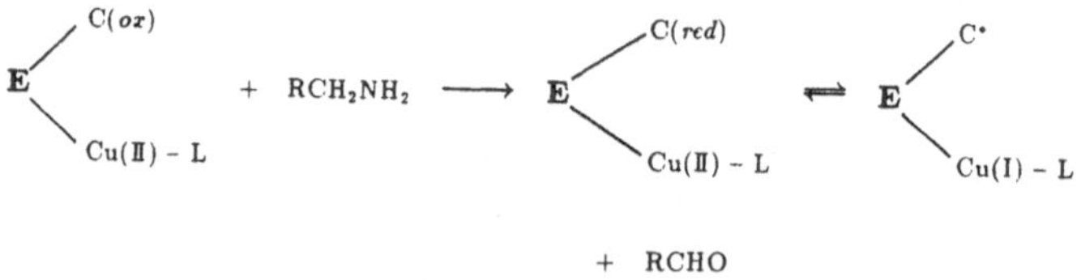

SCHEME 1
C = Organic cofactor
(PQQ); L = OH$_2$, CN$^-$

+ RCHO

species can be produced by adding CN⁻ to the substrate-reduced enzyme under
anaerobic conditions, i. e. independent of the order of addition of substrate or cyanide. Scheme I summarizes the proposed reaction. In order to
confirm that the disappearance of the Cu(II) EPR signal is associated with
reduction (and not with spin-coupling to produce an EPR nondetectable
state), the reaction was followed by CD spectroscopy, which permits the
direct observation of the Cu(II) d-d transitions [Dooley and Cote, 1985].
Fig. 2A shows the CD spectra of resting methylamine oxidase and its cyanide
complex. The Cu(II) d-d transitions are shifted $\cong$ 3000 cm⁻¹ to higher
energy, consistent with equational substitution by CN⁻. Methylamine effects
on the CD spectrum are displayed in Fig. 2B. A new species is formed that

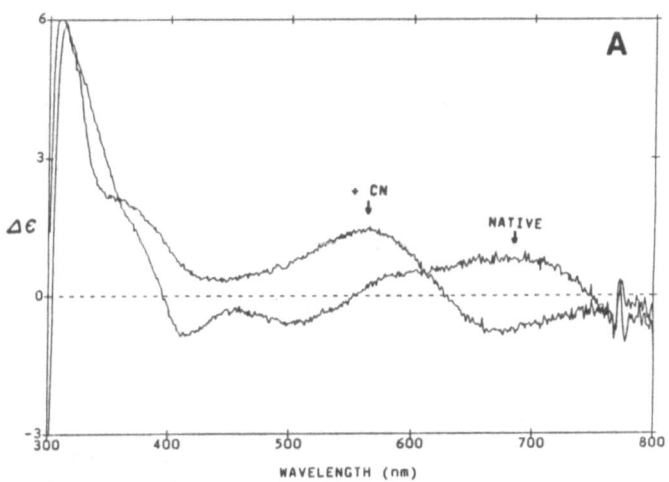

Figure 2. (A) CD spec-
trum of resting methyl-
amine oxidase and its
cyanide complex in 0.1M
potassium phosphate buf-
fer pH 7.0 at 20°C.
[CN⁻] = 50 mM.

(B) Methylamine titra-
tion of the methylamine
oxidase-cyanide complex.
Methylamine concentra-
tions are approximately
25 µM, 70 µM, 100 µM,
140 µM. Enzyme concen-
tration is 85 µM.

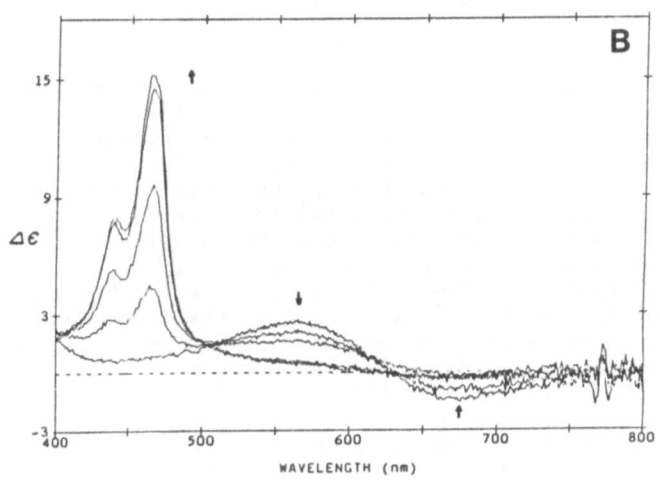

is very similar to the intermediates observed in plant [Finazzi-Agro, et al. 1984] and porcine kidney amine oxidase [Dooley, et al. 1987]. Moreover, the bands at 570 nm and 675 nm, assigned as Cu(II) d-d transitions, are bleached, thereby establishing that substrates reduce Cu(II) → Cu(I) in the amine oxidase-cyanide complex. Further evidence for Scheme I is provided by the observation that a Cu(I)-PQQ° state in methylamine oxidase can be generated using t-butylisocyanide (data not shown), another ligand that would stabilize Cu(I) relative to Cu(II) [Greenwood and Earnshaw 1984]. Not all ligands promote copper reduction, however. Addition of 1,10-phenanthroline to substrate-reduced pea seedling amine oxidase bleaches the distinctive absorption spectrum customarily observed for the substrate-reduced form (Figure 3). Thus there exist at least two forms of substrate-reduced pea

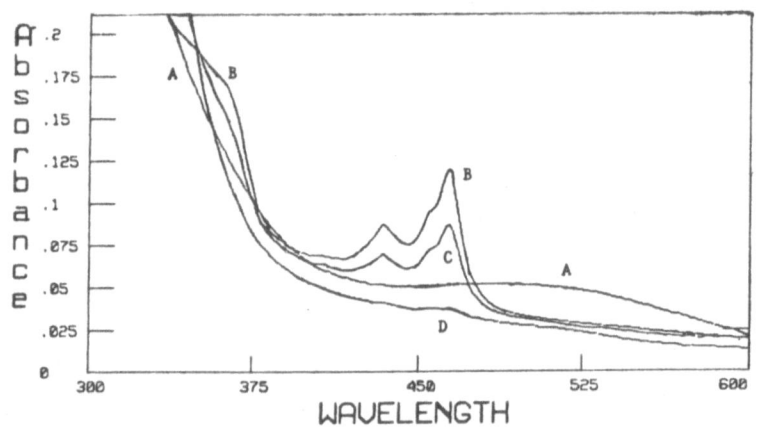

Figure 3. Effects of 1,10-phenanthroline on substrate-reduced pea seedling amine oxidase. (A) 1.5×10^{-5} M enzyme. (B) Anaerobically reduced with 2×10^{-4} M p-dimethylaminomethylbenzylamine. (C) Sample in (B) plus 4×10^{-4} M 1,10-phenanthroline. (D) after additional phenanthroline to give a final concentration of 1.2×10^{-3} M .

seedling amine oxidase, which can be interconverted by ligand substitution at the copper ions. This is also consistent with Scheme I. Other data suggest that Scheme I is incomplete, at least for certain amine oxidases. Reducing plant amine oxidases with substrates in the absence of cyanide still produces an intermediate with electronic absorption bands at 465, 430 and 360 nm [Bellelli, et al. 1985; Hill and Mann 1964] that does not display a semiquinone EPR spectrum. Similar results have been obtained for methylamine oxidase (data not shown). Further work is necessary to completely understand the reactions of amine oxidases with their substrates. It is nevertheless clear that the reactivity of the PQQ in these enzymes is very sensitive to the copper coordination environment. Ligand substitution is likely to involve the two coordinated H_2O molecules believed to be present in the Cu(II) sites [McCracken et al., 1987].

Resonance Raman spectroscopy of derivatized amine oxidases is an informative probe of the active site structure(s) around the PQQ [Moog, et al. 1986; Knowles, et al 1987; Williamson et al. 1987]. The phenylhydrazine derivatives of native and apo methylamine oxidase are compared in Fig. 4.

Based on the extinction coefficients, the reactions between the apo enzymes and hydrazines are not stoichiometric. We assume that this reflects reduction of the PQQ by dithionite, which would then be unreactive towards hydrazines. Closely similar vibrational frequencies in these spectra establish that the same product is formed. On the other hand, the shifts in some

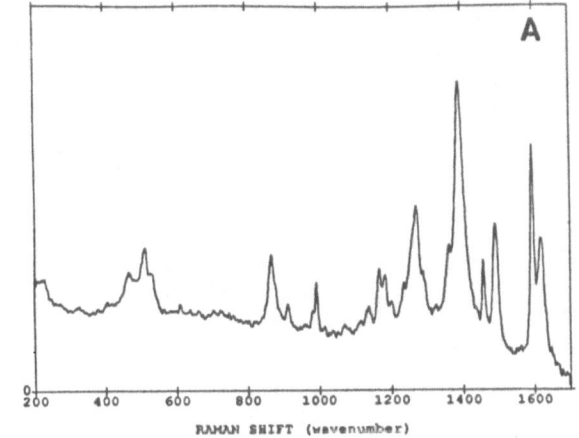

Figure 4. Resonance Raman spectra obtained with 457.9 nm excitation of the phenylhydrazine derivatives of native (A) and apo (B) methylamine oxidase.

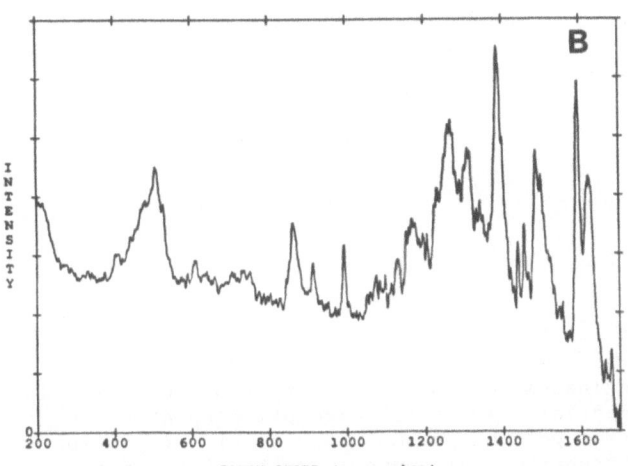

peaks and the variations in relative intensities indicate that the copper does alter the PQQ site. Comparing the 2,4-dinitrophenylhydrazine derivatives of apo (prepared as in [Suzuki, et al. 1983]) and native bovine plasma amine oxidase (Fig. 5) leads to the same conclusion. Suzuki's methodology appears to produce a metal-depleted enzyme that reacts stoichiometrically with hydrazines.

312

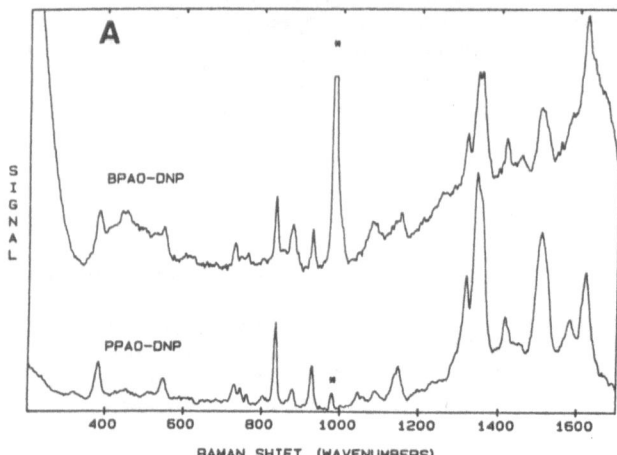

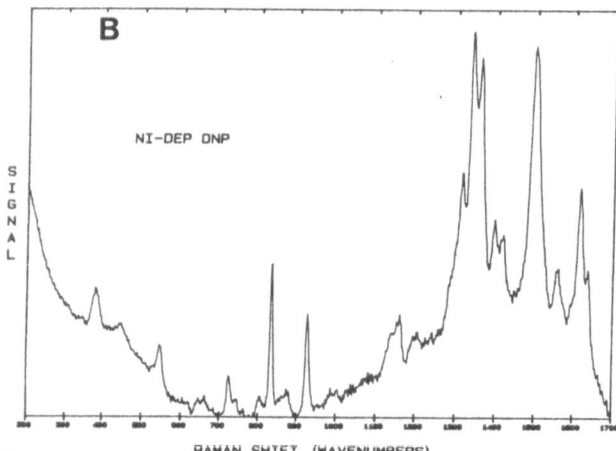

Figure 5. Resonance Raman spectra obtained with 457.9 nm excitation of the 2,4-dinitrophenylhydrazine derivatives of native bovine and porcine plasma amine oxidase (A). (B) metal-depleted bovine plasma amine oxidase.

2-HP reacts with amine oxidases to produce two spectroscopically distinct products (Fig. 6). Essentially identical absorption spectra are observed for amine oxidases from porcine plasma, porcine kidney, bovine plasma, pea seedling, and Arthrobacter, although the rate of conversion to the 520 nm form is variable. Very different resonance Raman spectra are observed with excitation at 457.9 nm or 514.5 nm (Fig. 7), establishing that the two products have different structures. Importantly, the same two products are formed with all the amine oxidases examined to date. Only the derivative with $\lambda_{max} \approx 415$ nm is observed in reactions of 2-HP with metal-depleted amine oxidases; subsequent Cu(II) addition produces the 520 nm form.

2HP DERIVATIVES

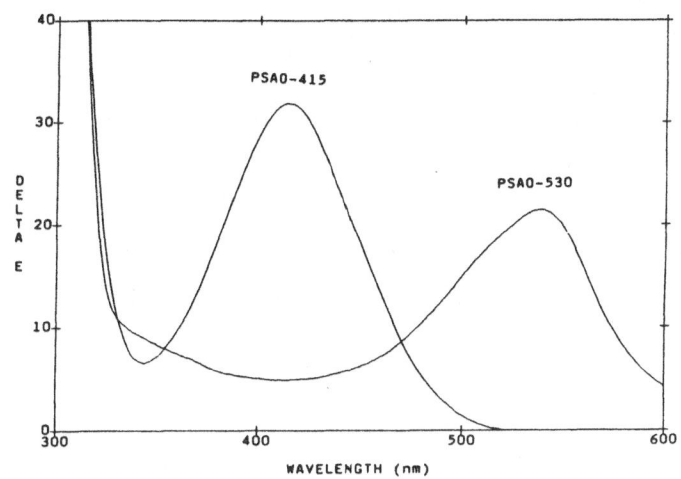

Figure 6. Optical absorption spectra of the 2-hydrazinopyridine derivatives of pea seedling amine oxidase in 0.1 M potassium phosphate buffer (pH 7.0). λ_{max} = 530 nm in this case.

Resonance Raman spectroscopy (Fig. 8) establishes that the final products are identical. Hence, the conversion of the initial 2-HP derivative to the 520 nm form requires copper. Although the absorption and resonance Raman spectra of the PQQ:2-HP derivative are quite similar to the spectra of the initial amine oxidase:2-HP derivatives, the spectra of the 520 nm forms are significantly different from the PQQ/Cu^{2+} derviative with 2-HP. Evidently free PQQ does not interact with Cu(II) in the same way as in the native enzymes. There are two plausible roles for copper in the 2-HP reaction: (1) Cu(II) may oxidize the initial 2-HP derivative; (2) derivatized PQQ may coordinate to Cu(II). Preliminary EPR experiments on the porcine plasma enzyme support the second possibility, as conversion of the 415 nm form to the 520 nm form apparently does not decrease the Cu(II) EPR intensity but does alter the g values.

Collectively the data are consistent with significant structural and chemical interactions between copper and PQQ in amine oxidases. Given the distances between the cofactors derived from NMR relaxation [Williams and Falk, 1986] and fluorescence quenching [Lamkin et al., 1988], copper coordination by the pyrrole nitrogen or a carboxylate group of PQQ remains a possibility. Alternatively, the interactions between copper and PQQ could be mediated by the polypeptide, especially since PQQ is known to be covalently bound to the protein [Lobenstein-Verbeek, et al. 1984]. If copper is the site for O$_2$ binding and reduction, then an efficient electron-transfer path from PQQ to Cu(II) is required.

314

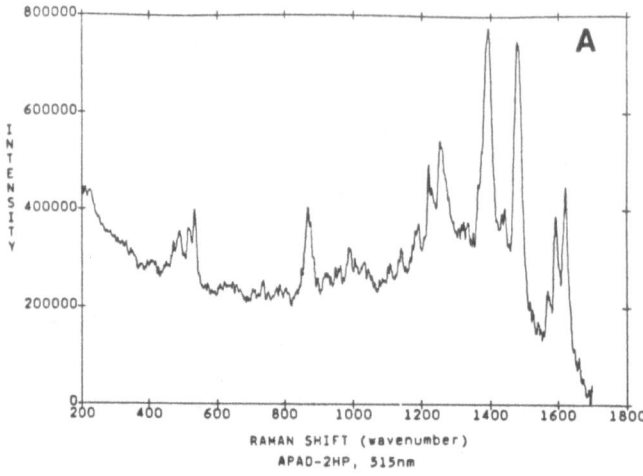

RAMAN SHIFT (wavenumber)

APAD-2HP, 515nm

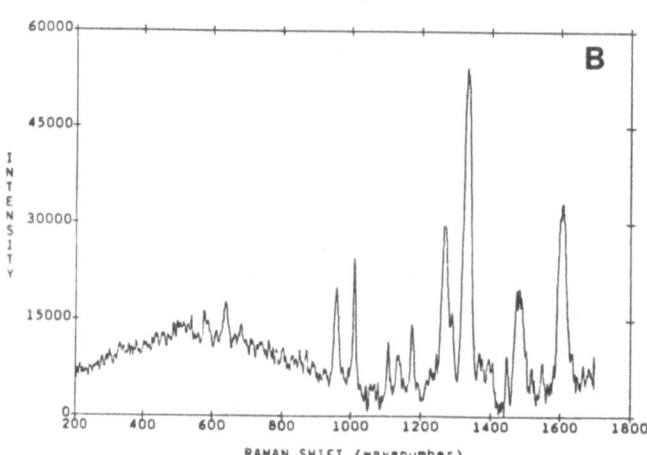

RAMAN SHIFT (wavenumber)

Figure 7. Resonance Raman spectra obtained with 457.9 nm (A) and 514.5 nm excitation (B) of the initial (A) and final (B) methylamine oxidase derivatives with 2-hydrazinopyridine.

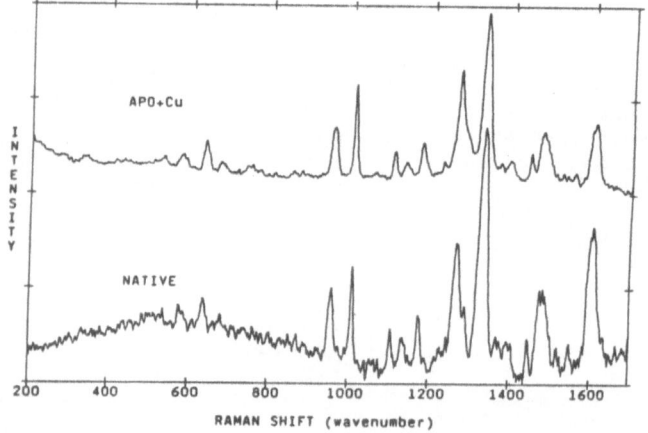

RAMAN SHIFT (wavenumber)

Figure 8. Resonace Raman spectra obtained with 514.5 nm excitation of the native and apo methylamine oxidase derivatives with 2-hydrazinopyridine (520 nm form).

REFERENCES

Barker R, Boden N, Cayley G, Charlton SC, Hanson R, Holmes MC, Kelly ID, Knowles PF, 1979. Biochem. J. 177: 289-302.

Bellelli A, Brunori M, Finazzi-Agro A, Floris G, Giartosi A, Rinaldi A, 1985. Biochem. J. 232: 923-926.

Dooley DM and Cote CE, 1984. J. Biol. Chem. 259: 2923-2926.

Dooley DM and Cote CE, 1985. Inorg. Chem. 24: 3996-4000.

Dooley DM and McGuirl MA, 1986. Inorg. Chim. Acta 123:4245-4248.

Dooley DM, McGuirl MA, Peisach J, McCracken J, 1987. FEBS Lett. 214: 274-278.

Farnum MF and Klinman JP, 1986. Biochemistry 25: 6028-6036.

Finazzi-Agro A, Rinaldi A, Floris G, Rotilio G, 1984. FEBS Lett. 176: 378-380.

Greenwood NN and Earnshaw A, 1984. Chemistry of the Elements. Pergamon, Oxford.

Hill JM and Mann PJC, 1964. Biochem. J. 91: 171-182.

Knowles PF, Pandeya KB, Rius FX, Spencer CM, Moog RS, McGuirl MA, Dooley DM, 1987. Biochem. J. 241: 603-608.

Knowles PF and Yadav KDS, 1984, in: Lontie R. Copper Proteins and Copper Enzymes, Vol. 2., CRC Press, Boca Raton, FL.

Lamkin MS, Williams TJ, Falk MC, 1988. Arch. Biochem. Biophys. 261: 72-79.

Lobenstein-Verbeek CL, Jongejan JA, Frank J, Duine JA, 1984. FEBS Lett. 170: 305-309.

McCracken J, Peisach J, Dooley DM, 1987 J. Am. Chem. Soc. 190: 4064-4072.

Mondovi B, 1985. Structure and Functions of Amine Oxidases. CRC Press, Boca Raton, FL.

Moog RS, McGuirl MA, Cote CE, Dooley DM, 1986. Proc. Natl. Acad. Sci. (USA) 83: 8435-8439.

Olsson B, Olsson J, Pettersson G, 1978. Eur. J. Biochem. 87: 1-8.

Rinaldi A, Giartosio A, Floris G, Medda R, Finazzi-Agro A, 1984. Biochem. Biophys. Res. Commun. 120: 242-249.

Rius FX, Knowles PF, Pettersson G, 1984. Biochem. J. 220: 767-772.

Summers MC, Markovic R, Klinman JP, 1979 Biochemistry 18: 1969-1979.

Suzuki S, Sakurai T, Nakahara A, Manabe T, Okuyama T, 1983. <u>Biochemistry</u> 22: 17630-1635.

Suzuki S, Sakurai T, Nakahara A, Manabe T, Okuyama T, 1986. <u>Biochemistry</u> 25: 338-341.

Tarini P, Sabatini S, Befani O, Chimenti F, Casanova C, Riccio PL, Mondovi B, 1982. <u>Anal. Biochem.</u> 125: 294-298.

Williams TJ and Falk MC, 1986 <u>J. Biol. Chem.</u> 261: 15949-15954.

Williamso,n PS, Moog RS, Dooley DM, Kagan HM, 1986. <u>J. Biol. Chem.</u> 261: 16302-16305.

Yadav KDS and Knowles PF, 1981. <u>Eur. J. Biochem.</u> 114: 139-144.

ENZYMOLOGY OF LYSYL OXIDASE

Herbert M. Kagan, Stephen N. Gacheru, Philip C. Trackman, Susan D. Calaman and Frederick T. Greenaway

Boston University School of Medicine, Boston, Massachusetts 02115 and Clark University, Worcester, Massachusetts 01610

KEY WORDS: Lysyl oxidase, crosslinkages, histidine, diaminocyclohexane

ABSTRACT

The pH dependency of steady-state kinetic parameters and the results of chemical modification of lysyl oxidase by diethylpyrocarbonate has identified an enzyme residue with a pK_a of 7.0 ± 0.1 and an enthalpy of ionization of 6.1 kcal mole^{-1}, consistent with an active site histidine residue. This residue could function as a general base in the abstraction of a proton from the substrate during lysyl oxidase catalysis. Evidence for complex formation between the amine substrate and PQQ in lysyl oxidase was derived from the similar perturbations induced by n-butylamine in the spectra of the enzyme and of PQQ under anaerobic conditions. A new avenue for the development of inhibitors of lysyl oxidase is suggested by the finding that the cis isomer but not the trans isomer of 1,2-diaminocyclohexane is a potent inhibitor of this enzyme, apparently by virtue of its interaction with PQQ at the active site.

INTRODUCTION

Lysyl oxidase (protein-lysine 6-oxidase; EC 1.4.3.13) is unique among the copper-dependent amine oxidases by catalyzing the oxidative deamination of lysine in soluble forms of elastin and collagen to peptidyl $\propto$-amino-adipic-δ-semialdehyde. The peptidyl aldehyde product can spontaneously condense with other residues of aminoadipic semialdehyde or with unmodified ϵ-amino groups eventually generating a variety of covalent crosslinkages which account for the insolubilization of these connective tissue proteins (Kagan, 1986). The finding that lysyl oxidase will also oxidize alkyl mono- and diamines has facilitated its assay as well as efforts to investigate its mechanism of action (Trackman et al., 1981). The critical role of lysyl oxidase in development is made evident by the severe defects in connective tissues seen in lysyl oxidase deficiency states, as in Menke's syndrome (Royce et al., 1980), in copper-deficiency (Siegel et al., 1970), and in lathyrism induced by the administration of inhibitors of lysyl oxidase such as β-aminopropionitrile (Levene, 1961). Such situations can result in deformed skeletal structure, loose, fragile skin, emphysema, and aneurysms, among other consequences. By the same token, the selective inhibition of lysyl oxidase by chemical agents has significant chemotherapeutic potential for the control of fibrotic disease.

It had long been suspected that lysyl oxidase contained a carbonyl as well as a copper cofactor. Evidence has been provided that the organic cofactor is pyrroloquinoline quinone in bovine aortic lysyl oxidase

317

J. A. Jongejan and J. A. Duine (eds.), PQQ and Quinoproteins, 317–326.

(Williamson et al., 1986), a conclusion which was independently upheld for the human placental enzyme (van der Meer and Duine, 1986). Steady-state kinetics indicated that lysyl oxidase follows a ping pong kinetic pattern (Williamson and Kagan, 1986), as follows:

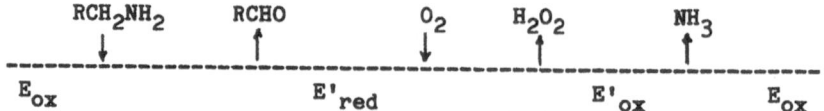

where E'_{red} represents a putative, aminated, substrate-reduced enzyme derivative presumably represented as the aminophenol form of PQQ, while E'_{ox} is presumed to be the 5-imino derivative of PQQ_{ox} (PQQ=NH). Consistent with this scheme, we have observed aldehyde production stoichiometric with active site concentration under anaerobic conditions (Williamson and Kagan, 1986). The catalytic cycle can thus be accounted for by the sum of two half-reactions, with one involving oxidation of the amine to the aldehyde concomitant with the reduction of the enzyme-bound carbonyl cofactor to the aminophenol form of PQQ. The second half-reaction would thus consist of the oxygen-dependent re-oxidation of the enzyme followed by the presumptively hydrolytic release of ammonia from 5-iminoPQQ. Amine oxidation is proposed to begin with the formation of a Schiff base between carbon 5 of PQQ and the amino group of the substrate, followed by the abstraction of the α-proton of the substrate to yield a transient, PQQ-linked, substrate-derived carbanion. Evidence for carbanion formation was provided by the observation that the electrophilic reagent, tetranitromethane, nitrates a catalytic intermediate derived from the substrate (Williamson and Kagan, 1987a). Moreover, both the k_{cat}/K_m and k_{cat} kinetic parameters exhibited kinetic isotope effects with α-deuterated butylamine or benzylamine substrates, indicative that α-proton abstraction is at least rate contributing in lysyl oxidase catalysis (Williamson and Kagan, 1987a,b).

In the present study, we have employed pH-dependent kinetic analyses and chemical modification of lysyl oxidase assessing for evidence of an enzyme residue which might function as a general base possibly involved in α-proton abstraction. We have also examined spectral features of lysyl oxidase complexes and model compounds for evidence of interaction between substrate or inhibitors with PQQ in this enzyme.

MATERIALS AND METHODS

Lysyl oxidase was isolated from calf aorta as described (Williams and Kagan, 1985). The product appeared as a single protein band of 32,000 Da by sodium dodecylsulfate polyacrylamide gel electrophoresis. The enzyme was assayed in 0.1 M sodium borate, 0.15 M sodium chloride, pH 8.0, against 125,000 cpm of an insoluble elastin substrate prepared from chick embryo aortas pulsed with L-[4,5-^{3}H]lysine as described (Kagan and Sullivan, 1982). Tritiated water formed by enzyme action was isolated by distillation and then quantified by liquid scintillation spectrometry. Lysyl oxidase was also assayed against alkyl mono- or diamine substrates by a peroxidase-coupled fluorescence method for the detection of H_2O_2 (Trackman et al., 1981). The Vmax values obtained at different pH values with this assay were shown to be independent of the differing ratios of the 0.02 M sodium borate or 0.02 M potassium phosphate buffer components used at each pH.

Steady state kinetic constants were obtained by entering velocity and

substrate concentration data into the Fortran program of Cleland   (Cleland,
1979)   which   applies a least squares fitting procedure to   the   Michaelis-
Menten equation to derive Vmax   and Km. pH–dependent kinetic   data were fit
to   the   following   equation:

$$Log\ Y\ =\ log\ \frac{C}{1\ +\ H/K_1\ +\ K_2/H}$$

where   Y   is either   Vmax   or Vmax/Km, C is   the   pH   independent
value for   Y,   and K_1 and K_2 are the   dissociation   constants   for   an
acid   or base, respectively (Cleland, 1979).

Modification   of   lysyl oxidase with diethylpyrocarbonate   (DEPC)   was
initiated   by   the addition   of   DEPC in ethanol to 0.3 mg ml^{-1} solutions
of enzyme in 0.2 M potassium phosphate, pH 7.0. The reaction mixtures   were
incubated   at 22° C, aliquots were removed at intervals and mixed with   20-
fold   molar   excesses of L–histidine to quench unreacted DEPC,   and   enzyme
activity   remaining was measured by the peroxidase–coupled assay for H_2O_2
release.   Histidine   modification   was quantified from the   increase   in
absorbance   at   240 nm (ϵ = 3200 M^{-1} cm^{-1}) (Miles, 1977).

RESULTS

<u>pH–Dependent</u>   <u>Kinetic</u> <u>Parameters</u>   The   pH–dependent profile of log   Vmax/Km
for n–hexylamine oxidation, shown in Figure 1, gave pK_a values of 7.0 ± 0.1
and   10.4 ± 0.1. The slope of the increase in the plot at   the   lower   pH
values   is 0.9, indicative of deprotonation governed by a single pK_a value
(Cleland, 1977). The corresponding analysis with ethanolamine as   substrate
resulted   in   first   and   second pK_a values of 7.0 ± 0.1 and   9.3   ±   0.2,
respectively.   The   second pK_a apparently reflects the   ionization   of   the
substrate   since   the pK_a values for the amino groups of n–hexylamine   and
ethanolamine are 10.56 and 9.5, respectively, in reasonable agreement   with
the second pK_a found in the kinetics experiments. It thus appears that   the
the protonated form of the substrate is the favored species for catalysis.

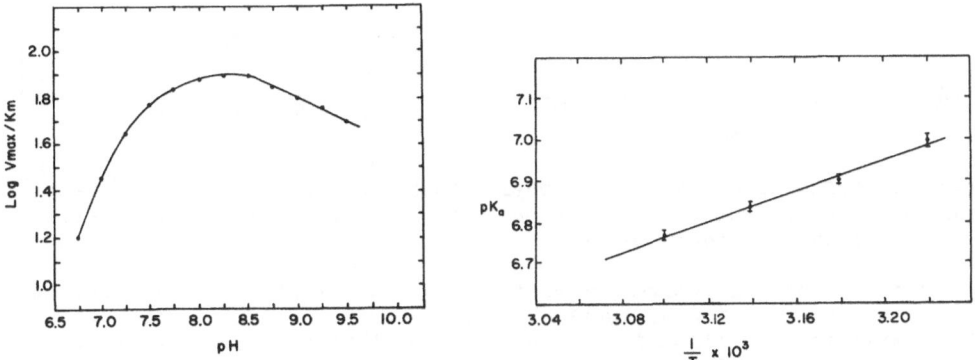

Figure 1. (Left) pH–Dependent Oxidation of n–Hexylamine by Lysyl Oxidase.
Figure 2. (Right) Van't Hoff Plot for the Temperature Dependency of pK_{a1}.

The substrate-independent pK_a of 7.0 ± 0.1 likely reflects the ionization of an enzyme residue which must be unprotonated for effective catalysis. The temperature dependency of this pK_a was assessed by performing steady-state kinetic analyses at different assay temperatures and assessing the lower pK_a value from plots of log Vmax/Km *versus* pH at each temperature. The enthalpy of ionization for this function was found to be 6.1 ± 0.1 kcal/mole, determined as described (Cleland, 1977), from the slope of a plot of pK_a against $1/T$, where the slope equals $\Delta H_{ion}/2.303R$, (Figure 2). *In toto*, the value of 7.0 for the substrate-independent, lower pK_a and the enthalpy of ionization of this function are most consistent with reported values for histidine residues (Cleland, 1977).

<u>Reaction of Lysyl Oxidase with Diethylpyrocarbonate</u> Incubation of lysyl oxidase with varying concentrations of DEPC at pH 7.0 inactivated the enzyme by a process which obeyed pseudo first order kinetics (Figure 3). A plot of k_{app} against DEPC concentration was linear (not shown), indicative that the modification stemmed from a bimolecular reaction between DEPC and an enzyme residue. The data of the experiment shown in Figure 3 were also plotted as ln k_{app} *versus* ln DEPC (Levy et al., 1963). The resulting plot was linear with a slope of 0.92, consistent with the reaction of one molecule of DEPC with one residue of the protein as inactivation occurs (Levy et al., 1963). As shown in Figure 4, the presence of 15 mM n-hexylamine in the chemical modification reaction mixture protected against the loss of activity caused by DEPC, indicative of modification at the active site. Separate experiments established that there was negligible direct reaction between n-hexylamine and DEPC under these conditions.

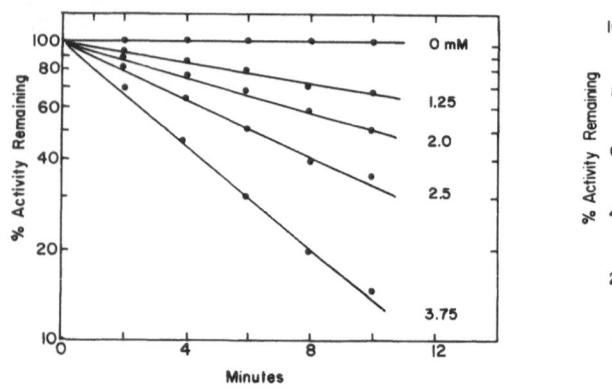
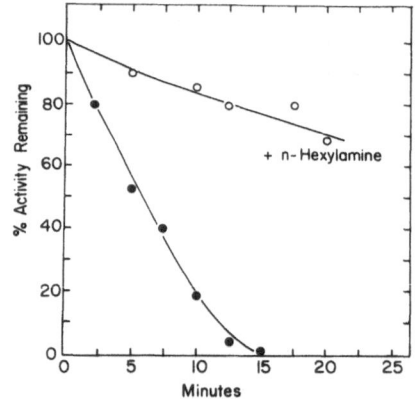

Figure 3. (Left) Inactivation of Lysyl Oxidase by Diethylpyrocarbonate
Figure 4. (Right) Protection by n-Hexylamine (15 mM) against Inactivation of Lysyl Oxidase by DEPC.

A difference spectrum exhibiting a maximum at 240 nm was generated during the inactivation of lysyl oxidase by DEPC, consistent with the N-carbethoxylation of histidine by this reagent (Miles, 1977). There was no significant change in the enzyme spectrum at 280 nm, arguing against tyrosine modification by DEPC. Histidine modification and enzyme inactivation were then quantified during the course of the modification by 3 mM DEPC. These parameters were linearly related, with the plot of this histidine per mole of lysyl oxidase at full inactivation (Figure 5). Further support for the relationship of histidine modification to enzyme inactivation stemmed from the finding that incubation of fully inactivated enzyme with 2.5 mM hydroxylamine resulted in the recovery of 71% of the enzyme activity, consistent with the hydroxylamine-mediated, nucleophilic displacement of a carbethoxy moiety from N-carbethoxyhistidine (Miles, 1977).

The pH dependency of the rate constant for inactivation (k_{inact}) of lysyl oxidase was determined as a function of pH to assess whether the histidine modified by DEPC correlating with enzyme inactivation exhibited the same pK_a as that catalytically functional residue titrating in the log Vmax/Km plot shown in Figure 1. As seen in Figure 6, k_{inact} increases between pH 5 and 9. Further, the experimentally determined values for k_{inact} agree closely with the theoretical titration curve for a single ionizing species with a pK_a of 6.9 ± 0.1, in good agreement with the pK_a of 7.0 ± 0.1 for the apparently functional residue as derived from the log Vmax/Km versus pH plot (see Figure 1).

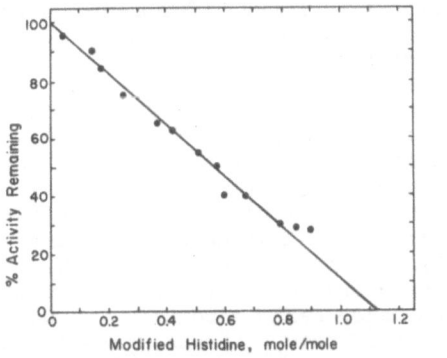 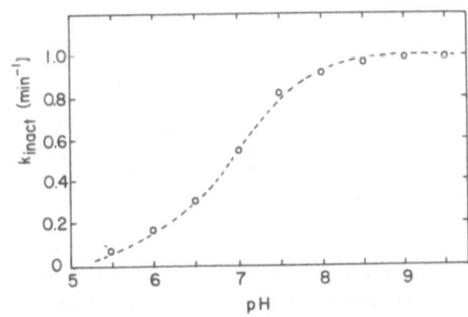

Figure 5. (Left) Enzyme Inactivation versus Histidine Modification.
Figure 6. (Right) pH-Dependency of k_{inact} for Inactivation by DEPC. Open circles, experimental data. The dashed line is a theoretical curve, assuming a pK_a of 6.9 for the species being modified.

Reaction of Lysyl Oxidase with Diamines and Cyclic Compounds In the course of these studies, we explored the pH-dependency of the oxidation of diamine substrates by lysyl oxidase. As shown in Figure 7, the pH optimum of Vmax occurs at lower values of pH while the Vmax decreases as the carbon chain length of alkyldiamines decreases from 6 to 3. In contrast, the pH optimum remains at 8.5 for n-hexylamine, n-butylamine and n-propylamine, while the Vmax for these monoamines decreases by only 5 to 10% as the carbon chain is shortened by one methylene unit (not shown). Ethylenediamine, the shortest diamine, is not catalytically oxidized but, in fact, enzyme which is pretreated with 10 mM ethylenediamine at 37° and then dialyzed exhaustively is inactive. Misiorowski and Werner (1978) had previously noted that 10 mM 1,3-diaminopropane caused partial, irreversible inactivation of chick bone lysyl oxidase and had observed that incubation of lysyl oxidase with other bifunctional molecules such as dithiothreitol also irreversibly inactivated the enzyme.

322

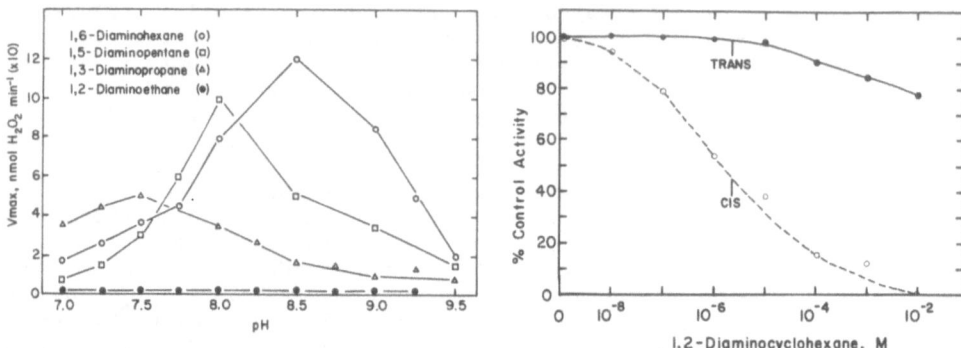

Figure 7. (Left) pH-Dependency of Diamine Oxidation by Lysyl Oxidase.
Figure 8. (Right) Inhibition of Elastin Oxidation by DACH Isomers.

We considered the possibility that the poor substrate capacity and inhibitory potential of shorter diamines reflects the bifunctional attack of both amino groups of such diamines on a component of the enzyme to form an inhibited complex and that such an interaction might exhibit steric specificity. Indeed, as shown (Figure 8), cis 1,2–diaminocyclohexane (cis-DACH) is a potent inhibitor of elastin oxidation with an IC_{50} of 1.3 uM while the trans isomer is at least 10^6 times less potent with an IC_{50} in excess of 0.1 M, clearly illustrating a sterically specific effect. Moreover, the inhibition of enzyme pre-incubated at 37° with cis-DACH was not reversible by dialysis. Steady-state kinetic analysis of the inhibition revealed that cis-DACH is competitive with the n-hexylamine substrate and has a K_I of 0.5 uM as an inhibitor of lysyl oxidase (not shown). We were unable to detect the release of H_2O_2 upon the incubation of catalytic quantities of lysyl oxidase with cis-DACH under assay conditions, indicating that productive oxidation with turnover of this compound did not occur.

In considering potential bifunctional targets of cis-DACH in the enzyme, the interaction of cis- and trans-DACH with authentic PQQ was explored. As shown, 1.2 mM cis-DACH induces marked changes in the spectrum of PQQ while little change was induced in the PQQ spectrum in the presence of 1.2 mM trans-DACH over the same time period (3 minutes) and under the same conditions (Figure 9) using either 50 mM sodium borate or 16 mM potassium phosphate, pH 8.0, as buffer. These spectral effects raise the possibility that PQQ is the site of attack by cis-DACH in the enzyme, possibly involving Schiff base formation between each of the amino functions of cis-DACH with each of the carbonyls in PQQ. Alternatively, cis-DACH might chelate copper at the active site. However, preincubation of lysyl oxidase with as much as 10 mM cis-DACH followed by dialysis against phosphate-buffered saline gave the same enzyme-bound copper content as did enzyme which was not exposed to cis-DACH but which otherwise was treated identically. We have also noted that ethylenediamine reacts with free PQQ to generate the same spectral changes in the spectrum of PQQ as induced by cis-DACH. Moreover, ethylenediamine is a reasonably potent, competitive inhibitor of lysyl oxidase with a K_I of 48 uM. It thus seems possible that such vicinal diamines inhibit lysyl oxidase by reacting with PQQ to form a stable, catalytically inert complex.

Studies in progress reveal that the spectral changes accompanying the reaction of cis-DACH or ethylenediamine with free PQQ are markedly suppressed in the absence of oxygen. Moreover, NMR spectroscopy of an aerobic solution of ethylenediamine and PQQ at a 0.8:1 molar ratio reveals the loss of two of the four aliphatic protons of ethylenediamine with the remaining two protons giving NMR signals at 8.69 and 8.73 ppm, indicative of aromatic protons. This is consistent with the oxidation of each carbon of ethylene-diamine in the PQQ-ethylenediamine complex. Thus, the oxygen dependency of the changes induced in the absorption spectrum of PQQ could indicate that one carbon of the bifunctional Schiff base may be oxidized, following which the reduced PQQ may be reoxidized by oxygen. The second carbon of ethylene-diamine may then also become oxidized by transfer of a pair of electrons to the PQQ moiety, potentially generating an unsaturated double bond between the two carbons of the PQQ-linked diamine.

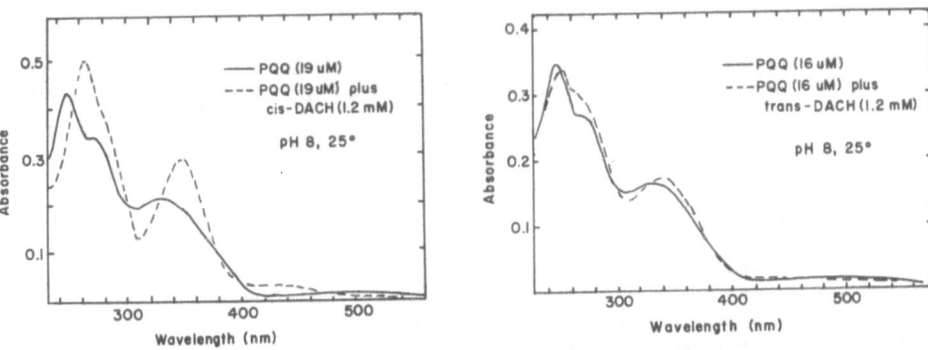

Figure 9. Interaction of cis-DACH (left) and trans-DACH (right) with PQQ.

Spectral Evidence for Enzyme-Substrate Complexes As noted, the mechanistic scheme for lysyl oxidase catalysis predicts the formation of a Schiff base between the substrate amino group and a reactive carbonyl of PQQ. As shown (Figure 10), introduction of n-butylamine under anaerobic conditions to lysyl oxidase in 50 mM sodium borate changes the enzyme absorption spectrum, yielding a difference spectrum (inset) with a maximum at 310 nm with lesser peaks noted at higher wavelengths. The corresponding difference spectrum of the effect of n-butylamine on the spectrum of PQQ also displays a peak at 310 nm (Figure 10). Spectral maxima in the 300-310 nm region of the PQQ spectrum have been correlated with the quinol and/or the amino-phenol, the 2-electron reduced forms of PQQ (Itoh et al., 1986). Little if any spectral changes were noted when n-butylamine was added under aerobic conditions to either the enzyme or PQQ. Moreover, introduction of oxygen caused the spectral perturbations seen anaerobically to largely disappear. These results thus support the conclusion that the amine substrate interacts with PQQ at the active site of lysyl oxidase in a manner that leads leads to the 2-electron reduction of PQQ, consistent with the results of model studies (Dekker et al., 1982; Itoh et al., 1986). Since product inhibition studies indicate that ammonia departs the enzyme after the binding of oxygen and in view of the available evidence that the aldehyde product is released in the absence of oxygen (Williamson and Kagan, 1986), it seems likely that the anaerobic reaction of the amine with the enzyme generates the reduced, aminophenol form of PQQ at the active site, consistent with the absorption peak at 300-310 seen in the anaerobic difference spectrum of the enzyme-butylamine complex in Figure 10.

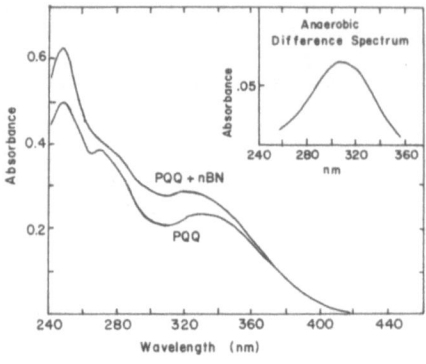 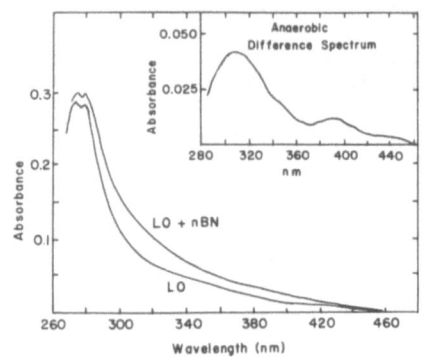

Figure 10. Changes in the Spectra of PQQ (left) and Lysyl Oxidase (right) Induced by n-Butylamine (0.01 M) under Anaerobic Conditions.

DISCUSSION

The analysis of the pH-dependent kinetics of lysyl oxidase identified a pK_a of 7.0 in the log Vmax/Km plot which apparently derives from a functional enzyme residue which must be unprotonated for effective catalysis. This pK_a is unlikely to simply reflect a pH-dependent conformational change since such effects often involve the ionization of more than one residue and are characterized by enthalpies of ionization significantly greater than that of 6.1 kcal mole^{-1} found for this ionization in lysyl oxidase (Cleland, 1977). The properties of this ionizing residue are consistent with those expected for peptidyl histidine (Cleland, 1977).

The inactivation of lysyl oxidase by DEPC correlated with the modification of one mole of active site histidine per mole of enzyme with a pK_a essentially the same as that of the enzyme base detected by steady-state kinetics. This argues against the possibility that other types of functional amino acid residues with different pK_a values are modified. It appears likely that the same functional histidine residue has been identified by both the steady-state kinetic and chemical modification results.

Histidine is clearly a reasonable candidate to serve as a general base catalyzing α-proton abstraction thus generating a substrate-derived carbanion intermediate in lysyl oxidase catalysis (Williamson and Kagan, 1977a). However, other possible catalytic roles for such an enzyme base should also be considered. For example, the formation of a Schiff base between the amine substrate and the PQQ cofactor would be greatly favored with an unprotonated, nucleophilic amine. Since the second pK_a in the log Vmax/Km plot indicates that the amine must be protonated to productively interact with the enzyme, it is possible that a general base residue of the enzyme may then deprotonate the incoming amine substrate thus rendering it nucleophilic. More precise identification of the role of the residue identified in the present study will await further detailed explorations of the structure-function relationships of lysyl oxidase.

Recent analyses of bovine plasma amine oxidase support the participation of a general base, yet to be chemically identified, catalyzing α-proton abstraction and tritium exchange in the sodium

borotritide-reduced enzyme-substrate complex (Farnum et al., 1986). In view
of the likely similarities in the mechanisms of action of lysyl oxidase,
plasma amine oxidase and diamine oxidase, each of which have been
identified as copper- and PQQ-dependent enzymes, it will be of considerable
interest to learn whether such similarities extend to the identities of
functional amino acid residues at the active site(s) of each enzyme.

The present results support the direct interaction of amine
substrates with PQQ in lysyl oxidase consistent with results with plasma
amine oxidase (Hartmann and Klinman, 1987). Studies in progress are
directed at characterizing the amine-PQQ linkage, assessing its
reducibility by sodium cyanoborohydride as would be expected of a Schiff
base. It also appears that cyclic, vicinal diamines can be potent
inhibitors of this enzyme likely by virtue of their interaction with PQQ.
Ongoing studies are probing the mechanism of these inhibitors, considering
the degree to which oxidation of the inhibitory cyclic and linear diamines
coupled with reduction of PQQ may lead to tightly and/or irreversibly bound
amine-PQQ complexes.

ACKNOWLEDGEMENTS

This work was supported by grants AM 18880 and HL 13262 to HMK from the
National Institutes of Health

REFERENCES

Cleland WW, 1979. Statistical analysis of enzyme kinetic data. Meth.
Enzymol. 63: 103-138.

Cleland, WW, 1977. Determining the chemical mechanisms of enzyme-catalyzed
reactions by kinetic studies. Advan. Enzymol. 45: 273-387.

Dekker RH, Duine JA, Frank J, Verweil J and Westerling J, 1982. Covalent
addition of H_2O, enzyme substrates and activators to pyrrolo-quinoline
quinone, the coenzyme of quinoproteins. Eur. J. Biochem. 125: 69-73.

Farnum M, Palcic M and Klinman JP, 1986. pH dependence of deuterium isotope
effects and tritium exchange in the bovine plasma amine oxidase reaction: a
role for single-base catalysis in amine oxidation and imine exchange.
Biochem. 25: 1898-1904.

Hartmann C and Klinman JP, 1987. Reductive trapping of substrate to bovine
plasma amine oxidase. J. Biol. Chem. 262: 962-965.

Itoh S, Kitamura Y, Ohshiro Y and Agawa T, 1986. Kinetics and mechanism of
the oxidative deamination of amines by coenzyme PQQ. Bull. Chem. Soc. Jpn.
58: 1907-1910.

Kagan HM, 1986. Characterization and regulation of lysyl oxidase. In:
Mecham RP. Biology of extracellular matrix. Academic Press, Orlando, FL,
pp. 322-398.

Kagan HM and Sullivan KA, 1982. Lysyl oxidase: preparation and role in
elastin biosynthesis. Methods Enzymol. 82A: 637-649.

Levene CI, 1961. Structural requirements for lathyrogenic agents. J. Exp. Med. 114: 295–310.

Levy HM, Leber PD and Ryan EM, 1963. Inactivation of myosin by 2,4-dinitrophenol and protection by adenosine triphosphate and other phosphate compounds. J. Biol. Chem. 238: 3654–3659.

Miles EW, 1977. Modification of histidyl residues in proteins by diethylpyrocarbonate. Meth. Enzymol. 47: 431–442.

Misiorowski RL and Werner MJ, 1978. Inhibition of lysyl oxidase by disulfhydryls, diamines and sulfhydryl-amines. Biochem. Biophys. Res. Commun. 85: 809–814.

Petterson G, 1985. Plasma amine oxidase. In: Mondovi B. Structure and functions of amine oxidases. CRC Press, Inc., Boca Raton, FL, pp 105–120.

Royce PM, Camakaris J and Danks, DM, 1980. Reduced lysyl oxidase activity in skin fibroblasts from patients with Menke's syndrome. Biochem. J. 192: 579–586.

Siegel RC, Pinnell, SR and Martin GR, 1970. Crosslinking of collagen and elastin. Properties of lysyl oxidase. Biochemistry 9: 4486–4492.

Trackman PC, Zoski CG and Kagan HM, 1981. Development of a peroxidase coupled fluorometric assay for lysyl oxidase. Anal. Biochem. 113: 336–342.

van der Meer RA and Duine JA, 1986. Covalently bound pyrroloquinoline quinone is the organic prosthetic group in human placental lysyl oxidase. Biochem. J. 239: 789–791.

Williams MA and Kagan HM, 1985. Assessment of lysyl oxidase variants by urea gel electrophoresis: evidence against disulfide isomers as bases of the heterogeneity. Anal. Biochem. 113: 336–342.

Williamson PR and Kagan HM, 1986. Reaction pathway of bovine aortic lysyl oxidase. J. Biol. Chem. 261: 9477–9482.

Williamson PR and Kagan HM, 1987a. Alpha proton abstraction and carbanion formation in the mechanism of action of lysyl oxidase. J. Biol. Chem. 262: 8196–8201.

Williamson PR and Kagan HM, 1987b. Electronegativity of aromatic amines as a basis for the development of ground state inhibitors of lysyl oxidase. J. Biol. Chem. 262: 14520–14524.

Williamson PR, Kittler JM, Thanassi JW and Kagan HM, 1986. Reactivity of a functional carbonyl moiety in bovine lysyl oxidase. Evidence against pyridoxal 5'-phosphate. Biochem. J. 235: 597–605.

Williamson PR, Moog RS, Dooley DM and Kagan, HM, 1986. Evidence for pyrroloquinoline quinone as the carbonyl cofactor in lysyl oxidase by absorption and resonance Raman spectroscopy. J. Biol. Chem. 261: 9477–9482.

LYSYL OXIDASE FROM THE YEAST PICHIA PASTORIS

S.S. Tur, P.M. Royce[*], and K. Lerch, Biochemisches Institut der
Universität Zürich, Winterthurerstrasse 190, CH-8057 Zürich, Switzerland.
[*] Stoffwechselabteilung, Universitäts-Kinderklinik, CH-8032 Zürich
Switzerland.

KEY WORDS   Amine oxidase, Lysyl oxidase, Pichia pastoris, Copper, PQQ

ABSTRACT

Benzylamine oxidase from Pichia pastoris is a copper containing quinoprotein with a molecular weight of 106 kdal. It catalyzes the oxidative
deamination of primary amines and diamines. Its most unusual feature is the
oxidation of ε-aminogroups in peptidyllysine residues. This reaction is
normally catalyzed by lysyl oxidase. The enzyme from P. pastoris shows an
astonishing similarity in substrate specificity to ly`syl oxidase with
respect to the amino acids juxtaposed to lysine. The fungal enzyme exhibits
k_{cat} values two to three orders of magnitude higher than those of
lysyl oxidase.

INTRODUCTION

Many yeasts are known to grow on amines as the sole nitriogen source,
thus requiring the presence of an amine oxidase (van Dijken & Bos, 1981).
In the yeast P. pastoris, two amine oxidases (metylamine and benzylamine
oxidase) have been isolated and partially characterized (Green et al.,
1983). Both enzymes are synthesized de novo when grown on butylamine as
sole nitrogen source. In this paper, the biosynthesis and some molecular
properties of benzylamine oxidase will be described. In addition, the
substrate specificity of this amine oxidase is compared to the one of
bovine aortic lysyl oxidase.

MATERIALS & METHODS

Purification. All steps were performed at 4°C. Cells of P. pastoris
were grown and harvested as previously described (Haywood & Large, 1981).
The cells (approx. 20 g wet weight) were resuspended in 1 vol of 0.1 M
potassium phosphate, pH 6.0 and disrupted by shaking vigourously with 1 vol
of glass beads (1 mm diam.) for 5 min. The supernatant was decanted and the
glass beads washed twice with the same buffer. All fractions were combined
and centrifuged at 20'000 g for 30 min. The supernatant was passed through
glass wool to remove lipids. The highly opalescent extract was applied to a
column (65 X 4.5 cm) of Sephadex G-25 equilibrated with 10 mM potassium
phosphate, pH 7.0. The most active fractions were pooled and applied to a
DEAE-Sepharose Cl-6B column (20 X 2.5 cm) previously equilibrated with 10
mM potassium phosphate, pH 7.0. After loading, the column was washed

J. A. Jongejan and J. A. Duine (eds.), PQQ and Quinoproteins, 327–333.
© 1989 by Kluwer Academic Publishers.

extensively with the same buffer before eluting the enzyme with a linear gradient of 0.01 - 1 M potassium phosphate, pH 7.0 (500 ml). The major peak of activity was eluted at 0.39 M potassium phosphate, pH 7.0. The active fractions were pooled and dialyzed against 10 mM potassium phosphate , pH 7.0. The enzyme was concentrated by rechromatography on a small column of DEAE-Sepharose Cl-6B (3 X 0.9 cm) equilibrated with 10 mM potassium phosphate, pH 7.0. The purity of the enzyme was assessed by SDS-PAGE according to Laemmli (1970) on a 7.5% polyacrylamide gel. The protein concentrations were determined by standard methods (Lowry et al., 1951). Copper and PQQ concentrations were measured as described earlier (Tur & Lerch, 1988).

Enzyme assays. The peroxidase coupled spectrophotometric assay as described by Haywood & Large (1981) was used for all substrates except collagen. Oxidation of collagen was measured in a tritium release assay with labeled substrate from chick calvaria as described earlier (Royce et al., 1980). Enzyme samples were incubated with 10^6 cpm labeled collagen for 10 hours at 37°C. Lysine containing model peptides were synthesized by solid phase peptide synthesis according to Merifield (1963).

RESULTS AND DISCUSSION

Benzylamine oxidase from P. pastoris is induced when the organism is grown in the presence of butylamine (Haywood & Large, 1981). In Fig. 1 is shown the kinetics of biosynthesis. A lag phase of approx. 1 h is followed by a rapid induction of the enzyme reaching 80% of maximal specific activity within 3 h. The specific activity leveled off after 24 h decreasing slowly thereafter (data not shown).

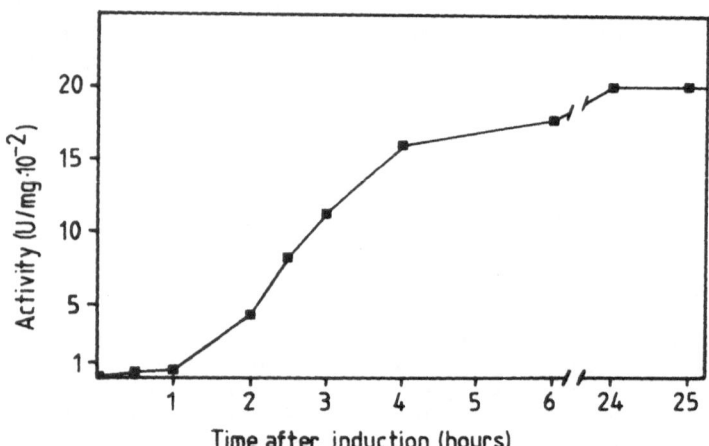

FIGURE 1. Induction of P. pastoris benzylamine oxidase with butylamine as sole nitrogen source.

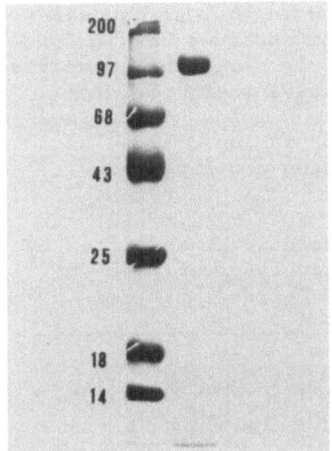

FIGURE 2. SDS polyacrylamide gel of purified benzylamine oxidase from P. pastoris. M_r of markers are indicated in kdal.

For the purification of benzylamine oxidase, cells were harvested 24 h after induction. Later harvesting led to significant losses of enzyme activity during purification, most likely due to presence of proteases. Starting with 20 g of packed P. pastoris cells (wet weight), approx. 2.5 mg benzylamine oxidase were obtained within 3 days. Fig. 2 shows an SDS polyacrylamide gel of purified benzylamine oxidase. The enzyme is a dimer (Green et al., 1983) consisting of two subunits with a molecular weight of 106 kdal each containing one copper and one PQQ (Tur & Lerch, 1988). This ratio is different from those found for other amine oxidases (Mondovi et al. 1987).

Benzylamine oxidase from P. pastoris oxidizes a large number of different substrates (Table 1.)

TABLE 1. Kinetic constants for the oxidation of various substrates by P. pastoris benzylamine oxidase a)

Substrate	k_{cat} (min^{-1})	K_M (mM)	k_{cat}/K_M (min^{-1}mM^{-1})
Butylamine	720	.235	3'100
Ornithine	870	.80	1'100
Lysine	1030	.091	10'600
Lysine methylester	210	.0047	44'600
N-α-acetyllysine	690	2.97	200
N-α-acetyllysine methylester	850	.013	65'400

a) Adapted from Tur & Lerch (1988).

In addition to simple aliphatic amines (e.g. butylamine), aromatic amines and diamines (Green et al., 1983) P. pastoris amine oxidase also uses ornithine, lysine and N- or C- protected lysine derivatives. A comparison of ornithine with lysine shows that the sidechain length influences the binding appreciably. Lysine is bound nearly 10-fold more strongly than

ornithine. The binding increases again about 20-fold in the C-protected
substrate lysine methylester. On the other hand, the observed binding of
the N-protected derivative N-α-acetyllysine is greatly diminished compared
to free lysine. Hence it is suggested that the negative charge of free
lysine hinders the formation of the enzyme-substrate complex. In Table 2
are shown the kinetic constants for P. pastoris benzylamine oxidase and
bovine lysyl oxidase using different lysine containing model peptides.

TABLE 2. Kinetic constants for the oxidation of synthetic hexapeptides by
P. pastoris benzylamine oxidase (a) and bovine aortic lysyl oxidase (b).

Substrate	k_{cat}	K_M	k_{cat}/K_M
	(min^{-1})	(mM)	$(min^{-1}mM^{-1})$
a)			
Ala-Lys-Ala-Tyr-Asp-Val	1054	.444	2'300
Ala-Lys-Glu-Tyr-Asp-Val	170	.523	300
Glu-Lys-Glu-Tyr-Asp-Val	–	–	–
His-Lys-His-Tyr-Asp-Val	740	.0024	308'000
Phe-Lys-Tyr-Tyr-Asp-Val	840	.35	2'400
b)			
Ala_2-Lys-Ala_2	.65	.082	7.9
Ala_2-Lys-Glu-Ala_2	.68	.76	.89
Ala_2-Lys-Tyr-Ala_2	1.2	.41	2.9

a) Adapted from Tur & Lerch (1988)
b) Adapted from Kagan et al. (1984)

The peptides are excellent substrates for the fungal enzyme thus establishing it as a lysyloxidase. Surprisingly, the k_{cat}/K_M values of P.
pastoris benzylamine oxidase are two to three orders of magnitude higher
than those of bovine lysyl oxidase. However, the two enzymes show a similar
pattern in substrate specificity with respect to the amino acids juxtaposed
to lysine. The highest k_{cat}/K_M values for both enzymes were found
for the peptide containing the sequence Ala-Lys-Ala. When a negative
charged amino acid such as glutamate is introduced at the C-terminal site
of the lysine, the k_{cat} drops to 13% and 11% of the initial value for
benzylamine oxidase and lysyl oxidase, respectively. With two negative
charges, the peptide becomes such a poor substrate, that accurate measurements were impossible. Introducing two positive charges (His-Lys-His)
increased the binding about 100-fold, whereas aromatic sidechains such as
Tyr and Phe juxtaposed to the susceptible lysine hardly affected the
kinetic parameters compared to Ala-Lys-Ala. The benzylamine oxidase like
lysyl oxidase prefers alanine-rich glutamate-free sequences, which are

commonly found to be the major crosslinking sites in elastin (Siegel, 1979). Aside these peptides, benzylamine oxidase was found to oxidize also elastin, collagen and histones (Tur & Lerch, 1988).

Like lysyl oxidase P. pastoris benzylamine oxidase is inhibited by β-aminopropionitrile. In Fig. 3 are shown Dixon-plots for the inhibition of benzylamine oxidase with three different substrates.

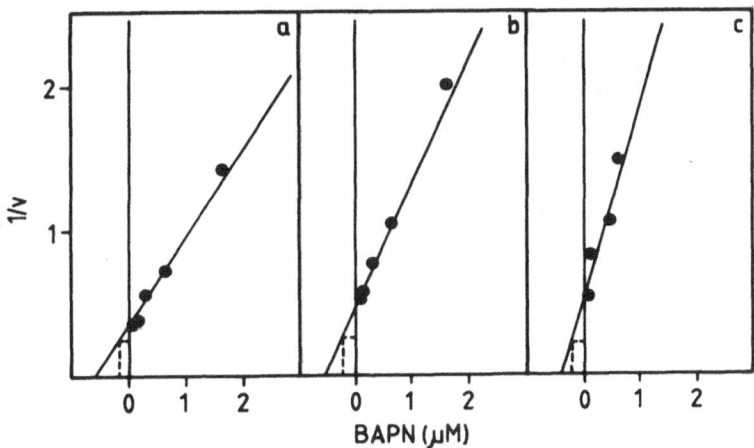

FIGURE 3. Dixon-plots for the determination of K_I values for benzylamine oxidase using β-aminopropionitrile as inhibitor and three different substrates. a) Lysine, b) N-α-acetyllysine methylester, c) Ala-Lys-Ala-Tyr-Asp-Val.

The K_I values are found to be 0.17 μM for lysine (a), 0.22 μM for N-α-acetyllysine methylester (b) and 0.216 μM for the hexapeptide Ala-Lys-Ala-Tyr-Asp-Val (c).

Fig. 4 shows the inhibition of benzylamine oxidase with β-amino-propionitrile using $(4,5-^3H)$-labeled collagen as substrate. The enzyme is inhibited to an extend of 50% at β-aminopropionitrile concentrations of approx. 100 μM, which is nearly 10-fold the higher than the one required for lysyloxidase (Siegel, 1979).

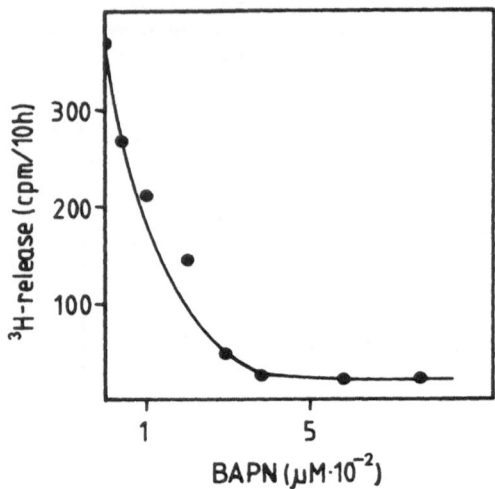

FIGURE 4. Inhibition of benzylamine oxidase by β-aminopropionitrile in a ^{3}H-release assay with labeled collagen.

In conclusion, the benzylamine oxidase exhibits an astonishing similarity to bovine aortic lysyl oxidase in many respects such as cofactors, substrate specificity and inhibition. These properties together with the high catalytic efficiency make the fungal enzyme highly attractive for structural and mechanistic studies.

ACKNOWLEDGEMENTS

This work was supported by the Swiss Science Foundation (Grants 3.236–0.89 and 3.861–0.86), The Geigy Jubiläumsstiftung, the Hartmann–Müller Stiftung and the Kanton of Zürich.

REFERENCES

Green J, Haywood GW, Large PJ, 1983. Serological differences between the multiple amine oxidases of yeast and comparison of the specificities of the purified enzymes from Candida utilis and Pichia pastoris. Biochem. J. 211: 481–493

Haywood GW, Large PJ, 1981. Microbial oxidation of amines. Bioch. J. 199: 187–201

Kagan HM, Sullivan KA, 1982. Lysyl oxidase: Preparation and role in elastin biosynthesis. Methods in Enzymol. 82: 637–650.

Kagan HM, Williams MA, Williamson PR and Anderson JM, 1984. Influence of sequence and charge on the specificity of lysyl oxidase toward protein and synthetic peptide substrates. J. Biol. Chem. 259: 11203-11207.

Laemmli UK, 1970. Cleavage of structural proteins during the assembly of the head of bacteriophage T4. Nature 227: 680-685.

Lowry OH, Rosebrough NJ, Farr AL and Randall RJ, 1951. Protein measurement with the folin phenol reagent. J. Biol. Chem. 193: 265-275.

Merrifield RB, 1963. Solid phase peptide synthesis. J. Am. Chem. Soc. 85: 2149-2154

Mondovi B, Morpurgo L, Agostinelli E, Befani O, McCracken J and Peisach J, 1987. A comparison of the local environment of Cu(II) in native and half-Cu-depleted bovine serum amine oxidase. Eur. J. Biochem. 168: 503-507

Royce PM, Camakaris J and Banks DM, 1980. Reduced lysyl oxidase activity in skin fibroblasts from patients with Menkes'-Syndrom. Biochem J. 192: 579-586

Siegel RC, 1979. Lysyl oxidase. Int Rev. Conn. Tiss. Res. 8: 73-118

Tur SS and Lerch K, 1988. Unprecedented lysyl oxidase activity of Pichia pastoris benzylamine oxidase. FEBS Letters, in press.

Van Dijken JP and Bos P, 1981. Utilization of amines by yeast. Arch. Microbiol. 128: 320-324

CHICK EMBRYO LYSYL OXIDASE: EVIDENCE FOR A PYRIDOXAL DERIVATIVE AS AN ESSENTIAL COFACTOR

C.I. Levene

Department of Pathology, University of Cambridge,
Tennis Court Road, Cambridge CB2 1QP, U.K.

Keywords: Lysyl oxidase, pyridoxal phosphate, lathyrism, collagen.

Lysyl oxidase catalyses the oxidative deamination of the ξ-amino group of specific lysine residues in elastin, and of lysine and hydroxylysine residues in collagen (Bailey et al., 1974; Gallop & Paz, 1975; Tanzer, 1976), to form the aldehyde precursors of the cross-links of these structural proteins, and is therefore directly responsible for the ultimate structural integrity of the connective tissues. Its existence was first demonstrated by Pinnell and Martin (1968). There is ample evidence to suggest that lysyl oxidase has a specific requirement for Cu^{2+} ions and molecular oxygen in vitro (Siegel et al. 1970) - animals raised on copper-deficient diets demonstrate cross-link defects in arterial elastin (O'Dell et al., 1961; Shields et al., 1962). It has also been demonstrated that in copper-deficient pigs, which exhibit these defects, there is a dramatic decrease in plasma amine oxidase activity (Blaschko et al., 1965). On the basis of these findings it was suggested (Page & Benditt, 1967) that lysyl oxidase may belong to that group of amine oxidases which are known to be copper-dependent and probably pyridoxal phosphate-dependent (Yasunobu & Yamada, 1963; Blaschko & Buffoni, 1965). Consistent with this hypothesis was the demonstration (Hill & Kim, 1967; Starcher, 1969) of a derangement of cross-link synthesis in aortic elastin in pyridoxine-deficient chicks. Indirect evidence for the presence of a pyridoxal cofactor has also come from the study of the inhibition characteristics of lysyl oxidase in vitro, which demonstrated the sensitivity of the enzyme to carbonyl reagents (Harris et al., 1974). Many of these carbonyl reagents have lathyrogenic activity in vivo, resulting in dramatically increased fragility of mesenchymal tissue (Levene, 1961b).

In view of the critical role of cross-linking in connective tissues, it is clearly important to establish the nature of the cofactors of this enzyme. The effects of copper deficiency on the connective tissues are well documented (Carnes, 1971) and, although vitamin B6 deficiency is rare in animals, there are situations where suboptimal dietary intake of this vitamin, particularly during human pregnancy, may well have an irreversible effect on the connective tissues (Levene & Murray, 1977), owing to inadequate cross-linking of both collagen and elastin.

More recent evidence resulting from studies in vivo and in vitro on lysyl oxidase from embryonic chick cartilage is here presented, suggesting that a form of vitamin B6 may play a catalytic role in the enzyme. For example, 24h after injection of 16 day chick embryos with G-^{3}H pyridoxine hydrochloride, some of this label appears in the epiphy-

335

J. A. Jongejan and J. A. Duine (eds.), PQQ and Quinoproteins, 335–344.
© 1989 by Kluwer Academic Publishers.

sial cartilage. Over 35% of this radioactivity appears in the form of
G-3H pyridoxal and a further 30% as other vitamin B6 compounds. Partial
purification of lysyl oxidase from the labelled epiphysial cartilage
reveals a single peak of radioactivity coinciding with a single peak
of enzyme activity (Fig. 1).

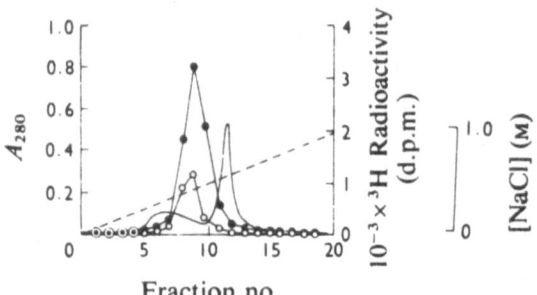

Fraction no.

Fig. 1. *Ion-exchange chromatography of lysyl oxidase from
17-day chick embryos injected with* [G-3H]*pyridoxine
hydrochloride 24h previously*
The 6M-urea eluate from collagen–Sepharose 4B was
chromatographed on DEAE-cellulose with a linear
gradient of NaCl (----). Elution profiles for lysyl
oxidase activity (●), for incorporation of radioactivity ( ) and for material absorbing at 280nm
(——) are shown. The fraction size was 10ml.

Figure 1

On dialysis against phosphate buffered saline, 75% of this radioactivity is non-diffusible. After incubation with isonicotinic acid
hydrazide (INAH), a carbonyl reagent that inhibits chick embryo lysyl
oxidase both in vivo and in vitro, a further 70% of the radioactivity
is lost, with a roughly corresponding loss of enzyme activity. This
suggests that a form of vitamin B6 is required as a cofactor of lysyl
oxidase, (Murray & Levene, 1977).

The reason for studying INAH instead of the classical lathyrogen β-
aminopropionitrile (BAPN) (Levene & Gross, 1959) was as follows: one's
main interest in lathyrism stems from its usefulness as a tool with
which to study mechanisms by which collagen, once laid down, may be
mobilised, but studies on the lathyrogenic nitriles, such as BAPN, have
consistently failed to elucidate such mechanisms during the past 30 years.
During an investigation into the nature of compounds possessing lathyrogenic activity, using a chick embryo assay, it was repeatedly noted
that INAH, a compound effective against tuberculosis, possessed marked
lathyrogenic activity. This study discussed its mode of action, comparing it with BAPN, as well as the reversal in vivo of the INAH effect
by pyridoxal (Levene, 1961a).

When applied to the chorio-allantoic membrane of the chick embryo, INAH was shown to produce an increase in the fragility of the embryo and in the amount of collagen which was extractable from the bones with cold IM NaCl. The administration of pyridoxal reversed these phenomena almost completely; its effect on a stoichiometric basis exceeded that of all other aldehydes tested. (Levene, 1961a) (Fig. 2).

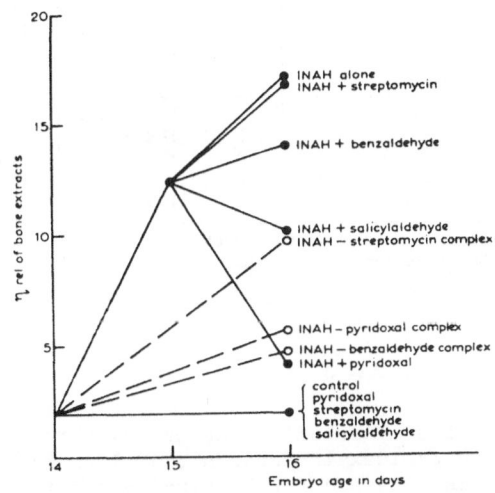

Comparison of (a) the effects of various aldehydes on INAH-treated embryos and (b) the effects of INAH with those of various aldehyde-INAH complexes.

Figure 2

The effect of INAH differed from that of BAPN in that the latter was of greater magnitude, and was not affected by pyridoxal; moreover, BAPN caused skeletal deformities, whilst INAH, even at 12 times the concentration, produced no deformities. The aldehyde group of pyridoxal was shown to be necessary for its interaction with INAH.

Using an assay developed in the chick embryo to detect lathyrogens by the increase which they produce in salt solubility of connective tissue collagen, a number of compounds tested were found to inhibit collagen cross-linking in vivo. The assay is based on the finding by Gross (1958) that a linear relationship exists between the relative viscosity of extractable collagen and the actual amount extracted. The lathyrogenic compounds detected by this method included (Table 1):

Variation in Relative Viscosity of Bone Extracts in Embryos Treated with Various Lathyrogens under Constant Conditions

Compound*	No. of extracts observed	η rel of 1 M sodium chloride extracts	
		Standard deviation	Coefficient of variation
			per cent
Aminoacetonitrile	3	45.2 ± 3.8	8.4
Methylene aminoacetonitrile	3	44.9 ± 2.8	6.3
β-Aminopropionitrile	3	35.5 ± 4.1	11.6
Thiosemicarbazide	3	34.9 ± 3.6	10.3
Semicarbazide	5	22.1 ± 1.4	6.4
Acetone semicarbazone	3	25.1 ± 3.0	12.0
Isoniazid	9	15.5 ± 1.2	7.7
Nicotinic acid hydrazide	3	17.6 ± 0.1	0.8
Benzhydrazide	3	17.3 ± 0.8	4.6
Cyanoacetic acid hydrazide	3	14.8 ± 2.0	13.5
Hydrazine hydrate	4	14.7 ± 2.7	18.4

Weighted mean of coefficient of variation = 9.00 per cent.
* 0.054 mM injected at 14 days and examined 2 days later.

Table   1

It will be noted that semicarbazide was observed to be a lathyrogen. However, when the pyridoxal hydrazones of INAH and the semi-
carbazones of pyridoxal were tested for lathyrogenic activity, they
were found to be greatly diminished when compared to the parent lathyrogens (Table 2).

Effect of Schiff-Base Formation on Activity of Various Lathyrogens

Group	Compound	Observed melting point	η rel of bone extracts from embryos treated with 0.054 mM of compound	Melting point of parent compound	η rel of bone extracts from embryos treated with .054 mM of parent compound
		°C.			
Semicarbazones	Pyridoxal semicarbazone	235	3.7	*Semicarbazide*	
	Salicylaldehyde semicarbazone	228	9.1	175°–185°C dec.	24.6
	Benzaldehyde semicarbazone	207	4.1		
Hydrazones of isonicotinic acid hydrazide	Pyridoxal hydrazone of INAH	256	5.6	*Isonicotinic acid hydrazide (INAH)* 171°C	
	Benzaldehyde hydrazone of INAH	193	4.6		15.1
	Salicylaldehyde hydrazone of INAH	248	died		
	Streptomycin hydrazone of INAH‡	230 dec.	9.7		
	Pyridoxal hydrazone of nicotinic acid hydrazide	125	5.0	*Nicotinic acid hydrazide* 158°C	17.8

Table   2

These lathyrogenic compounds fell into 4 groups - nitriles,
ureides, and 2 new groups - hydrazides and hydrazines, forming a
spectrum of diminishing potency in their ability to produce solubility
of collagen and skeletal deformities in the chick embryo (Table 3).

339

Classification of compounds found to be lathyrogenic in the chick embryo based on the salt-extractibility of collagen from their bones.

Lathyrogenic compound (0.054mmol/egg)	Relative viscosity of bone extracts
Normal control	2
I. *Organic nitriles*	
Methylene aminoacetonitrile	44
Aminoacetonitrile	41
β-Aminopropionitrile	36
II. *Ureides*	
Semicarbazide	25
Acetone semicarbazone	22
III. *Hydrazides*	
Nicotinic acid hydrazide	18
Benzhydrazide	18
Isonicotinic acid hydrazide	15
Cyanoacetic acid hydrazide	14
p-Nitrobenzhydrazide	14
Glycine hydrazide	11
γ-L-Glutamylhydrazide	11
IV. *Hydrazines*	
Hydrazine hydrate	13
Unsym. dimethylhydrazine	5
Sym. dimethylhydrazine	5

Table 3

Pyridoxal reversed to a great degree the effects on collagen solubility and skeletal deformity produced by the ureide and hydrazide lathyrogens, though not by the nitriles; this reversal was considered to be due to the formation of a Schiff's complex between the aldehyde and the terminal amine of the lathyrogen (Levene, 1961b). Further evidence supporting this view emerged from the following study: weanling rats which had been acurely intoxicated with the lathyrogenic compounds aminoacetonitrile and methylene aminoacetonitrile were shown by Karnovsky (1960) to suffer from a marked and constant decrease in the glycogen content of the cells of the epiphyseal plate as well as an impaired ability of these cells to synthesise glycogen from glucose-1 -phosphate. He also demonstrated that these agents produce a marked glycostatic effect on the liver of starved rats, this being abolished by adrenalectomy; he therefore concluded that the liver effect is probably due to non-specific stress; the cause of the glycogen changes in the cells of the epiphyseal plate, however, remained a matter of speculation. A histochemical and quantitative assessment was consequently made of the effect of various lathyrogenic compounds on the glycogen content of the livers of 16 day old chick embryos treated 2 days earlier. There was no obvious correlation between their effect on glycogen and on the solubility of bone collagen. However, INAH and pyridoxal individually, lowered the glycogen content of the liver, but injection of pyridoxal 24h after INAH restored the glycogen level to normal, suggesting a mutual antagonism, possibly by Schiff base formation (Levene, 1962).

340

The next problem was to demonstrate if possible, that normal purified collagen in solution actually possessed carbonyl groups which were
either absent or masked in lathyritic collagen; this was achieved by
reacting purified normal guinea pig skin collagen in solution with 2:4
dinitrophenyl-hydrazine, a carbonyl-blocking agent, followed by exhaus-
tive dialysis and eventual lyophilisation. The same was done to collagen
extracted and purified from BAPN-treated guinea-pigs. Normal collagen
was found to have a pale lemon yellow colour, whereas lathyritic collagen was white. In solution the normal collagen showed a shift of spectral absorption following 2:4 dinitrophenylhydrazine treatment (Fig. 3a),
whereas the lathyritic collagen apparently showed none (Fig. 3b), (Levene,
1962).

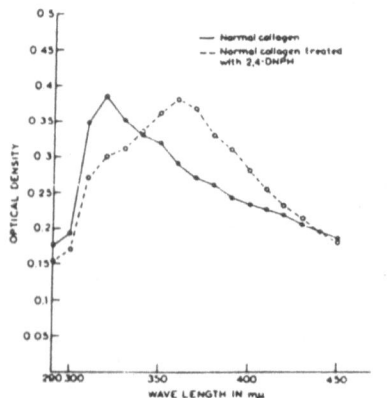

Spectral absorption curves of acetic acid solutions of normal, guinea pig
before and after treatment with 2,4-dinitrophenylhydrazine; examined with visible

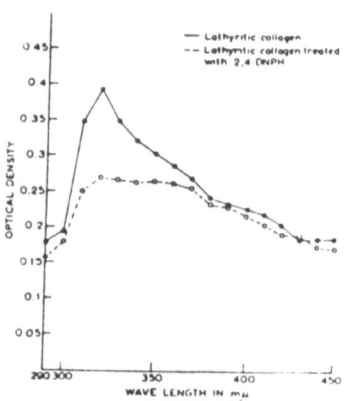

8. Spectral absorption curves of acetic acid solutions of collagen from guinea
pigs treated with BAPN; before and after treatment of collagen with 2,4-dinitrophenyl-
hydrazine; examined with visible light

Fig. 3a   Fig. 3b

Mention has already been made of the finding that purified chick
embryo lysyl oxidase co-chromatographed with tritium label derived
from ^{3}H-pyridoxine which had been injected in vivo into normal fertile
eggs from which the lysyl oxidase was later isolated and purified.

Arem and Misiorowski (1976) have shown that lysyl oxidase is
inhibited by INAH both in vivo and in vitro; many carbonyl agents inhibit the enzyme (Harris et al., 1974). We subsequently studied the
effect on one day old chicks, of vitamin B6 deficiency lasting 11 days;
the activity of lysyl oxidase was measured in the aortas and epiphyseal
cartilages of chicks raised on pyridoxine-deficient and control diets.
Lysyl oxidase activity was found to have diminished in both tissues in
the deficient chicks. Deficient chicks given 150 μg of pyridoxine 14h
before being killed showed a marked increase in lysyl oxidase levels in
both tissues. It was concluded that a pyridoxine derivative as well as
copper were essential cofactors for lysyl oxidase. (Murray et al., 1978).

Further evidence in favour of a pyridoxal derivative as cofactor
emerged from the work of Bird & Levene (1982). This was based on the

fact that both crude and partially purified preparations of embryonic chick aortic lysyl oxidase tend to gradually lose enzymic activity when illuminated, or when the urea extractant is removed by dialysis. Full activity was restored to such preparations by dialysis versus pyridoxal 5'-phosphate prior to assay (Table 4).

illuminated enzyme	79 ± 14 (33%)[+]
illuminated enzyme + 10^{-5}M PLP	208 ± 14 (88%)
illuminated enzyme + 10^{-5}M PLP + 10^{-5}M $CuCl_2$	194 ± 17 (82%)
dialyzed enzyme	72 ± 9 (31%)
dialyzed enzyme + 10^{-5}M PLP	228 ± 32 (97%)
dialyzed enzyme + 10^{-5}M PLP + 10^{-5}M $CuCl_2$	284 ± 10 (120%)

Table 4

Upon treatment with potassium cyanide or semicarbazide, purified embryonic chick aortic lysyl oxidase gives rise to fluorescent derivatives. The fluorescence spectrum of the semicarbazide adduct closely resembled that of pyridoxal phosphate semicarbazone. A preliminary ultraviolet/visible spectrum of chick aortic lysyl oxidase is also presented; this shows features which add to the existing evidence (Fig. 4).

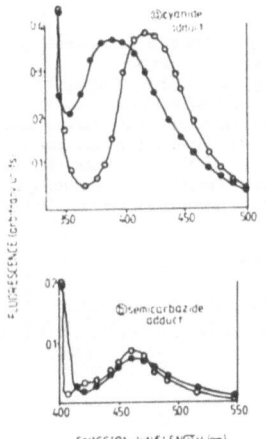

Fluorescence emission spectra of chromophores isolated from highly-purified chick aortic lysyl oxidase (●—●): (a), after treatment with potassium cyanide (13); (b), after treatment with semicarbazide (14). The fluorescence emission spectra of similar derivatives made from pyridoxal 5'-phosphate are shown for comparison (O—O). Excitation wavelengths were 325 nm (a) and 380nm (b).

Figure 4

342

A subsequent study of the effect of injection of 4-deoxypyridoxine, a pyridoxal analogue, into 13 day old chick embryos demonstrated an increase in the solubility of leg bone and cartilage collagen and a diminution in the salt and urea extractable lysyl oxidase, giving further support to the role of a pyridoxal derivative in the cross-linking of collagen (Bird & Levene, 1983).

Carrington et al (1984) then confirmed that INAH markedly decreased the liver content of pyridoxal phosphate; this took approx. 6h, whereas the inhibition of lysyl oxidase and the increase in collagen solubility occurred more slowly. A reversal of these effects of INAH was obtained by the injection 24h later of a stoichiometric amount of pyridoxal. Treatment of chick embryos with BAPN, an irreversible inhibitor of lysyl oxidase in vitro, caused inhibition of the enzyme, which began to recover within 24h but which was not affected by the administration of pyridoxal; with INAH inhibition, however, lysyl oxidase activity showed no sign of recovery by 48h (Fig. 5).

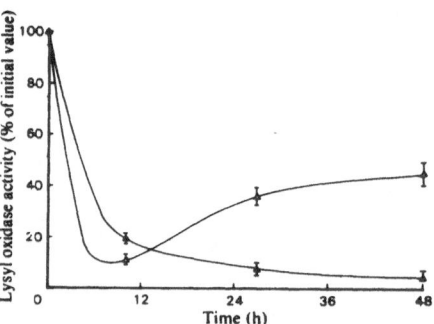

Inhibition of lysyl oxidase by isoniazid and BAPN in 14-day chick embryos
Isoniazid (7.4 mg/egg) and BAPN (1 mg/egg) were injected into eggs which had been incubated for 14 days. The leg bones were removed at intervals, and PBS extracts were assayed for lysyl oxidase activity. Results are means±half the difference of two experiments. Curves represent changes in lysyl oxidase activity after treatment with BAPN (△) or isoniazid (▲).

Figure 5

It was proposed that INAH may cause the inhibition of lysyl oxidase by competing for its obligatory cofactor, pyridoxal phosphate (Levene & Carrington, 1985).

Acknowledgements

The author gratefully acknowledges the Medical Research Council of Great Britain, the Arthritis and Rheumatism Council and the British Heart Foundation, without whose financial support these studies would not have been possible.

Figure 2: Reproduced from the Journal of Experimental Medicine, (1961) 113, 795-810 by Copyright permission of the Rockefeller University Press. Figures 3a & b: Reproduced from the Journal of Experimental Medicine, (1962) 116, 119-130 by Copyright permission of the Rockefeller University Press. Figure 4: Reproduced by permission of the Academic Press from Biochem. Biophys. Res. Commun. (1982) 108, 1172-1180. Figure 5: Reproduced by permission from the Biochemical Journal (1984) 221, 837-843. Copyright (c) The Biochemical Society, London. Table 1: Reproduced from "Drug Toxicity", a Symposium, ed. J.W. Gorrod (1978) pp 269-284 by Copyright permission of Taylor & Francis Ltd., London. Table 2: Reproduced from the Journal of Experimental Medicine, (1961) 114, 295-310 by Copyright permission of the Rockefeller University Press. Table 4: Reproduced by permission of the Academic Press from Biochem. Biophys. Res. Commun. (1982) 108, 1172-1180.

References

Arem, A.J. and Misiorowski, R. 1976. J. Med. (Westbury, N.Y.) 7: 239-248.

Bailey, A.J., Robins, S.P. and Balian, G. 1974. Nature (London) 251: 105-109.

Bird, T.A. and Levene, C.I. 1982. Biochem. Biophys. Res. Commun. 108: 1172-1180.

Bird, T.A. and Levene, C.I. 1983. Biochem. J. 210: 633-638.

Blaschko, H. and Buffoni, F. 1965. Proc. R. Soc. London, Series B 163: 45-60.

Blaschko, H., Buffoni, F., Weissman, N., Carnes, W.H. and Coulson, W.F. 1965. Biochem. J. 96: 40.

Carnes, W.H. 1971. Fed. Proc. Fed. Am. Soc. Exp. Biol. 30: 995-1000.

Carrington, M.J., Bird, T.A. and Levene, C.I. 1984. Biochem. J. 221: 837-843.

Gallop, P.M. and Paz, M. 1975. Physiol. Rev. 55: 418-487.

Gross, J. 1958. J. Exp. Med. 107: 265-277.

Harris, E.D., Gonnerman, W.A., Savage, J.E. and O'Dell, B.L. 1974. Biochim. Biophys. Acta 341: 332-344.

Hill, C.H. and Kim, C.S. 1967. Biochem. Biophys. Res. Commun. 27: 94-99.

Karnovsky, M.J. 1960. Lab. Invest. 9: 639-655.

Levene, C.I. 1961a. J. Exp. Med. 113: 795-810.

Levene, C.I. 1961b. J. Exp. Med. 114: 295-310.

Levene, C.I. 1962a. Brit. J. Exp. Path. 43: 596-599.

Levene, C.I. 1962b. J. Exp. Med. 116: 119-130.

Levene, C.I. and Carrington, M.J. 1985. Biochem. J. 232: 293-296.

Levene, C.I. and Gross, J. 1959. J. Exp. Med. 110: 771-790.

Levene, C.I. and Murray, J.C. 1977. Lancet. i: 628-630

344

Murray, J.C. and Levene, C.I. 1977. Biochem. J. <u>167</u>: 463-467.

Murray, J.C., Fraser, D.R. and Levene, C.I. 1978. Exp. & Mol. Pathol. <u>28</u>: 301-308.

O'Dell, B.L., Hardwick, B.C., Reynolds, G. and Savage, J.E. 1961: Proc. Soc. Exp. Biol. & Med. <u>108</u>: 402-405.

Page, R.C. and Benditt, E.D. 1967. Biochemistry. <u>6</u>: 1142-1148.

Pinnell, S.R. and Martin, G.R. 1968. Proc. Natl. Acad. Sci. U.S.A. <u>61</u>: 708-716.

Shields, G.S., Coulson, W.F., Kimball, P.A., Carnes, W.A., Cartwright, G.E. and Wintrobe, M.M. 1962. Amer. J. Pathol. <u>41</u>: 603-617.

Siegel, R.C., Pinnell, S.R. and Martin, G.R. 1970. Biochemistry <u>9</u>: 4486-4492.

Starcher, B.C. 1969. Proc. Soc. Exp. Biol. Med. <u>132</u>: 379-382.

Tanzer, M.L. 1976. in Biochemistry of Collagen (Ramachandran, G.N. and Reddi, H. eds), Plenum Press, New York, U.S.A. 137-162.

Yasunobu, K.T. and Yamada, H. 1963. in Symposium on the Chemical and Biological Aspects of Pyrodoxal Catalysis (Snell, E.E., ed), Pergamon Press, New York, U.S.A., 453-465.

BOVINE SERUM AMINE OXIDASE: BENZYLHYDRAZINE REACTION

L. Morpurgo, E. Agostinelli, J. Muccigrosso, B. Mondovi', and *L. Avigliano.

CNR Centre of Molecular Biology and Dept. of Biochemical Sciences, University "La Sapienza", 00185 Rome, Italy; *STBB Dept. University of L'Aquila, 67100 L'Aquila, Italy.

Key words: amine oxidase, copper, inhibitors, benzylhydrazine.

INTRODUCTION

Bovine serum amine oxidase (BSAO) catalyzes the oxidative deamination of primary amines by O_2 with formation of the aldehyde, NH_3 and H_2O_2 . The dimeric enzyme (m.w. 90,000/ subunit) contains two Cu(II) ions and one carbonyl group titratable by hydrazines (Pettersson, 1975), which was shown to belong to a pyrrolo quinoline quinone (PQQ) cofactor (Ameyama et al., 1984; Lobestein-Verbeek et al., 1984). The inactive hydrazine adducts show an intense absorption band with position and intensity dependent on the particular hydrazine and on the existence of hydrazone-azo tautomeric equilibria (Morpurgo et al., 1988). In the phenylhydrazine derivatives the azo-tautomer is stabilized by conjugation of PQQ and phenyl π-systems through the bridging −N=N− group. These derivatives, in which the PQQ moyety is in a "reduced" hydroquinonic conformation, miss the 480 nm absorption band of the native enzyme, assigned as the $n \rightarrow \pi^*$ transition of polycyclic quinones. The BSAO-hydrazine adducts are generally stable, except a few ones, which decompose with time, while the enzyme recovers its native properties. Benzylhydrazine (BHy) and hydrazine (Hy) belong to the latter group (Hucko-Haas & Reed, 1970 ; Morpurgo et al., submitted).

RESULTS

BHy reacted with BSAO producing, both in air and in anaerobiosis, an intense absorption band at 405 nm, while the protein absorption band at 480 nm was decreased (Fig. 1,a). In anaerobiosis the 405 nm band rapidly decayed and a species with maximum absorbance at 355 nm was formed (Fig. 1,c). On opening the cuvette to air, a third species was obtained with maximum absorbance at 335 nm and a shoulder at 410 nm (Fig. 1,b). The latter species could also be obtained by reacting BSAO with Hy, as confirmed by CD and EPR spectra. No change of the EPR integrated intensity and no recovery of the 480 nm band intensity was seen at all stages of the reaction. All native BSAO properties were recovered only after a very slow (8 hours) decay of the Hy-adduct.

345

J. A. Jongejan and J. A. Duine (eds.), PQQ and Quinoproteins, 345–347.
© 1989 by Kluwer Academic Publishers.

In the presence of Cu-binding inhibitors only the formation of the 355 nm absorbing species and the subsequent decay to the 335-410 nm absorbing one were seen in air. Both reactions were relatively fast with N_3^-, a reversible inhibitor of the enzymatic activity, and much slower with N,N-diethyldithiocar-bamate (DDC), a bidentate irreversible inhibitor. At less than saturating (6mM) DDC concentrations the 405 nm band was in part formed, its intensity being proportional to the residual activity of the sample. The BSAO-DDC initial spectral properties were recovered on standing.

The 355 nm absorbing species became absolutely stable in BSAO-DDC samples from which one Cu ion was removed (Morpurgo et al., 1987).

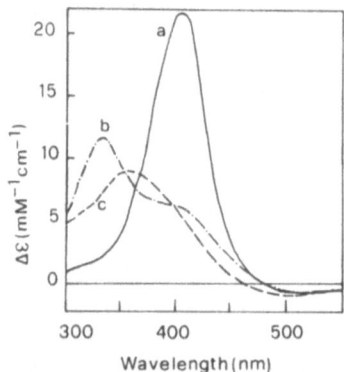

FIGURE 1. Differential optical spectra of BHY-reacted BSAO. 20 μM BSAO was reacted with 100 μM BHY: a) immediately after BHy addition in air or anaerobiosis; b) after 15 min in air; c) after 15 min in anaerobiosis.

DISCUSSION

The assignment of the BSAO adducts responsible for the three spectra of Fig. 1 is shown in Scheme 1. The mechanism proposed for BHy reaction is very similar to the ping pong mechanism outlined by Pettersson (1985) for the enzymatic reaction, which involves the transfer of $-NH_2$ from the substrate to the reduced enzyme via the formation of the Schiff-base of the cofactor, isomerization to the Schiff-base of the aldehyde, and hydrolytic release of the aldehyde in the first part of the process. In the second part, the enzyme is reoxidized by O_2 with release of H_2O_2 and NH_3, hydrolyzed from the imino adduct. The outstanding difference of the present reaction is the inability of the benzaldehyde hydrazone and the BSAO-Hy adduct to hydrolyze. The latter one is actually formed from BSAO and Hy. As a consequence the release of product follows O_2 reaction and cycling is almost prevented by the very slow oxidation of the Hy-adduct.

Both Cu-binding anions produced a similar change of the

reaction mechanism, i.e. the stabilization of the benzaldehyde hydrazone adduct. DDC produced a substantial decrease of the reaction rate, but was unable to stop the reaction as did the Cu removal. This suggests that Cu is not directly involved in O_2 binding, but only in the stabilization of a conformer more reactive toward O_2 with possible formation of a hydroperoxy derivative as in the autoxidation of phenylhydrazones (Bellamy & Guthrie, 1965).

BHy is a mechanism based inhibitor of BSAO as it is of dopamine - β - hydroxylase, a 2 Cu : 1 PQQ enzyme, which catalyzes a different reaction with a mechanism involving Cu reduction, unlike BSAO (Fitzpatrick & Villafranca, 1986).

SCHEME 1

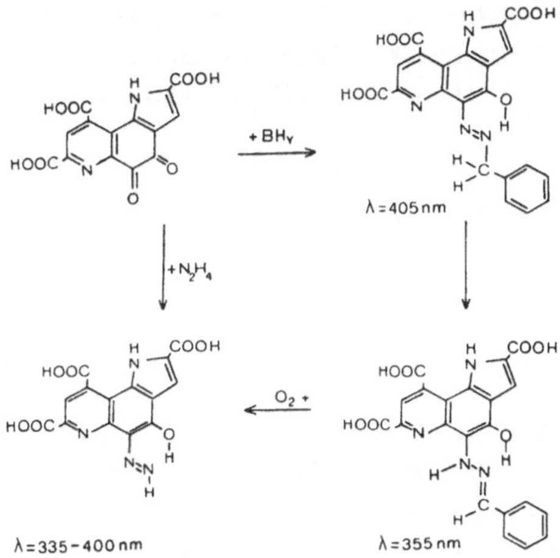

The partial support of Ministero P.I. is acknowledged.

REFERENCES

Ameyama M., Hayashi M., Matsushita K., Shinagawa E. & Adachi O. (1984) Agric. Biol. Chem. 48,561-565.

Bellamy A.J. & Guthrie R.D. (1965) J. Chem. Soc. 2788-2795.

Fitzpatrick P.F. & Villafranca J.J. (1986) J. Biol. Chem. 261, 4510- 4518.

Hucko-Haas J.E. & Reed D.J. (1970) Biochem. Biophys. Res. Commun. 38, 396-400.

Lobenstein-Verbeek C.L., Jongejan J.A., Frank J. & Duine J.A. (1984) FEBS Lett. 170,305-309.

Morpurgo L., Agostinelli E., Befani O. & Mondovi' B. (1987) Biochem. J. 248, 865-870.

Morpurgo L., Befani O., Sabatini S., Mondovi' B., Artico M., Corelli F., Massa S., Stefancich G. & Avigliano L. (1988) Biochem. J., in press.

Morpurgo L., Agostinelli E., Muccigrosso J., Martini F., Mondovi' B. & Avigliano L., submitted.

Pettersson G. (1985) in "Structure and Functions of Amine Oxidases" (Mondovi' B., ed.) pp. 105-120, CRC Press, Boca Raton, Florida.

PRIMARY STRUCTURE OF A PQQ
CONTAINING PEPTIDE FROM PORCINE KIDNEY DIAMINE OXIDASE.

Robert A. van der Meer[1], Pieter D. van Wassenaar[2], Johannes H. van Brouwershaven[2] and Johannis A. Duine[1]

[1]Department of Microbiology & Enzymology, Delft University of Technology, Julianalaan 67, 2628 BC Delft, The Netherlands. [2]Unilever Research Laboratory, Olivier van Noortlaan 120, 3133 AT Vlaardingen, The Netherlands.

Diamine oxidase (porcine kidney), Amine oxidases (copper-containing), PQQ-hydrazone, Primary structure, Tryptic peptide

ABSTRACT

After treating porcine kidney diamine oxidase (PKDAO, EC 1.4.3.6) with the inhibitor 2,4-dinitrophenylhydrazine (DNPH), the enzyme was subjected to proteolysis with trypsine. A peptide was found in the hydrolysate which contained the C(5) hydrazone of PQQ and DNPH. The peptide was purified to homogeneity after which the amino acid sequence was determined. It appeared to consist of 11 amino acids with PQQ bound to, most probably, a lysyl residue. Possibilities for insertion in or attachment of PQQ to the protein chain are discussed.

INTRODUCTION

Copper-containing amine oxidases (EC 1.4.3.6) have long been recognized to play key roles in cellular processes. Until recently, however, the study of these enzymes has been seriously hampered by lack of knowledge on the nature of the organic cofactor. Application of the so-called "hydrazine method" enabled us to detect and quantify PQQ as covalently-bound cofactor in these enzymes.

How are the cofactors distributed among the subunits and do they interact with each other? In order to obtain insight into these aspects, it should be known how PQQ is bound to the protein chain(s). Since PQQ in porcine kidney diamine oxidase (PKDAO) reacts with 2,4-dinitrophenylhydrazine (DNPH) to the stable C(5)-hydrazone [van der Meer et al., 1986], it was reasoned that hydrolysis of the derivatized enzyme with a specific protease might yield a PQQ-DNPH peptide with adequate stability and detectability for further analysis. As described here, tryptic hydrolysis yielded such a peptide.

J. A. Jongejan and J. A. Duine (eds.), PQQ and Quinoproteins, 348–350.

MATERIALS AND METHODS

Enzyme isolation and derivatization Isolation of homogeneous PKDAO and derivatization of its PQQ with DNPH to the hydrazone were performed as described previously [van der Meer et al., 1986].

Proteolysis of derivatized enzyme Proteolysis of derivatized enzyme (184 mg) with trypsin (2 mg; Boehringer Mannheim) was carried out in 15 ml 0.02 M Tris, pH 8.0, by incubation for 1 h at 37 °C. The reaction was stopped by adding sodium dodecyl sulphate (SDS) to a final concentration of 0.1 %.

Peptide purification The trypsine digested derivatized PKDAO was applied to a column of Tris-Acryl GF05 (1.6 cm x 28 cm) equilibrated with 0.1 M Tris, pH 8.0, containing 0.1 % SDS. The yellow coloured PQQ-DNPH peptide (λ_{max} = 445 nm) eluted with a K_{av} of 2. Further purification was achieved by applying the pooled yellow fractions (brought to pH 2.0 with 2 M HCl) to a Sep-Pak C_{18} cartridge, and eluting the yellow compound as described for the PQQ-DNPH hydrazone itself [van der Meer et al., 1987].

Sequencing of the peptide Sequence analysis of the peptide was performed with a gas-phase sequenator (Applied Biosystems, model 470A) using 25 % TFA for the conversion step. The resulting phenylthiohydantoin (PTH) amino acids were analysed on-line by reversed-phase HPLC on a PTH analyser (Applied Biosystems, model 120A) with a PTH C_{18} column of 2.1 mm x 220 mm (Applied Biosystems).

Amino acid composition Amino acid analysis of the peptide was carried out using a Waters PICO.TAG amino acid analysis system.

RESULTS

Derivatization of homogeneous PKDAO (1.1 µmol enzyme, having a specific activity of 1.1 U/mg protein) with DNPH under an oxygen atmosphere resulted in the formation of the hydrazone in a 91 % yield.

The procedure used to prepare the peptide gave a product that showed a certain spectral and chromatographic similarity to the C(5) hydrazone of PQQ and DNPH. The yield was rather low (0.1 %), as compared with the nearly quantitative yield of PQQ-DNPH itself. Polyacrylamide gel electrophoresis in the presence of SDS showed that the preparation is homogeneous and that the peptide has a M_r of approximately 1400 Da.

Amino acid sequencing of the isolated tryptic peptide gave the following sequence:

```
1              5        8        10
His-Ser-Asp-Ala-Val-Phe-Thr-X-Asn-Tyr
```

Reversed phase HPLC of the fraction containing amino acid number 8 revealed the presence of a compound more or less similar to the C(5) hydrazone of PQQ and DNPH (slightly different retention time and absorption spectrum). Hydrolysis in 2 N HCl gave the genuine C(5) hydrazone of PQQ and DNPH (albeit in a low yield, due to the instability of the hydrazone under this condition) and an amino acid determined to be norleucine (with the PICO.TAG system). Total hydrolysis of the peptide revealed the following composition: amino acid (residues/mol): Asp (2.2), Thr (1.5), Ser (0.8), Gly (0.1), Ala (1.0), Val (0.9), Nor-Leu (1.0), Tyr (0.9), Phe (0.8), His (0.8), Arg (1.0).

DISCUSSION

The procedure used to isolate a PQQ-containing peptide from PKDAO resulted in a homogeneous preparation in a yield of only 0.1 %. This low yield can be attributed to an as yet uncomprehensible instability of the PQQ-DNPH hydrazone in the peptide effected by the presence of trypsine, an instability which has also been observed for the hydrazone itself [unpublished results].

The results of the amino acid sequencing, combined with the amino acid composition and the fact that it is a tryptic peptide, make it reasonable to propose that amino acid number 11 is an arginine residue. Although the amino acid at position 8 was found to be norleucine, this is an unnatural amino acid and it has no site for attachment to PQQ with a bond scissable by pronase or 2 N HCl. In view of the carbon skeleton and the conditions of detachment, we tentatively propose that X is a lysyl residue linked via its ε-NH$_2$ group to a carboxylic acid group of PQQ.

PQQ can either be attached to or inserted into the peptide chain, by participation with one or with two carboxylic acid groups, the latter possibility assumes that the second link is scissable under the degradative conditions of the Edman procedure (this assumption seems somewhat unlikely). Since the effect of trypsin on PQQ-protein bonds are unknown, even the possibility exists that there is a link to another part of the protein chain or to the protein chain of the other subunit. Dual crosslinking of PQQ in one and the same protein chain has been suggested in the case of methylamine dehydrogenase from *Pseudomonas* AM1 [Ishii *et al.*, 1983]. The suggestion that the two subunits of PKDAO are crosslinked via PQQ is, however, very unlikely since only the protein band of the monomer is observed on electrophoresis of PKDAO in an SDS system and it is not reasonable to assume scission of these bonds by SDS.

Further work on the nature of the amino acid and its bond with PQQ is in progress.

REFERENCES

Ishii, Y., Hase, T., Fukumori, Y., Matsubara, H., and Tobari, J. (1983) J. Biochem. *93*, 107–119.
Van der Meer, R.A., Jongejan, J.A., Frank Jzn., J., and Duine, J.A. (1986) FEBS Lett. *206*, 111–114.
Van der Meer, R.A., Jongejan, J.A., and Duine, J.A. (1987) FEBS Lett. *221*, 299–304.
Van der Meer, R.A., Jongejan, J.A., and Duine, J.A. (1988) FEBS Lett. *231*, 303–307.

PQQ AND QUINOPROTEINS: AN IMPORTANT NOVEL FIELD IN ENZYMOLOGY

J.A. DUINE, Department of Microbiology & Enzymology, Delft University of Technology, Julianalaan 67, 2628 BC Delft, The Netherlands

INTRODUCTION

In view of the landmark attained with the first International Symposium on PQQ and Quinoproteins, it seemed a good opportunity to use the present contribution to evaluate the relevance of this topic for enzymology and related scientific disciplines. To this end, a short historical overview is given, the balance is drawn of the currently known quinoproteins, a comparision is made with well known cofactors, and finally some conclusions and prospects are presented. After a long and hidden past, the topic has reached the status of maturity and all requirements are met to become a fully-fledged part of enzymology.

THE HISTORY OF PQQ

Non-phosphorylative sugar dissimilation by bacteria has already been known for a long time. Interest in the fifties into the biochemistry of it revealed that the enzymes involved in the primary oxidation step were NAD(P)-independent and that their activity was sometimes restored by addition of an unknown compound (see e.g. ref. (1)). One of these enzymes, glucose dehydrogenase, was further investigated by Hauge, culminating in a paper (2) in which the unusual properties of the cofactor were demonstrated. This work was not continued, progress only achieved after an impetus came from a different topic in microbial biochemistry, namely the dissimilation of Cl-compounds. Bacterial conversion of methanol became popular in the sixties, stimulated by the interest in single cell protein. Methanol dehydrogenase also appeared to be a curious enzyme and the proposition was made that its cofactor could be a pteridine (3). This view was confirmed in the following years as the group of Forrest proposed several pteridine derivatives (e.g. in ref. (4)). Therefore, the finding that the cofactor was neither a pterine nor a flavin but an o-quinone with 2 nitrogen's and 3 hydrogen's (5, 6) was a real breakthrough. Structure elucidation was then rapidly performed by two research groups (7, 8). Taking the discovery of the cofactor as the start and the final structure elucidation as the end, the whole period took about 20 years. Reasons for this long time could be the disinterest of biochemists during the past decennia in enzymes, and that of enzymologists in the outsiders in which PQQ was originally detected.

Already during the structure elucidation of PQQ, it became clear that several other bacterial dehydrogenases were quinoproteins. The strategy was rather easy, those enzymes where the properties did not agree with those of a flavoprotein were checked by denaturing the protein and assaying the supernatant with a quinoprotein apo-enzyme (9), while a further check was possible with HPLC (10).

After it appeared that so many bacterial dehydrogenases were quinoproteins, a logical step was to look in other classes of enzymes and in eukaryotes. This appeared to be a difficult enterprise since all cofactors which could possibly be PQQ were tightly bound to the protein in these cases. Although protein hydrolysis lies at hand in such a situation, it was already known that this would not be feasible since PQQ reacts with certain amino acids to a number of compounds which were unknown and could

351

J.A. Jongejan and J.A. Duine (eds.), PQQ and Quinoproteins, 351–360.
© 1989 by Kluwer Academic Publishers.

not be identified at that time. Therefore, an approach was chosen which prevented these unwanted reactions, resulting in the so-called hydrazine method (11). After the successful demonstration of the presence of PQQ in copper-containing amine oxidases, the question arose whether other metallo-enzymes could be quinoproteins. Although some of them have been well studied with a number of sophisticated techniques like EXAFS, NMR, ESR, this did not lead to the conclusion that an organic cofactor might be present. Therefore, the mechanistic views were solely based on a metallo-ion as the cofactor, althought theses were difficultly to reconcile with other enzymic data. With the discovery of PQQ in these enzymes, much more satisfactory explanations can be given now.

Not only the combination PQQ/metal ion appears to be successfully employed in Nature, but also the combination PQQ/other organic cofactor, as demonstrated in the case of atypical PLP-containing enzymes found in the groups of amino acid decarboxylases and amino transferases (12). Several of these enzymes play an essential role in cellular metabolism. Thus, after a history of about 30 years, PQQ has come of age as a cofactor, present from microbes to man and functioning in several classes of enzymes.

In retrospect, the question can be posed why has the cofactor been overlooked for such a long time in a number of well-known mammalian enzymes? An important reason might be that in all these cases, PQQ is covalently bound and its reactivity with free amino acids precludes its isolation in identifiable form. Another reason is the tendency which exists to use the absorption spectrum of an enzyme for identification of its cofactor. This not only has led to the erroneous conclusion of a pteridine in methanol dehydrogenase, but also to that of PLP in copper-containing amine oxidases. In addition, it is responsible for the failure to observe PQQ in certain metallo- as well as certain pyridoxo-proteins. Finally, most of us are unaware of the fact that cofactor identification has been accomplished in only an astonishing low number of cases (13) (as revealed by a compilation of enzymes classfied by the Enzyme Nomenclature Commission). Therefore, the history of PQQ shows that cofactor enzymology could be more rewarding than is generally believed.

MICROBIAL QUINOPROTEINS

Dehydrogenases. A substantial number of bacterial NAD(P)-independent dehydrogenases have been established to be quinoproteins, as shown in Table 1 . The establishment is, however not always unambiguous. In order to indicate this, the following criteria were used for the compilation: genuine quinoproteins are those for which a quantitative determination of PQQ has been reported or where reconstitution has been achived of its apo-enzyme; other enzymes, suspected to be quinoproteins, are rubricated under the heading "Miscellaneous". Curiously, quinoprotein dehydrogenases have been only detected in bacteria, to be more precise in Gram-negative species, where they are located in the periplasm. However, a number of exceptions exist: methanol dehydrogenase from Nocardia spec. 239 since the bacterium is a Gram-positive while the enzyme requires NAD for activity (14) and the complex of which it forms part should, therefore, be localized at the cytoplasmic side of the membrane; methanol dehydrogenase from the strict anaerobe Clostridium autotrophicum since the bacterium is also a Gram-positive and the enzyme has several uncommon properties (15). The enzymes release PQQ on denaturation, but not methylamine dehydrogenase, an enzyme where the cofactor occurs in covalently bound form but can be detached in the form of PQQ-phenylhydrazone (16).

Not unexpectedly, the substrates for these periplasmic quinoprotein dehydrogenases are in their non-phosphorylated form, which could indicate that the enzymes are particularly well suited for conversions at the outside of the bacterial cytoplasmic membrane. Although this may be benificial with respect to the bio-energetics of the organisms (e.g. transport systems are not necessary and substrate oxidation leads to external proton production), the necessity to assemble the holo-enzyme in the periplasmic space (assuming that the common mechanisms for protein transport are operating) seems less economic because a waste by escape from cofactor to the extracellular fluid is unavoidable in such a process. This is indeed what has been found: bacteria with periplasmic quinoprotein dehydrogenases excrete substantial amounts of PQQ into their culture medium (17). Another unattractive consequence of the outside location are the provisions which should exist in some cases for effective product transfer to internal systems, e.g. for those quinoproteins which produce toxic aldehydes that have to be safely transferred to either dissimilation and/or assimilation pathways.

As is shown in Table 1, several bacterial quinoproteins are involved in the dissimilation of C1-compounds. However, the enzymes are not specifically designed for this task since they have a broad substrate specificity. Their involvement in conversion of higher alcohols or amines has indeed been found in certain facultative methylotrophs. In addition, related types of enzymes have been discovered in non-methylotrophic bacteria, e.g. quinoprotein alcohol dehydrogenase in Pseudomonas aeruginosa and putida spec. (18, 19). On the other hand, quite different types also occur, e.g. the haemoquinoprotein alcohol dehydrogenase from Ps. testosteroni (20) and perhaps from Rhodopseudomonas acidophila (21) and Acetic acid bacteria (22). Most probabaly, a similar situation holds for amine dehydrogenases since several types are known (EC 1.4.99.3; 1.4.99.4) as well as a haem-containing amine dehydrogenase (23-25).

From the discovery of quinoprotein dehydrogenases, it appears now that certain bacteria have a whole set of different enzymes (with respect to the nature of the cofactor) catalyzing the same reaction (26). For instance, Ps. aeruginosa grown on ethanol contains an NAD-dependent as well as a quinoprotein alcohol dehydrogenase (27). The benefit for a bacterium containing such a set of dehydrogenases might be related to the bioenergetical flexibility acquired with such a variety. Support for this view comes from the observation that some organisms use quinoprotein glucose dehydrogenase to oxidize glucose as an auxiliary energy source (28). Since a dissimilation pathway, using one or more quinoproteins, frequently has to compete with another route (without quinoproteins) in the same organism, possibilities to manipulate preferent induction or specifically designed inhibitors are necessary to reveal the significance of the enzymes. Insight into this could also have relevance for practical purposes since the multiplicity of conversion routes is especially pronounced in microbes performing incomplete oxidations (e.g. bacteria applied in the manufacturing of vinegar, gluconic acid, etc.).

Oxidases. The only bacterial quinoprotein oxidase which has been established sofar is methylamine oxidase (29) from the Gram-positive methylotroph, Arthrobacter P1 (however, a similar enzyme may exist in certain Escherichia coli strains (30)). From the absence of a periplasm, it is obvious that the organism does not have the methylamine dehydrogenase mentioned before, though the oxidase might be evolutionary related to the methylamine dehydrogenase/amicyanine couple. In view of its properties, methylamine oxidase is a so-called copper-containing amine oxidases (EC 1.4.3.6), enzymes having Cu(II) ions and covalently-bound PQQ as cofactors and widely distributed in yeast's, fungi, plants, and mammals. Amines are also oxidized via flavoprotein oxidases (EC 1.4.3.4). In some cases, the two types of enzymes have overlapping substrate specificities. Comparative enzymology is necessary to reveal the special features of each type of enzyme.

The combination, covalently-bound PQQ and Cu ions, seems to be versatile with respect to catalytic possibilities of oxidation since it also operates in fungal galactose oxidase. Although this enzyme has been studied for decennia, mechanistic views were problematic since they had to be based on a single Cu ion as cofactor. With the recent finding of PQQ as an additional cofactor (van der Meer, Jongejan and Duine, to be published elsewhere), the different enzyme redox-forms could be assigned and a mechanism proposed, consistent with the spectroscopic and mechanistic data reported in the literature.

Amino acid decarboxylases. Recently, glutamate decarboxylase from E. coli was found to contain PLP as well as covalently bound PQQ (Groen, van der Meer and Duine, unpublished results). From this finding, the conclusion can be drawn that this bacterium synthesizes covalently bound PQQ but not free PQQ (common laboratory strains produce quinoprotein glucose dehydrogenase apo-enzyme (31)). Since holo- and apo-glucose dehydrogenase seem randomly distributed among genera (and sometimes species) of Enterobacteriacea (32), the ability to produce free PQQ is apparently not crucial and easily lost or acquired.

Miscellaneous. Nitrile hydratase is a bacterial iron protein with an organic prosthetic group which might be PQQ or a related compound (33). Nitroalkane oxidase is a fungal enzyme containing a flavin as well as a PQQ-like compound (34). The discovery of PQQ in dopa decarboxylase from pig kidney (12) and in glutamate decarboxylase from E. coli indicates that other atypical microbial PLP-enzymes might be quinoproteins too. In this respect, certain bacterial amino acid decarboxylases and ω-aminotransferases are interesting, as revealed on PLP-depletion (35).

Table 1. Microbial quinoproteins

Enzyme	Organism	References*
DEHYDROGENASES		
Methanol dehydrogenase (EC 1.1.99.8)	Gram-negative methylotrophic bacteria	(36)
Methylamine dehydrogenase (EC 1.4.99.3)	Gram-negative methylotrophic bacteria	(16)
Alcohol dehydrogenase	Ps. aeruginosa, Ps. putida	(18, 19)
Quinohaemoprotein alcohol dehydrogenase	Ps. testosteroni	(20)
Quinate dehydrogenase	Several Gram-negative bacteria	(37)
Glucose dehydrogenase (membrane bound)	Many Gram-negative bacteria	(38, 39)
Glucose dehydrogenase (soluble; EC 1.1.99.17)	Acinetobacter calcoaceticus LMD 79.41.	(40, 41)
Polyvinylalcohol dehydrogenase	Ps. spec.	(42)

Table 1, continued

OXIDASES
Methylamine oxidase	Arthrobacter P1	(29)
Galactose oxidase	Fungi	(unpubl.)
(EC 1.1.3.9)		

DECARBOXYLASES
Glutamate decarboxylase	Escherichia coli	(unpubl.)
(EC 4.1.1.15)		

MISCELLANEOUS
Methanol dehydrogenase	Clostridium thermoautotrophicum	(15)
Methanol dehydrogenase	Nocardia spec. 239	(14)
Glucose dehydrogenase	Zymomonas mobilis	(43)
Alcohol dehydrogenases	Rhodops. acidophila, Acetobact. spec.	(21, 22)
Amine dehydrogenases	Ps. spec.	(23–25)
(e.g. EC 1.4.99.4)		
Glycerol dehydrogenase	Gluconobacter industrius	(36)
(Form)aldehyde dehydrogenases	Acetob. spec., Hyphomicrobium spec.	(21,
(e.g. EC 1.2.99.3)	Rhodops. acidophila	36, 44)
Polyethyleneglycol dehydrogenase	Flavobacterium spec.	(36)
Lactate dehydrogenase	Propionibacterium pentosaceum	(36)
Tryptophan side chain	Ps. spec.	(36)
oxidase (dehydrogenase)		
Amine oxidase	Aspergillus spec., yeasts	(45, 46)
Nitroalkane oxidase	Fusarium oxysporum	(34)
Lysyl oxidase	Pichia pastoris	(47)
Nitrile hydratase	Brevibacterium spec.	(33)
Atypical pyridoxoprotein	Microbia	(12)
amino acid decarboxylases		
and transaminases		

----

*For several enzymes, reference (36) refers to the original report.

MAMMALIAN AND PLANT QUINOPROTEINS

Oxidoreductases. As far is known at the moment, eukaryotes have no quinoprotein dehydrogenases (although it has been claimed (48) that dog liver choline dehydrogenase is a quinoprotein, direct evidence for this has not been provided and others consider this enzyme as a flavoprotein (49)). On the other hand, many, if not all, copper-containing amine oxidases from this group of organisms are quinoproteins (Table 2). Covalently-bound PQQ with Cu-ions also occur in hydroxylases (mono-oxygenases), as demonstrated for dopamine β-hydroxylase (50). The hitherto uncharacterized peptidyl glycine mono-oxygenase, is very similar to this enzyme since it contains Cu ions and needs ascorbic acid as a source of reduction equivalents (51). Similarly, membrane-bound methane mono-oxygenase contains Cu ions and its activity in vitro could be demonstrated in the presence of ascorbic acid (52). Effective combinations also exist for other metal ions, as has been found for soybean lipoxygenase-1, a dioxygenase where an Fe ion and covalently-bound PQQ form an electron relay system, operating in stereoselective removal of reduction equivalents and insertion of dioxygen (53). Several types of lipoxygenases have been found in mammalian systems but as these have not been characterized, it is unclear whether they are quinoproteins.

Amino acid decarboxylases and &-amino transaminases. Dopa decarboxylase
from pig kidney is a pyridoxo-quinoprotein (12). The covalently bound PQQ
has been proposed to be the cofactor that is active in the decarboxylation
step while PLP functions as acceptor for the product after which
hydrolysis occurs (54). Studies on model systems consisting of PQQ and
basic amino acids show that PQQ is able to decarboxylate and to convert
ω-amino groups into an aldehyde group (55). Thus, it can be expected that
PQQ will be present in several other members of this types of enzymes.

Table 2. Plant and mammalian quinoproteins

Enzyme	Organism	References
OXIDASES		
Plasma amine oxidase (EC 1.4.3.6)	Bovine serum	(56)
Diamine oxidase (EC 1.4.3.6)	Pig kidney	(57)
Lysyl oxidase (EC 1.4.3.13)	Human placenta and arteria, chicken cartillage	(58, 59)
MONO-OXYGENASES		
Dopamine β-hydroxylase (EC 1.14.17.1)	Bovine medulla	(50)
DIOXYGENASES		
Lipoxygenase-1 (EC 1.13.11.12)	Soybean	(53)
DECARBOXYLASES		
Dopa decarboxylase (EC 4.1.1.28)	Pig kidney	(12)
MISCELLANEOUS		
Amine oxidases	Plants and mammals	(60, 61)
Lipoxygenases	Plants and mammals	
Peptidyl glycine mono-oxygenase	Mammals	(51)
Atypical pyridoxoprotein amino acid decarboxylases and transaminases	Plants and mammals	(12)

COMPARISON WITH OTHER COFACTORS AND ENZYMES

PQQ and other cofactors. With respect to redox behaviour, PQQ is
comparable to flavins since it shows 2-electron as well as 1-electron
redox behaviour, the latter in consecutive steps. In accordance with this,
just as flavoproteins, quinoproteins occur in the class of
respiratory-chain coupled dehydrogenases as well as oxidases. However,
sofar it has been found that all quinoprotein oxidases contain a
functional metal ion (in this context, the excellent chelating properties
of PQQ should be mentioned (62, 63)).
PQQ has also similarity to PLP since both contain a reactive carbonyl
group, are non-nucleotide cofactors and are able to decarboxylate amino

acids. The similarity goes even further with the pyruvoyl group cofactor in view of covalent bonding to the proteins.

Tyrosine and glutamic acid are the precursors of bacterial PQQ biosynthesis (64-66). It has been speculated that the biosynthesis process could proceed on a protein matrix, just by fusion, cyclisation, oxidation and hydroxylation of the precursors, and finally the excision of the cofactor (13). The discovery that E. coli produces covalently PQQ (in glutamate decarboxylase), but not free PQQ (as is also apparent from its glucose dehydrogenase, occurring in the apo-form), suggests that free PQQ is not used in the assemblage of enzymes with covalently PQQ but that the cofactor is made in situ. The recent 3-dimmensional structure elucidation of methylamine dehydrogenase (Vellieux, Huitema, Groendijk, Kalk, Frank, Jongejan, Duine, Drenth and Hol, unpublished results), supports this view. It appears that, although the cofactor is detached as the PQQ-phenylhydrazone (16), its structure in the active site, is the so-called pro-PQQ, consisting of the indole moiety of PQQ connected with a glutamic acid residue, covalently bound at 2 positions in the protein chain. Although it remains to be seen whether this is an exception or that all quinoproteins from which covalently bound PQQ has been extracted contain in fact pro-PQQ, the observation that free PQQ has not been detected sofar in mammalian systems and free intermediates do not occur in the biosynthesis route (67), are in accordance with an in situ synthesis. If this view is correct, PQQ will not be a vitamin and effects upon its administration to animals or plant cultures (68-70) should be due to secondary phenomena. Finally, it is interesting to note that several enzymes have been found, using a tyrosyl radical in their protein chain as cofactor (71). This compound could be seen as the evolutionary predecessor of PQQ with pro-PQQ as an intermediate. In this reasoning, free PQQ, sofar only detected in a number of bacteria, is the final stage in the evolution of the amino acid cofactor.

Quinoproteins, a special role? Before PQQ was discovered, the well known cofactors seemed well equipped to perform a catalytic role for all enzymic reactions. Although the different types of reaction pose heavy demands with respect to catalytic versatility, the protein part of an enzyme is able to transform the properties of the cofactor into the desired direction. This is illustrated for instance in the case of flavoproteins where the redox potentials span the wide range required for the different tasks performed by these enzymes. Therefore, at first sight PQQ seems superfluous, reflected by the fact that the reactions catalyzed by quinoproteins find their counterpart in groups of enzymes using a different cofactor but catalyzing the same reaction and the normal behaviour of quinoprotein dehydrogenases, that is they use all the common electron acceptors of the respiratory chain (cytochromes b and c, ubiquinones, and copper proteins), just like flavoproteins. In this view, PQQ could be a remnant from evolution, its presence explained from the inability of organisms to remove the superfluous genetic burden. On the other hand, it could also be reasoned that quinoproteins have specific properties, not yet discovered or still awaiting exploitment in the evolutionary history. It will be clear that comparative enzymology of enzymes catalyzing the same reaction with different cofactors could throw light on these points, as discussed elsewhere (26).

CONCLUDING REMARKS AND PERSPECTIVES

-From the many already established quinoproteins, it is clear that PQQ belongs to the group of the important cofactors since it is versatile and has a wide distribution, quinoproteins with covalently bound PQQ presumably being omnipresent in view of the vital role of some of these enzymes in cellular metabolism.

-The reactivity and the tentative ideas on the mechanism of biosynthesis seem to preclude a role of PQQ or a PQQ-like compound as a vitamin in mammals. On the other hand, recent reports suggest physiological effects of PQQ on administration to plants and animals. Therefore, studies are necessary to discriminate between indirect and direct effects. In addition, the development of reliable analytical procedures to detect PQQ in all its forms is crucial in order to obtain insight into distribution and the way of biosynthesis in these organisms.

-The presence of PQQ in so many enzymes involved in synthesis or conversion of bioregulators (Table 2) gives new opportunities to develop inhibitors regulating important physiological processes in higher organisms. In addition, it points to the necessity to reconsider the explanations given for the effects of certain inhibitors ascribed to the blocking of enzymes using a certain cofactor (for instance the presumed effect of certain drugs with carbonyl group reactive moities on PLP-enzymes).

-Several well-known enzymes with an already established cofactor appear now to be quinoproteins. This implicates that mechanistic views on e.g. certain metallo-enzymes, frequently based on the participation of only a metal ion, have to be revisited. The same comment applies to the pyridoxo-quinoproteins. In order to be able to propose novel mechanisms, it will be necessary to obtain insight into the interaction between the cofactors. The discovery of pro-PQQ indicates, however, that first the real structure of the cofactor in the active site should be determined.

REFERENCES

1) Szymona, M. and Doudoroff, M., (1960) J. Gen. Microbiol. 22, 167-183.
2) Hauge, J.G. (1964) J. Biol. Chem. 239, 3630-3639.
3) Anthony, C. and Zatman, L.J. (1967) Biochem. J. 104, 960-969.
4) Urushibara, T., Forrest, H.S., Hoare, D.S. and Patel, R.N. (1971) Biochem. J. 125, 141-146.
5) Duine, J.A. Frank, J. and Westerling, J. (1978) Biochim. Biophys. Acta 524, 277-287.
6) Westerling, J., Duine, J.A. and Frank, J. (1979) Biochem. Biophys. Res. Commun. 87, 719-724.
7) Salisbury, S.A., Forrest, H.S., Cruse, W.B.T. and Kennard, O. (1979) Nature 280, 843-844.
8) Duine, J.A., Frank, J. and Verwiel, P.E.J. (1980) Eur. J. Biochem. 108, 187-192.
9) Duine, J.A., Frank, J. and van Zeeland, J.K. (1979) FEBS Lett. 108, 443-446.
10) Duine, J.A. and Frank, J. (1980) Biochem. J. 187, 221-226.
11) R.A. van der Meer, J.A. Jongejan and J.A. Duine, this Proceedings.
12) Groen, B.W., van der Meer, R.A. and Duine, J.A. (1988) FEBS Lett. 237, 98-102.
13) Duine, J.A. and Jongejan, J.A. (1989) Vitamins and Hormones 46, in press.

14) Duine, J.A., Frank, J. and Berkhout, M.P.J. (1984) FEBS Lett. 168, 217–221.
15) D.K. Winters and L.G. Ljungdahl, this Proceedings.
16) van der Meer, R.A., Jongejan, J.A. and Duine, J.A. (1987) FEBS Lett. 221, 299–304.
17) Van Kleef, M.A.G. and Duine, J.A. (1989) Appl. Environ. Microbiol., in press.
18) Groen, B.W., Frank, J. and Duine, J.A. (1984) Biochem. J. 223, 921–924.
19) H. Goerisch and M. Rupp, this Proceedings.
20) Groen, B.W., van Kleef, M.A.G. and Duine, J.A. (1986) Biochem. J. 234, 611–615.
21) K. Yamanaka, this Proceedings.
22) Adachi, O., Shinagawa, E., Matsushita, K. and Ameyama, M. (1982) Agric. Biol. Chem. 46, 2859–2863.
23) Niimura, Y., Omori, T. and Minoda, Y. (1986) Agric. Biol. Chem. 50, 1445–1451.
24) Nozaki, M. (1987) Meth. Enzymol. 142, 650–655.
25) Shinagawa, E., Matsushita, K., Nakashima, K., Adachi, O. and Ameyama, M. (1988) Agric. Biol. Chem. 52, 2255–2263.
26) Duine, J.A. In: The Roots of Modern Biochemistry (Kleinkauf, H., von Dohren, H. and Jaenicke, L., eds.) Walter de Gruyter & Co., (1988), Berlin, New York, pp. 671–682.
27) Groeneveld, A., Dijkstra, M. and Duine, J.A. (1984) FEMS Microbiol. Lett. 25, 311–314.
28) O.M. Neijssel, R.W.J. Hommes, P.W. Postma and D.W. Tempest. this Proceedings.
29) van Iersel, J., van der Meer, R.A. and Duine, J.A. (1986) Eur. J. Biochem. 161, 415–419.
30) Parrott, S., Jones, S. and Cooper, R.A. (1987) J. Gen. Microbiol. 133, 347–351.
31) Hommes, R.W.J., Postma, P.W., Neijssel, O.M., Tempest, D.W., Dokter, P. and Duine, J.A. (1984) FEMS Microbiol. Lett. 24, 329–333.
32) Neijssel, O.M. (1987) Microbiol. Sci. 4, 87–90.
33) Nagasawa, T. and Yamada, H. (1987) Biochem. Biophys. Res. Commun. 147, 701–709.
34) K. Tanizawa, T. Moriya, T. Kido, H. Tanaka and K. Soda, this Proceedings.
35) Soda, K. In: Transaminases (Christen, P. and Metzler, D.E., eds.), Wiley & Sons, (1985), New York, pp. 421–430.
36) Duine, J.A., Frank, J. and Jongejan, J.A. (1987) Adv. Enzymol. 59, 169–212.
37) van Kleef, M.A.G. and Duine, J.A. (1988) Archiv. Microbiol. 150, 32–36.
38) A-M. Cleton-Jansen, N. Goosen, K. Vink and P. van de Putte, this Proceedings.
39) K. Matsushita, E. Shinagawa, O. Adachi and M. Ameyama, this Proceedings.
40) Dokter, P., Frank, J. and Duine, J.A. (1986) Biochem. J. 239, 163–167.
41) Geiger, O. and Goerisch, H. (1986) Biochemistry 25, 6043–6048.
42) Shimao, M., Ninomiya, K., Kuno, O., Kato, N. and Sakazawa, C. (1986) Appl. Environ. Microbiol. 51, 268–275.
43) M. Strohdeicher, S. Bringer-Meyer, B. Neusz, R.A. van der Meer, J.A. Duine and H. Sahm, this Proceedings.

360

44) F.P. Kesseler, I. Baduns ans A.C. Schwartz, this Proceedings.
45) Adachi, O. and Yamada, H. (1969) Agric. Biol. Chem. 1707-1714.
46) Haywood, G.W. and large, P.J. (1981) Biochem. J. 199, 187-192.
47) S.S. Tur, P.M. Royce and K. Lerch, this Proceedings.
48) Ameyama, M., Shinagawa, E., Matsushita, K., Takimoto, K., Nakashima, K. and Adachi, O. (1985) Agric. Biol. Chem. 49, 3623-3626.
49) Chi-Shui, L. and Ru-Dan, W. (1986) J. Prot. Chem. 5, 193-200.
50) van der Meer, R.A., Jongejan, J.A. and Duine, J.A. (1988) FEBS Lett. 231, 303-307.
51) Eipper, B., Mains, R.E. and Glembotski, C.C. (1983) Proc. Natl. Acad. Sci. USA 80, 5144-5148.
52) Tonge, G.M., Harrison, D.E.F. and Higgins, I.J. (1977) Biochem. J. 161, 333-344.
53) van der Meer, R.A. and Duine, J.A. (1988) FEBS Lett. 235, 194-200.
54) Duine, J.A. and Jongejan, J.A. (1989) Annu. Rev. Biochem. 58, in press.
55) M.A.G. van Kleef, J.A. Jongejan and J.A. Duine, this Proceedings.
56) Lobenstein-Verbeek, C.L., Jongejan, J.A., Frank, J. and Duine, J.A. (1984) FEBS Lett. 170, 305-309.
57) van der Meer, R.A., Jongejan, J.A., Frank, J. and Duine, J.A. (1986) FEBS Lett. 206, 111-114.
58) van der Meer, R.A. and Duine, J.A. (1986) Biochem. J. 239, 789-791.
59) Williamson, P.R., Moog, R.S., Dooley, D.M. and Kagan, H.M. (1986) J. Biol. Chem. 261, 9477-9482.
60) A. Finazzi-Agro, this Proceedings.
61) Glatz, Z., Kovar, J., Machola, L. and Pec, P. (1987) Biochem. J. 242, 603-606.
62) Jongejan, J.A., van der Meer, R.A., van Zuylen, G.A. and Duine, J.A. (1987) Rec. Trav. Chim. Pays-Bas 106, 365.
63) S. Suzuki, T. Sakurai, S. Itoh and Y. Oshiro, this Proceedings.
64) van Kleef, M.A.G. and Duine, J.A. (1988) FEBS Lett. 237, 91-97.
65) Houck, D.R., Hanners, J.L. and Unkefer, C.J. (1988) J. Am. Chem. Soc. 110, 6920-6921.
66) Houck, D.R., Hanners, J.L., Unkefer, C.J., van Kleef, M.A.G. and Duine, J.A., this Proceedings.
67) van Kleef, M.A.G. and Duine, J.A. (1988) BioFactors 1, in press.
68) R. Rucker, J. Killgore, L. Duich, N. Romero-Chapman, C. Smidt and D. Tinker, this Proceedings.
69) T. Matsumoto, O. Suzuki, H. Hayakwama, S. Ogiso, N. Hayakawa, Y. Nimura, I. Takahashi and S. Shionoya, this Proceedings.
70) Xiong, L.B., Sekiya, J. and Shimose, N. (1988) Agric. Biol. Chem. 52, 1065-1066.
71) Prince, R.C. (1988) Trends Biochem. Sci. 13, 286-288.

AUTHOR INDEX

SUBJECT INDEX

Acetogens, *35*
Albumin, PQQ binding to, *145*
Alcohol dehydrogenase,
see also: ethanol dehydrogenase,
methanol dehydrogenase
  biomimetic reaction, *199*
  from *P. testosteroni*, *205*
  from *P. aeruginosa*, *354*
Amicyanin, *2, 4, 269*
Amine dehydrogenase, *355*
Amine oxidases
  from bovine serum, *345, 356*
  from *P. pastoris*, *327*
  from *Aspergillus* sp., *355*
Amine oxidase
  active site structure of, *283*
  biomimetic reactions, *196*
  copper-containing, *279, 283,*
  289, 297, 345, 348
  hydrazine treated, *345*
  mechanistic probes of, *297*
  models of, *253*
  PQQ as cofactor of, *159, 297*
  role of copper in, *289*
  spectroscopic properties of, *280*
Aminoguanidine
  reaction with PQQ, *239*
Azurin, *5*
Autoreduction, *7*
Azo-compound, *117*
Bioenergetics, *57*
Biosynthesis of PQQ
  in *A. calcoaceticus*, *169, 190*
  in *Hyphomicrobium* X, *177*
  in *K. aerogenes*, *187*
  in *Methylobacterium* AM1, *177*
  in *M. organophilum*, *190*
  genes involved in, *169*
  glutamate as precursor of, *180*
  mutants of, *169, 187, 190*
  tyrosine as precursor of, *180*

363